Environmental Earth Sciences

Series Editor

James W. LaMoreaux, Tuscaloosa, AL, USA

Environmental Earth Sciences encompass mulitdisciplinary studies of the Earth's atmosphere, biosphere, hydrosphere, lithosphere and pedosphere and humanity's interaction with them. This book series aims to provide a forum for this diverse range of studies, reporting on the very latest results and documenting our emerging understanding of the Earth's system and our place in it. The type of material published traditionally includes:

- proceedings that are peer-reviewed and published in association with a conference;
- post-proceedings consisting of thoroughly revised final papers; and
- research monographs that may be based on individual research projects.

The Environmental Earth Sciences series also includes various other publications, including:

- tutorials or collections of lectures for advanced courses;
- contemporary surveys that offer an objective summary of a current topic of interest; and
- emerging areas of research directed at a broad community of practitioners.

More information about this series at https://link.springer.com/bookseries/8394

Hüseyin Gökçekuş · Youssef Kassem
Editors

Climate Change, Natural Resources and Sustainable Environmental Management

Springer

Editors
Hüseyin Gökçekuş
Department of Civil Engineering, Civil and Environmental Engineering Faculty
Near East University
Cyprus, Turkey

Youssef Kassem
Department of Mechanical Engineering, Engineering Faculty
Near East University
Cyprus, Turkey

ISSN 2199-9155 ISSN 2199-9163 (electronic)
Environmental Earth Sciences
ISBN 978-3-031-04374-1 ISBN 978-3-031-04375-8 (eBook)
https://doi.org/10.1007/978-3-031-04375-8

This Springer imprint is published by the registered company Springer Nature Switzerland AG
The registered company address is: Gewerbestrasse 11, 6330 Cham, Switzerland

Contents

Climate Change Impact on Water Structures

Zekâi Şen[1,2(✉)]

[1] Engineering and Natural Sciences Faculty, Istanbul Medipol University, Beykoz, 34815 Istanbul, Turkey
zsen@medipol.edu.tr

[2] Center of Excellence for Climate Change Research/Department of Meteorology, King Abdulaziz University, PO Box 80234, Jeddah 21589, Saudi Arabia
zsen@kau.edu.sa

Abstract. The literature is full of global warming and climate change impact studies on the environment, ecosystem, and different components of hydrological cycle. Unfortunately, these impacts on engineering structure design, maintenance, and operation and management studies are rather rare. Climate change impact started to play significant role since the last three decades almost in every aspects of life especially on meteorological and climatological events and their impacts on water resources, which are managed by engineering structures. Its effects on hydro-meteorological records are searched with objective methodologies quantitatively, but the same is not valid for engineering water structures performances among which are dams, weirs, reservoirs, culverts, channels, bridges, wells, highways and their side drainages, levees, etc. This paper provides the review of the necessary adaptation, combat and mitigation activities against the climate change and variability for protection, construction or augmentation of the engineering water structures design capacity. Land use practices and geomorphological changes also trigger the climate change impacts on the engineering water structures. The main aim of this paper is to present the impact of such changes on the engineering water structure capacity, operation and maintenance.

Keywords: Arid region · Climate change · Engineering · Risk · Structure · System · Water

1 Introduction

Water resources structures are major social engineering units that are essential for individuals, societies, countries and humanity, in general. Any societal development is based on the water resources system availability and adaptation to natural hazards (droughts and floods) and anthropogenically induced variations (including climate change), which can be regarded as one of the accumulating variability [1–6]. The potential impact of climate change on the hydrologic regime is a crucial question for water resources engineering structures and contemporary hydrology and water resources management.

Hitz and Smith [7] review of global impact studies could not find clear relationship between changes in water supply and increases in global mean temperature. They concluded that higher magnitudes of climate change are likely to increase stress on water resources. This is due in part to the fact that current water resource infrastructure is

H. Gökçekuş and Y. Kassem (Eds.): NRSEM 2021, EESCI, pp. 1–10, 2022.
https://doi.org/10.1007/978-3-031-04375-8_1

generally designed for today's climate. Results from global studies in this sector are highly inconsistent with some studies quite sensitive to the climate model and mode of aggregation [1, 8] and others showing little net global impact [9–11].

The IPCC [4, 5] reports of climate impact studies suggest large differences in the vulnerability of water resource systems to climate variables. Isolated single-reservoir systems in arid and semiarid areas are extremely sensitive. They lack the flexibility for adaptation to climate impacts that could vary from decrease in reservoir yields even more than 50% at one extreme to increased seasonal flooding at the other potential regional impacts of climate change could include increased frequency and magnitude of droughts and floods, and long-term changes in mean renewable water supplies through changes in precipitation, temperature, humidity, wind intensity, snowpack accumulation duration, nature and extent of vegetation, soil moisture, and runoff [12]. It is necessary to maintain availability and construction of engineering water structures according to climate change expectations and to manage existing infrastructure accordingly.

Climate change is one of the pressures facing water resources and their management over the next several years and decades [13, 14]. Herein, pressure means on water supply and demand sides. Especially, supply side is directly related to climate change, after all precipitation and runoff are the main sources of surface water engineering storage structures and groundwater storage and recharge to aquifers. On the demand side pressure, there are social affairs such as population increase, extravagant life style, more consumption in almost all aspects of modern life, land use, energy generation, etc. On the other hand, Kundzewicz and Somlyody [15] provided a discussion on climate impact uncertainties analyses on water resources, water management planning, design and adaptation. A detailed account has been given by Kundzewicz et al. [16] on the implications of projected climate change for freshwater resources and their management. Climate change effects are the focus of many scientific, engineering, economic, social, cultural, and global nuisances, and these effects awaits cost-effective remedial solutions.

One-third of the developing world will face severe water shortages in the 21th century even though large amounts of water will continue to flood annually to sea from arid regions [17]. Many responses to current climatic variability would not in themselves be an enough response to climate change. For example, a changing climate would alter the design standard of engineering water structures, such as channels or a defense walls. It could alter the effectiveness of building codes based on designing against specified return period events (such as the 10-year return period gust). Finally, it could alter the area exposed to a potential hazard, meaning that previously assumed development to be "safe" was now located in a risk area.

The main purpose of this study is to direct researchers' and engineers' attention to climate change impacts on the water structures and to their operation and management under such impacts. Additionally, engineers are advised to revise the basic equations and procedures by considering climate change impacts.

2 Climate and Human Activities

The climate is the long-term average feature of short-term meteorological events, and therefore, it represents comparatively more stable behavior than daily, weekly, monthly, annual and decadal behaviors, and according to the World Meteorology Organization

(WMO) the long term is defined as 30 years or more. Such natural changes have taken place temporarily and spatially. The following points can be counted among such natural changes.

1) During the long-life history of earth depending on the fluctuation of carbon dioxide (CO_2) fluctuations there has been global scale climate changes through dry, wet and ice periods,
2) The revolution of the earth axis like a ball along different axis angle, which is today at 23.5° away from the vertical and perpendicular to the earth's orbit plane around the sun. This change of Earth's obliquity oscillation between 22.1 and 24.5° has about 41,000-year cycle.
3) Plate-tectonic movement explains the global distribution of geological phenomena. Before billion years ago there were different land pieces intact from each other. For instance, China and Arabian Peninsula were in the southern hemisphere, but they are presently in the northern hemisphere. Such plate or continental movements also caused to natural climate changes,
4) Rather local climate changes took place as a result of volcanic eruptions due to the release of gasses, dusts into the atmosphere, which gave rise to regional chemical composition change,

There were human population growth, trade and land use activities without impacting the atmospheric composition at all hence, there were no significantly noticed global climate change except re. However, with the advancement of technology after the industrial revolution, the global atmosphere started to be loaded with CO_2 as a result of fossil fuel burn for energy and therefore, greenhouse gases (GHG) started to be concentrated in

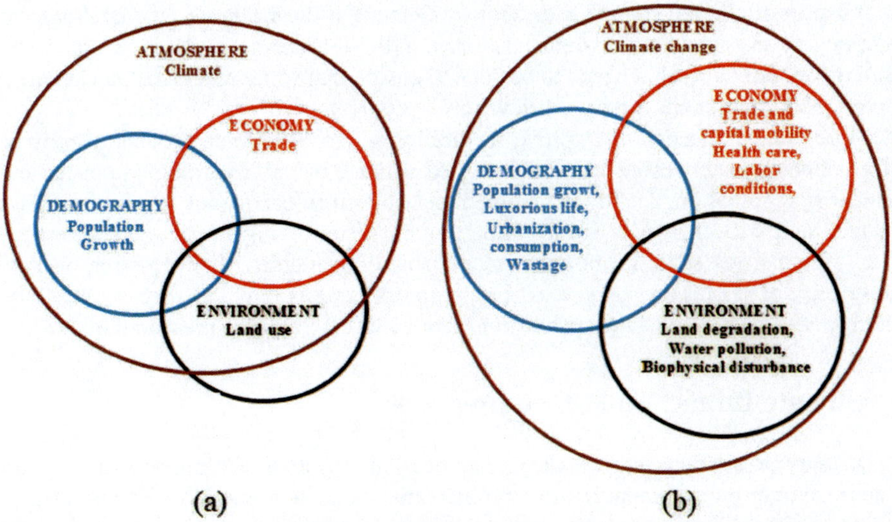

Fig. 1. Atmospheric and industrialization, a) Prior to climate change, b) Posterior to climate change

the atmospheric cover of the world. Figure 1b shows some new types of human activities within the demography, economy and environment circles. Comparison of Fig. 1a and 1b indicates the extensive additional activities increase.

3 Climate Change

Climate change has the potential to either aggravate or alleviate water situation in an area. On balance, however, the impacts are likely to be adverse because the existing water infrastructure and consumptions are based on past climate and hydrology records. It is difficult to plan for and justify expensive new projects when the magnitude, timing, and even the direction of the changes at the basin and regional levels are not well known. Narrowing the range of uncertainty for improved water planning, operation and management depends on a better understanding of the following points.

1) The processes governing global and regional climates,
2) The links between climate and hydrology,
3) The impacts of the climate on unmanaged ecosystems,
4) The impacts of ecosystem change on the quantity and quality of water,
5) The impacts of increased atmospheric CO_2 on vegetation and runoff.

Unlike the structural supply-side approach, demand management that introduces additional incentives to conserve and opportunities to reallocate supplies as conditions change does not require long lead times, large financial commitments, or accurate information about the future climate. Integrated management of existing supplies and infrastructure at the river basin and watershed levels offers a potentially cost-effective means of increasing reliable supplies and resolving water conflicts in many regions. While the prospect of climate change adds another element of uncertainty to the challenge of matching future supplies with demands through different storages (dams, weirs, dikes, etc.), it does not alter what needs to be done to ensure that water resources management and distribution systems are managed wisely.

More work is needed to improve the ability of global climate models to provide information on water-resources availability, to evaluate overall hydrologic impacts, and to identify regional impacts. Information about how storm frequency and intensity has changed and will change is vitally important for determining impacts on water and water systems, yet such information is not reliably available. More research on how the severity of storms and other extreme hydrologic events might change is necessary. Anthropogenic climate change projections from 1860 to 2100 are presented by [18].

4 Climate Impact and Structures

If necessary precautions are not taken from now on future climate impacts may render water resources management in more problematic state due to the following points.

1) The intensity, frequency pattern and amount of precipitation are bound to change especially in the sub-tropical climate belt of world, and accordingly necessary

precautions must be established for adaptation from now on based on the models' future predictions [2, 19].

2) In any region of significant water resources, the results of General Atmospheric Ocean Circulation Model (GAOCM) model results coupled with a set of scenarios must be downscaled to regional or better local scales [20, 21]. For practical studies, water resources experts without background information about the results of such models, the major recommendation is to base their water structure design on about 15% temporal rainfall decrease,
3) The spatial (regional or local) precipitation occurrences take place also under uncertainty, but increase is expected in mountainous high elevation and at mid-latitude places, especially in winter seasons. Hence, for summer seasons water resources necessitate optimum management practices under the climate change principles,
4) Global warming impact appears in temperature, evaporation and evapotranspiration increases, which reduces the surface water flow and runoff, and consequently renewability rates of surface and groundwater resources may decrease to certain extend. Climate change impact water resources management must take into consideration preference water supply from most evaporation water impoundments and recharge of surface water as much as possible into groundwater reservoirs,
5) Climate change is bound to cause floods and flash floods with abundant water volumes, and these water amounts must be directed in a beneficial way either to surface reservoir impoundments or groundwater recharge. Furthermore, the necessary precautions must be taken in terms of dams, levees, bends, channels, weirs, ditches, pipes as engineering water structures,
6) Global warming is expected to shift tropical and sub-tropical climate belts towards Polar Regions northerly and southerly and because of this the drought and dry periods are bound to increase especially in the sub-tropical regions. Due to decrease of precipitation and increase in evapotranspiration trigger drought potentiality in intensity, frequency, duration and spatial extent. The droughts start with meteorological events followed by hydrological types and agricultural types all of which cause social disturbances. The water engineering should be ready for action with the start of meteorological drought, alerted with the start of hydrological drought and hazard warnings towards the end of hydrological drought prior to the start of agricultural drought with consciousness of climate change impacts. Although there is no human interference to meteorological drought, but hydrological drought situation needs careful water resources administration and management so that agricultural drought is hindered,
7) As for the climate change water resources management practices differ in arid and semi-arid regions than humid places. Areas next to the arid and semi-arid regions are subject to desertification and such occurrences must be dealt with special treatments such as afforestation and reforestation studies and applications,
8) Snow and its melt are important water resources in many parts of the world especially at high mountainous elevations and mid-latitudes. The climate change impact causes reduction in snow fall and global warming starts early melting than historical cases. These events necessitate water resources management rearrangement to

cope with such effects to impound and manage melt snow water for better beneficial uses. Especially, in countries, where hydroelectricity generation is favorable, there might be reductions in power generation,

9) Global warming in general and climate change impacts affect water quality variations due to either natural or human activities. About the natural types the most spelled one is the sea water rise and salt water intrusion to the coastal aquifers. With the over pumping of groundwater aquifers, the saline or brine water at the base of aquifer may rise above due to up coning effect, and hence, fresh water layers in the aquifer become saline. On the other hand, due to air pollution acid rains may alter the quality of surface and subsequently groundwater resources. Such events cause increase of water cost, because of extra water treatment plant works,
10) Among the water resources impacts are the heat waves and islands in cities, where water demand is the maximum and sea surface waves that cause erosion and sedimentation at coastal areas.

There are a set of water engineering structures that are prone to the impact of climate change presently and soon. The following items explain the necessary precautions and adaptation possibilities.

1) Dams, (Fig. 2a): Decrease in the precipitation and especially in the form of snow will cause decrease in the surface water (runoff), and therefore, surface dams are bound to store less water volumes, and as consequence of such a situation the hydropower generation amount is bound to decrease at 10%–15% in the eastern part of Turkey. On the other hand, dry and drought periods are bound to carry more sediment loads behind the dams, which may fill the dead storage and consequently cause decrease in the active storage,
2) Subsurface dams, (Fig. 2b): These are very convenient for arid, desertic and dry climate regions. As a result of climate change impact in some regions their number will increase for the purpose of avoidance of evaporation losses and loss of surface water runoff amounts into the seas or desert areas,
3) Wells, (Fig. 2c): At regions of surface water reductions will trigger direct use of groundwater abstractions by wells, which is expected to cause groundwater table drops and as its consequent the water will be abstracted from deeper levels from the earth surface and this will cause economic losses due to energy consumption and technologically use of more powerful submersible pumps,
4) Drainage basins, (Fig. 2d): In many humid and semi humid regions and countries the surface water flow is expected to reduce and increase in floods and especially flash floods. On the other hand, in arid and semi-arid regions of the world such as the Arabian Peninsula, the precipitation increases are predictable actually and according to General Circulation (Climate) model studies. Hence, small scale surface dams and subsurface dams' number are bound to increase,
5) Culverts, (Fig. 2e): These are water passage tunnels beneath roads and highways to reduce the washing away effect of surface water. At regions of climate change impact precipitation increase or even at other regions due to the floods and especially flash floods their present dimensions may not be able to cope with surface flow volumes and velocities, and therefore their control or design updates are concern of engineers,

6) Channels or canals, (Fig. 2f): These are the natural and engineering made water conveyance structures, which will be subject to climate change impact. Their dimensions must also be controlled with the GCM output results so far as the surface flow discharges are concerned,
7) Pipes (Fig. 2g): Especially water distribution systems have majority of cases pipes for avoiding extra water losses due to the evaporation. The climate change is temperature increase, and hence, not to lose more water available surface water must be transported through pipes. Especially, in a country like Turkey with different climate regions, in the future regional water transfers may come into agenda, and in this case in dimensioning the pipe networks the climate change impact must be taken into consideration.

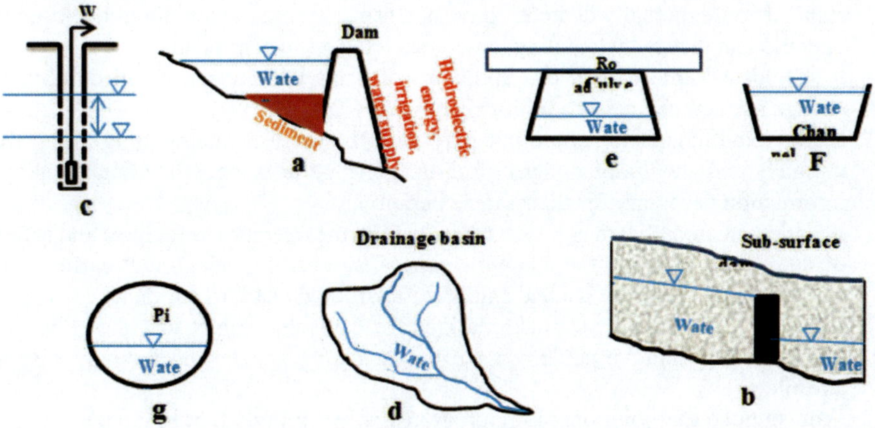

Fig. 2. Climate change and water structures

5 Recommendations

The climate change impact and variation effects must be reduced to the minimum by adaptation activities prior to any further treatment. Adaptation works help to minimize possible future dangers. The following recommendations can be applied in any suitable location, but this does not mean that there is no more risk in the future.

1) Whenever possible with a collective understanding of climate change, in the presence of border and transboundary waters within the states themselves, it will be necessary to promote peaceful cooperation among the relevant states through sustainable river basin management or other approaches and to develop a balanced interaction between different types of use at all levels,
2) In order to increase the groundwater resource locally, it can be ensured that the sea water can be mixed within the permissible limits near the shore,
3) Carrying out studies for the protection and sustainability of aquatic ecosystems under the light of climate change scenarios,

4) Developing strategic plans for the management and protection of water resources by using climate change model scenario data,
5) Identifying and implementing technologies that may be suitable for water harvesting in order to benefit from the increasing flood and sudden flood waters,
6) Adapting to climate change by considering different water supply storage facilities, sustainable groundwater withdrawal, inter-basin water transfer, water storage, water reuse and sea water treatment facilities, which may be more flexible for a medium-term future,
7) Hazards (risk) may play a role among the parameters affecting dam safety in terms of water and amounts of flood and drought. Many different effects of climate change and variability contribute to the emergence of risks related to the amount of floods,
8) Especially in the design of medium and large-scale dams, floods that may occur suddenly and gradually decreasing or increasing water amounts should be used in operation studies by calculating the climate change scenario data,
9) It should be kept in mind that spillways of dams that are not based on climate change impact may cause additional dangers,
10) Experts and engineers should not play with the safety of dams by ignoring the untimely and unwarranted signs of climate change or neglect the climate change information necessary for their safer operation,
11) Another situation that may pose a threat to both the operation principles and safety of dams is to examine the situations where snow and glaciers melt earlier than expected due to climate change and bring additional water to our dams,
12) In future it is necessary to make designs for dams that can absorb the effects of climate change and the sudden changes in the hydrological cycle that may arise as a result,
13) Dam singular and joint operation programs are very important in increasing dam safety by minimizing the effects of climate change,
14) It is true that dams do not cause greenhouse effect and thus climate change by producing carbon dioxide, but they are not allowed to lose their safety to some extent by being exposed to climate change due to the increase in GHG effect,
15) Necessary investments should be made to reduce the impact of climate change in structural (storage, control, transmission) and non-structural (demand and water basin management, service delivery, etc.) approaches regarding water management,
16) Supply-oriented adaptation techniques: Examples of supply-oriented adaptation include increasing the measures taken against drought, flood and flash flood modification or expansion of water supply and distribution infrastructure for consumers,
17) Demand-driven adaptation techniques: These include water demand management (e.g. sparing water consumption in irrigation), changing water allocations, and non-structural drought, flood and flood management measures (land use),
18) Conducting studies and developing strategic plans for the protection of aquatic ecosystems against the effects of climate change,
19) Encouraging the climate resistant water infrastructure and starting local applied studies on this subject,
20) Establishment of inter-basin water transport and distribution network from basins with excess water to basins that may be lacking,

21) Agricultural lands and pastures in the lower parts of the basins, dams, ponds, highways, railways, ports, bridges and so on. In order to minimize the effects of floods caused by floods, attention should be paid to the installation of side channels, retaining walls and gutters,
22) Elimination or minimization of flood-induced sweeping material (sediment) density currents and damages occurring in the main stream, side streams and rivulets,
23) By using the results of the climate change scenario data, it can be decided where, in what scale and area and what kind of adaptation studies will be carried out,
24) The emergence of arid and wet circuits directly affects droughts and flood events and their damages. In order to reduce these damages, the issues should be closely examined, and adaptation solutions should be produced,
25) It should be emphasized that flood and flood risk maps are prepared according to the climate change model results and updated every 5 or 10 years,
26) How many days after the beginning month of the snow melting will shift should be calculated in the light of the data of the scenarios and the dam operations should be updated and arranged accordingly,
27) Investigating rainfall harvesting possibilities and transforming them into hydrology studies and flow harvesting and eventually groundwater feeding.

6 Conclusion

There are many studies concerning climate change impacts on industry, agriculture, water resources, waste, buildings, transportation, economy, social integration and many environmental problems, the engineering water structures behaviors neither well documented nor examined fully in the literature. For humanity the most precious commodity is water and for water to serve humanity the basic corner stones are water structures. Additionally, numerous studies are available against climate change impact mitigation and adaptation studies. This paper concentrates on the human activities, i.e., anthropogenic climate change direct effect possibilities on eater structures such as surface and underground dams, culverts, groundwater resources and wells, channels or canals, pipes and drainage basins.

References

1. Arnell, N.W.: Climate change and global water resources. Glob. Environ. Change **9**, 31–49 (1999)
2. Döll, P.: Impact of climate change and variability on irrigation requirements: a global perspective. Clim. Change **54**, 269–293 (2002)
3. IPCC: Impacts: adaptation and vulnerability. In: McCarthy, J.J., Canziani, O.F., Leary, N.A., Dokken, D.J., White, K.S. (eds.) Contribution of Working Group II to the Third Assessment Report of the Intergovernmental Panel on Climate Change. Cambridge University Press, Cambridge (2001). 1000 pp
4. IPCC: Impacts, adaptation and vulnerability. In: Parry, M.O., Canziani, O., Palutikof, J., van der Linden, P., Hanson, C. (eds.) Contribution of Working Group II to the Fourth Assessment Report of the IPCC. Cambridge University Press, Cambridge (2007). 976 pp

5. IPCC: The physical science basis'. In: Contribution of Working Group I to the Fifth Assessment Report of the Intergovernmental Panel on Climate Change. Cambridge University Press, Cambridge (2013)
6. IPCC: Climate change 2014: impacts, adaptation, and vulnerability. Part B: regional aspects. In: Barros, V.R., et al. (eds.) Contribution of Working Group II to the Fifth Assessment Report of the Intergovernmental Panel on Climate Change. Cambridge University Press, Cambridge, New York (2014). 688 pp
7. Hitz, S., Smith, J.: Estimating global impacts from climate change. Glob. Environ. Change **14**, 201–218 (2004)
8. Arnell, N.W.: Climate change and global water resources: SRES emissions and socio-economic scenarios. Glob. Int. Change **14**, 31–52 (2004)
9. Vörösmarty, C.J., Green, P., Salisbury, J., Lammers, R.B.: Global water resources; vulnerability from climate change and population growth. Science **289**, 284–288 (2000)
10. Döll, P., Siebert, S.: A digital global map of irrigated areas. Int. Comm. Irrig. Drain. J. **49**, 55–66 (2000)
11. Alcamo, J., et al.: Global estimates of water withdrawals and availability under current and future business-as-usual conditions. Hydrol. Sci. J. **48**, 339–348 (2003)
12. Solomon, S., et al. (eds.): Climate Change 2007: The Physical Science Basis. Cambridge University Press, Cambridge (2007)
13. Gleick, P.H.: The World's Water. The Biennial Report on Freshwater Resources. Island Press, Washington, D.C. (1998)
14. Gleick, P.H.: Water planning and management under climate change. Water Resour. Update, 25–32 (1998)
15. Kundziewicz, Z.W., Somlyody, L.: Climatic change impact on water resources in a systems perspective. Water Resour. Manage **11**, 407–435 (1997)
16. Kundzewicz, Z.W., et al.: The implications of projected climate change for freshwater resources and their management. Hydrol. Sci. J. **53**(1), 3–10 (2008)
17. Keller, A., Sakthivadivel, R., Seckler, D.: Water scarcity and the role of storage in development, p. 16. IWMI, Colombo (2000)
18. Jones, R.N.: Incorporating agency into climate change risk assessments. Clim. Change **67**, 13–36 (2004)
19. Şen, Z., Mohorji, A.M., Almazroui, M.: Engineering risk assessment on water structures under climate change effects. Arab. J. Geosci. **10**(23), 1–9 (2017). https://doi.org/10.1007/s12517-017-3275-7
20. Santer, B.D., Wigley, T.M.L., Barnett, T.P., Anyamba, E.: Detection of climate change. In: Houghton, J.T., Meira Filho, L.G., Callander, B.A., Harris, N., Kattenberg, A., Maskell, M. (eds.) Contribution of Working Group I to the Second Assessment Report on the Intergovernmental Panel on Climate Change, pp. 407–443. Cambridge University Press, New York (1996)
21. Arnell, N.W.: Effects of IPCC SRES* emissions scenarios on river runoff: a global perspective. Hydro. Earth Syst. Sci **7**(5), 619–641 (2003)

"Darwinism" and the Future of a Globalized World

Walter W. Kofler[1,2(✉)]

[1] I.M. Sechenov Moscow State Medical University (Sechenov University), Mokhovaya 11 bld 4, 103009 Moscow, Russia
walter.kofler@ias-he.com
[2] Medical University of Innsbruck, Innsbruck, Austria

Abstract. The given political situation is dominated on a loss of predictability and of the validity of signed contracts. Classic Darwinist can justify this with the evolutionary principle of natural selection that the fittest will/ should survive. But Darwin relativized this position for the understanding of the evolution of humans as social beings. He introduced "sympathy distinct from love". Both principles seem to exclude each another as mechanics and electromagnetism seemed to be incompatible. Einstein developed the technique of hypothetic-deductive theories of principle and could interlink both with the Relativity Theories. This technique is used to interlink both positions of Darwin. The joint basis of both evolutionary principles is surprisingly WINWIN. Its principles are demonstrated on the metaphor of chess. It is explained why also the genetically based evolution can be understood as a special case of WINWIN. Therefore evolution on the basis of "sympathy" is understandable as the next step of evolution with natural selection as the precursor principle. All the demands of precursor levels have to be taken further on in consideration. "Sympathy" is never sufficient to deal adequately with the challenges of the globalized world. Responsibility would be a possible solution.

Keywords: Evolution · Survival of the fittest · Sympathy distinct of love · WINWIN · Responsibility

1 Introduction

Our world is a ball - This is a fact. But how can we handle this with responsibility for global peace and welfare in ecology, economy and culture? The analysis of the starting point is disillusioning: The predictability of political processes is reducing especially because of the loss of the power of signed agreements. Given contracts seem negligible if the assumption is given to be able to make more national or personal win thanks to physical, economic or political power. The consequences on the global economy are not to oversee, especially for the "less powerful" economies. But the economists predict for the long term also a loss for all, even the "short term winner". Similar consequences are to expect in ecology and sociocultural short term effects. This demonstrates: Balanced agreements are in the interests of each single country - any responsibility for ecosystems

H. Gökçekuş and Y. Kassem (Eds.): NRSEM 2021, EESCI, pp. 11–17, 2022.
https://doi.org/10.1007/978-3-031-04375-8_2

and any social-cultural dynamic too. It makes capacities free for future oriented wins and creative alternatives. Individual wins are to expect just in short term calculation.

What could be feared if agreements would be recalled in the interest of unilateral protectionism which is relevant for the ecological or economic stability of the world? This could cause a loss of confidence in the predictability of all instruments which should prevent the individual against future risks. The effects for health are known since decades: The so called "sense of coherence" is accepted as one of the best confirmed factors for health promotion and salutogenesis [1].

World-Famous Darwin 1 and overlocked Darwin 2

Classic Darwinists can support the focus on the individual surplus: "I follow just Darwin's principle of evolution: The fittest should survive!". But was this really the position of Darwin? You can come to this conclusion if you read only his first main book: "On the Origin of Species by Means of Natural Selection, or the Preservation of Favoured Races in the Struggle for Life" [2]. But Darwin published 1871 a second fundamental work: "The Descent of Man" [3]. The most relevant part of this book - in my understanding - was overlocked: Darwin relativized his concept of natural selection based on self-oriented application of power for the evolutionary progress to humans as social beings: "No tribe could hold together if murder, robbery, treachery, etc. were common; consequently such crimes within the limits of the same tribe are branded with everlasting infamy. He proposed for the evolution of the primate to the person an additional evolutionary principle: The "sympathy - distinct from love: "A human who possessed no trace of such instincts would be an unnatural monster". Darwin predicted a further evolutionary process to an eco-socio-cultural-sustainable word: "As man advances in civilization, and small tribes are united into larger communities, the simplest reason would tell each individual that he ought to extend his social instincts and sympathies to all the members of the same nation, though personally unknown to him. This point being once reached, there is only an artificial barrier to prevent his sympathies extending to the men of all nations and races... Sympathy beyond the confines of man, that is, humanity to the lower animals, seems to be one of the latest moral acquisitions.... but to the humblest living creature".

2 Epistemological Incompatibility

But we have to see the different epistemological positions of the principle of evolution thanks to natural selection and thanks to the principle of sympathy: They seem to exclude each other - similar as Newton's theory of mechanics and Maxwell's theory of electromagnetism seemed to exclude each another.

Basics

Einstein developed a technique to link such indispensable but theoretically not conclusive frames (1949). He named it "theories of principles". It is known in philosophy as "hypothetic-deductive" [4]. Einstein concluded from the fact that electromagnetism and mechanics are two parts from the same scientific discipline (physics) that joint principles must "behind" both parts of physics. Otherwise they would not be parts of the same.

Therefore it was conclusive to invent an idea of a hypothetic universe "behind" the fundamental entities of the universe which were known 1905 and 1915: Only gravitation, electromagnetic fields, quantum and "solid bodies including the electron" were known. Therefore he hypothesized an universe without electromagnetic fields and without solid bodies but consisting of physical entities with the potential for the occurrence of electromagnetic fields and solid bodies: The energetical field. The attributed principles allowed him the conclusion of $E = mc^2$ - which fits as well to the formulas for mechanics as to the formulas of electromagnetism - if electromagnetic fields are handled "as they would have a mass equivalent to Planck's quantum multiplied with their frequency". Then he could neglect the energetical field - as a scaffold can be removed after finishing the house. So the Special Relativity Theory is indispensable for physics including cosmology. But the standard model of cosmology does not cover energetical field. The invention of the energetical field was only a "helping construction". The SRT does falsify neither "Newton" nor "Maxwell". Only their applicability is restricted on their classic problems.

Now it is stay of knowledge that electromagnetic fields are the ancestors of atoms and electrons. So it is conclusive to interpret Einstein's "behind" as "evolutionary earlier".

Application to Link Health Aspects and Evolution

Kofler used the hypothetic-deductive technique to develop such a theory of principles for medicine and other health related scientific disciplines [e.g. 5]. The applicability should be demonstrated on the integration of "Darwin 1" (the evolutionary model on the basis of natural selection) and of "Darwin 2" (the principle for the next evolutionary step on the basis of "sympathy") within one extended view of bio-socio evolution. Darwin developed his theories just for permanent multicellular. He assumed an ancient precursor cell for animals and plants - similar as Einstein assumed the energetical field. But Darwin did not make a hypothetic proposal about its nature [2]. Darwin accepted the assumption of just one evolutionary process for inanimate and animates [6]. But he avoided operational statements about that: "What manner the mental powers were first developed in the lowest organisms, is as hopeless an enquiry as how life itself first originated. These are problems of the distant future, if they are even to be solved by man." and: "It is mere rubbish thinking, at present, of origin of life; one might as well think of origin of matter [7].

Starting point for the attempt to link "Darwin 1" and "Darwin 2" is the hypothetic assumption of a world without as well multicellular which guide their activities to survive thanks to the use of natural selection as without of humans which use sympathy as tool for an evolutionary progress. Both proposals of Darwin can be integrated into an extended view of biological and socio-cultural evolution thanks to the WINWIN-concept. This means that the prerequisite for one's own gain is that the other also gains.

Chess – A Metaphor for a Comprehensive Understanding of Evolution

The model can be explained with the metaphor of chess [8]. This game is based on agreements about the attribution of meaning to structures: Such agreements restrict the freedom of the users: One type of agreement deals with the restrictions about the directly observable structures. It has to be accepted that the field is just 8 to 8 squares in black and white, and there are black and white characteristic figures: famers, horses, the king etc. The other type of agreement is not directly observable: The rules for the allowed processes. So a horse has to be moved two steps forward and one step to the side or vice versa. Therefore a chess-player can predict what kind of movements has to be expected. Why the players are willing to accept the logically not deducible rules? Because of the possibility of two types of win:

- To have the classic (evolutionary old) chance to win the game. But the looser is not losing his life: The falling of the king is just a symbol for the death.
- To have to emergent option for a new type of fun thanks to the unpredictable and individual use of the now possible options for individual creative movements.

But this is only possible if there is a partner who knows and accepts the consents. Therefore the rules must be distributed to others - better not only to family members. The prerequisite for the persistence of emergent win was to share the chance to win with others. The prerequisite is WINWIN.

The Power of the Metaphor

This explains fundamental aspects of the evolutionary progress:

The New Emergent Level Is Based Just on the Characteristics of the Precursors: Wood and stone can be used adequately to their nature for many applications. To use for figurers is just one. There would be an endless number of possible figures from wood or stone. But the joint agreement focus just on the six. These six are relevant only because of the attribution of meaning in agreement. The same is to observe in live: Only 20 amino acids are essential for live - 20 from nearly endless theoretically possible ones. There are aspects which can predict thanks to observation: the different structures obviously from the forms and the process consents conclusively from the systematic analysis of observations of plays.

There Are Aspects Which Are in Principle Hidden for the Outside Observer: In which way and why the individual player will use the allowed movements. Usually to win the game. But sometimes the father makes a "wrong" step: Maybe to let win the son. Then the son has pleasure and is willing to learn chess as source for pleasure for his life: So emergent new applications can be created for new evolutionary levels but again based just on the evolutionary precursor. There is no need on influence from "outside": The creators of chess were able to invent the game just on the basis of their creativity. This should focus the interest on two indispensable and independent steps for the evolutionary progress.

The Creation and Realization of the Emergent: Just two maharajas should have been created chess. So individual wishes, fears and options and the special environmental possibilities are the starting points. They created the rules not with the intention to start an evolutionary process: The personal surplus was the reason: To share the knowledge was the prerequisite to have interesting partners to play and to have fun. The worldwide distribution was also not based on the intention to push an evolutionary process: Again individuals were interested to have an individual surplus. So the idea is spread out unconsciously and with extreme speed in comparison to evolution on the basis of genes and reproduction. Dawkins has seen the need of an evolutionary principle beside the genes. He proposed the idea of Meme [8]. So the evolutionary process can be understood often as an unintended process thanks to intended actions.

3 The Compatibility with the Classic Evolutionary Process Based on Genes and Biological Reproduction

This seems to be in contradiction to the evolutionary process of multicellular. But the experiences with chess helps to understand the obvious differences as the expression of a very complex follow up of emergent steps just on the basis of the different ancestor levels.

From One Level to the Next Cause Restrictions. The emergent new has to be based on the options which are given by the previous evolutionary level and therefore restricted on the compatibility of the intentions of the precursors. The options are also restricted by influences from the given environments: The applicability of wood and stone had to be respected for the figures of chess and their possible movements. This has to be respected even in the case of the father who is interested to support the development of the personality of the son thanks to stimulating the son to like chess. The restrictions of the precursors and the environments are easy to understand and not relevant in the case of father and son. But this is changing with the number of levels of precursors which have to be integrated.

Survival Despite Massive Limitation Leads to "Machine-Like Processes". As higher the related living being is as longer is the cascade of precursors with their specific intentions and prerequisites up to the level of the single fertilized cell with its rules thanks to the genetic code. The high evoluted individual has neither real information about the meaning of the own genes and the genes of potential sexual partners nor from the demands of the levels which are based on the level of the single cellular. The individuals can only estimate about that from obvious characteristics. Therefore the individual influence is restricted to the selection of sexual partners according to better genes. And the genetic code is based on immaterial structures which can be influenced physically and chemically. Not a surprise that the emergence of new needs so long time in high animals but is so quick e.g. in virus and microbes.

4 Speculations for the Future: From Darwin's Sympathy to Responsibility Reproduction

The proposed extended view of the biological and social evolution thanks to combining "Darwin 1 and Darwin 2" allows additional assumptions of possible further evolutionary steps "after Darwin2". Darwin 2 deals just with the evolutionary process of humans which are interested on their individual surplus. "Sympathy" is not "altruism". It can be seen as close to game theory: Sympathy is a tool for individual, but not immediate surplus. But is this sufficient within the given situation? The knowledge is known since "Darwin" that evolutionary process is not predictable but permanent running. Same visionaries have recognized even in the 20th century that this process can be influenced especially thanks to activities of humans. But our generation is the first in which this knowledge is widespread. First political consequences have been to integrate ecosocial market economy. But this was focused only on sustainability with the intention that the next generations of humans will have also the needed resources.

We have learned that there is no more automatic feedback of processes which are never compatible with the further needed cascade of demands up to the stability of natural ecosystems. The pictures of Chinese farmers demonstrate this interdependency: They have to pollinate the blossoms of their apple trees with brushes because of the lack of bees. UN's Global Assessment Report 2018 is giving alarm that 1 million species are endangered in their persistence - and with them the human mankind [9].

What kind of positive effects could be expected in the 21st century if the principles of WINWIN would be applied just on logic arguments to extend "sympathy beyond the confines of man, that is, humanity to the lower animals, seems to be one of the latest moral acquisitions....and to the humblest living creature"- as Darwin predicted as ecological oriented visionary.

But the actual knowledge about the processes within a more and more globalized world makes it obvious: self-oriented sympathy even including to nature conservation is never enough. The activities of mankind influence very complex eco-socio-cultural nets. An option would be to use the principle of WINWIN as starting point for the next evolutionary step: From sympathy to responsibility. But the experiences of chess demonstrate: We should not expect a worldwide spreading out if this is based just on orders of the political, economic and cultural authorities. The solution could be the personal experience that the individual decision to act and to abstain from given options in a weighting valuation process according to the individual responsibility is an individual surplus, an instrument to be part of community and society in a dynamic cultural and ecological setting. This should experience as a resource for sense of coherence, additional to the options of "sympathy" and the needed "survival". There are same processes especially of so many young people all over the world which give hope. But the political situation is dominated actually from a loss on confidence even on the level of "Darwin 1". But is there really another option then responsibility?

References

1. Mittelmark, M.B., et al. (eds.): The Handbook of Salutogenesis. Springer, Cham (2017). https://doi.org/10.1007/978-3-319-04600-6
2. Darwin, C.: On the Origin of Species by Means of Natural Selection, or the Preservation of Favoured Races in the Struggle for Life. Murray, London (1859)
3. Darwin, Ch.: The Descent of Man, and Selection in Relation to Sex. Part 1 IV Moral Sense, pp. 112, 117, 122. Murray, London, Edition Appleton and Comp, New York (1871)
4. Einstein, A.: Autobiographical notes. In: Schilpp, J.A. (ed.) Einstein Philosopher - Scientist, Library of Living Philosophers, VII, pp 1–36. Open Court, La Salle (1949)
5. Kofler, W.: The relevance of Sechenov for the development of the theory of an "Extended view" of a human person as a social being. In: Sechenov Honor Lectures 2004, Russian Academy of Scienes, Moscow, pp. 3–68 (2005)

6. Darwin, Ch.: The Descent of Man, I, Chapter III, p. 66. Murray, London (1875)
7. Darwin, Ch.: Letter to J. D. Hooker. In: Burkhardt, F., Smith, S. (eds.) The Correspondence of Charles Darwin 1863, vol. 11, p. 278. Cambridge University Press, Cambridge (1999)
8. Dawkins, R.: The Selfish Gene. Oxford University Press, New York (1976)
9. Kofler, W.: Pfizer ends Alzheimer-research: an emergency signal for medicine and politics. Her. Int. Acad. Sci. Russ. Sect. **1**, 56–62 (2018)

Water Related Sectors and Risks in Adaptation to Climate Change

Mahnaz Gümrükçüoğlu Yiğit(✉)

Environmental Engineering Department, Sakarya University, Sakarya, Turkey
mahnaz@sakarya.edu.tr

Abstract. Global warming causes an increase in precipitation distribution, soil moisture, river and groundwater changes and evaporation, and has negative effects on sectors such as water supply-sanitation, ecosystems and biodiversity, agriculture, health, land use and forestry, urban settlements and infrastructure. Climate change projections show that the availability and accessibility of water resources will be directly affected. Turkey will primarily be affected by the decrease in water resources and drought, which will occur with the increase in temperature and decrease in precipitation. According to the pessimistic scenario, effects such as decrease in river flows and groundwater, increase in evaporation, decrease in soil moisture will occur. If the total water used in different sectors in Turkey is compared to the freshwater source, the water use index is approximately 26% and it is in the position of the country experiencing water stress. There is also a 10% decrease in precipitation and an increase in heavy precipitation in Sakarya province and the sectors that will be affected by these change are primarily water resources, agriculture, urban infrastructure, biodiversity, public health and industry. It is important and necessary to reduce the impact of climate change on water resources and improve the capacity to adapt to changes in order to strengthen water security, reduce disaster risk, and increase the resilience of ecosystems and economies. In this study, within the framework of adaptation studies to climate changes, sectoral effects, risks, fragility and opportunities were evaluated both in general and in Sakarya province, and local solution proposals were presented.

Keywords: Climate changes · Water resources · Risks · Climate changes adaptation · Turkey · Sakarya

1 Introduction

The global temperature has increased by ~1 °C over the past century and it is predicted that there will be an increase of between 2.5 and 5.0 °C by the end of this century; consequently, it is expected that there will be worldwide changes in climate patterns, such as an increase in the intensity and frequency of heat waves, changes in the water cycle and precipitation patterns, melting glaciers, an increase in the number of disastrous storms and floods, and an decrease in biodiversity [1]. The decrease in precipitation and the increase in temperatures, which will continue to increase according to future projections, is causing a decrease in water resources, an increase in evaporation, a decrease in soil

H. Gökçekuş and Y. Kassem (Eds.): NRSEM 2021, EESCI, pp. 18–27, 2022.
https://doi.org/10.1007/978-3-031-04375-8_3

moisture, and deterioration in water quality. These effects result in negative impacts on sectors such as agriculture, water resources, ecosystems, health, forestry, industry, and urban infrastructure.

The main risks related to the effects on water resources resulting from climate change are lack of access to safe drinking water and the resulting health effects, decreased agricultural productivity and rural income loss because of the scarcity of irrigation water, loss of terrestrial and inland water ecosystems, decrease in biodiversity along with ecosystem services, economic costs associated with floods and droughts, and higher costs in urban settlements and for industrial production [2]. In the present study, within the framework of adaptation studies on the effects of climate change, the affectability; risks and adaptation activities in the drinking and utility water; and effects on agricultural and industrial sectors, which are important sectors for our country and especially for our region, were assessed in both Sakarya province and Turkey, and suggestions for local and general solutions were drafted.

2 Water Resources of Turkey Under the Impact of Climate Change

According to different models and scenarios of Representative Concentration Pathway (RCP)4.5 and RCP8.5, temperatures will continue to increase in Turkey, which is located in the semi-arid climate zone, until the end of the century. It is predicted that this increase will be ~2 °C according to the most optimistic scenario and ~6 °C according to the most pessimistic scenario. By the end of the century, a 60-mm average decrease in total precipitation is expected according to both scenarios (RCP4.5 and RCP8.5). All scenarios predict that effects such as a decrease in river flows and groundwater, an increase in evaporation, and a decrease in soil moisture will occur. For example, in 2020, a decrease in precipitation of 11.5% was observed in the Aegean and Mediterranean regions compared with the average decrease over the past 30 years. Table 1 shows the temporal change in precipitation within the region and indicates that the amount of precipitation has decreased over time [3]. Moreover, increasing temperatures leads to less snowfall, smaller snow-covered areas, and an earlier snow melt. This situation increases water stress, especially in summer months. Effects such as the loss of surface waters, increase in summer temperatures, decrease in winter precipitation, frequency of droughts, deterioration of the soil, erosion on the coasts, overflow, and floods threaten the existence of our water resources [4].

Table 1. Regional Precipitation and differences of Turkey [3]

Regional precipitation 01 October 2019–30 September 202					
Region	Average precipitation 2019–2020 (mm)	Average precipitation 1981–2010 (mm)	Average precipitation 2018–2019 (mm)	Rate of change	
				According to 1981–2010	According to 2018–2019
Marmara	558.4	662.3	676.4	−15.7	−17.4
Aegean	508.1	592.2	663.3	−14.2	−23.4
Mediterranean	726.9	666.5	899.5	9.1	19.2
Central Anatolia	363.0	406.5	433.4	−10.7	16.2
Blacksea	633.9	696.5	710.5	−9.0	10.8
East Anatolia	541.7	558.3	656.1	−3.0	17.4
Southeast Anatolia	629.4	532.2	886.7	18.3	29.0

Even though a general decrease in precipitation is predicted, an increase in heavy precipitation activity is expected in the northeast part of the country. Because of this increase, river and lake levels will rise, flow rates of the stream will increase, and there will be flood risks. Increases in the number of floods could cause an interruption in the water supply and deteriorate water quality. In addition, there will be less water to infiltrate the soil, thus less groundwater recharge.

Evaporation will tend to increase, except in the eastern part of the country. As temperatures increase, evaporation in lakes and dams will increase, groundwater levels will decrease, and soil will become dry from evapotranspiration. In such cases, groundwater storage will be adversely affected, droughts will be common, agricultural and industrial production will decrease, and forests will be damaged. It has been observed that the water needs of some river basins have already exceeded the potential of the resources [5].

In addition to these negative effects of climate change, as the use of water in other sectors, especially agriculture and industry, increases, issues such as acquiring, storing, improving the quality of, conserving, and determining the amount of water use become more important [6]. With its available resources, Turkey is not a water-rich country on a global scale. The amount of usable water per capita in Turkey is ~1340 m^3/year, and it is estimated that it will become a water-poor country from the effects of climate change and population growth (Table 2). The amount of usable water per capita will decrease to 1069 m^3/year by 2050, as shown in Fig. 1. As of 2018, when 55 Bm^3 water used industrially is compared to the renewable freshwater resource, Turkey's water exploitation index was ~26%. According to this ratio, Turkey is a country experiencing water stress.

Table 2. Water usage amounts of Turkey [7, 8]

Turkey population 2021 (TÜİK)	83 614 377
Renewable water potential	234 billion m^3
Usable water availability of Turkey	112 billion m^3
Annually actively used water in Turkey	54 billion m^3
Gross water availability per person	2799 m^3
Available water for per person	1340 m^3
Actual water availability for per person	646 m^3

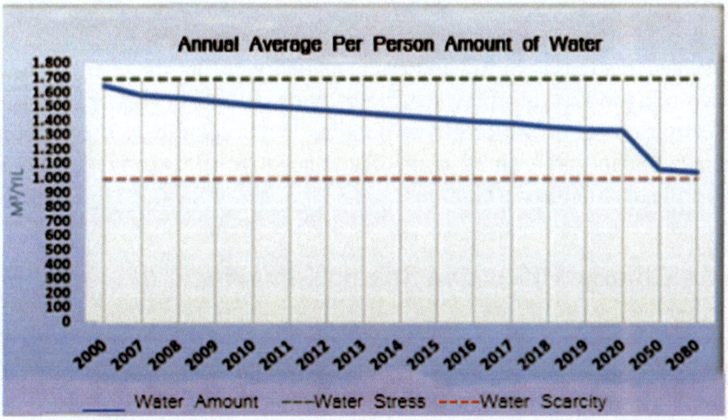

Fig. 1. Annual average amount of water for per person [9]

The agricultural sector is one that will be most affected by water scarcity in Turkey. Climate-related risks within this sector in addition to water scarcity can be listed as deterioration in soil structure, salinization, pests harmful to plants, increase in pathogens, and decreasing soil moisture. Within the agricultural sector, these risks will lead to problems such as an increase in irrigation need and demand because of high evapotranspiration, irrigation problems, a decrease in soil fertility from sedimentation, a decrease in harvests because of flooding, an increase in the effects of agricultural pests, and changes in planting and harvesting times [9]. The water used in irrigation is 74% of the total amount of water used throughout the country (Fig. 2). In addition, because surface irrigation methods are used in ~60% of these irrigations, the rate of water loss is high at ~35–60% [8]. This water loss adversely affects the sustainability of water resources while dealing with the effects of changing climatic conditions.

Water used for drinking constitutes 15% of the total use; the remaining 11% is for industrial use. Surface water supplies 48% for drinking, domestic, and industrial water and groundwater supplies the remaining 52% [9]. In addition, there is a 35% loss between the water drawn from the water sources and the water distributed by the municipalities.

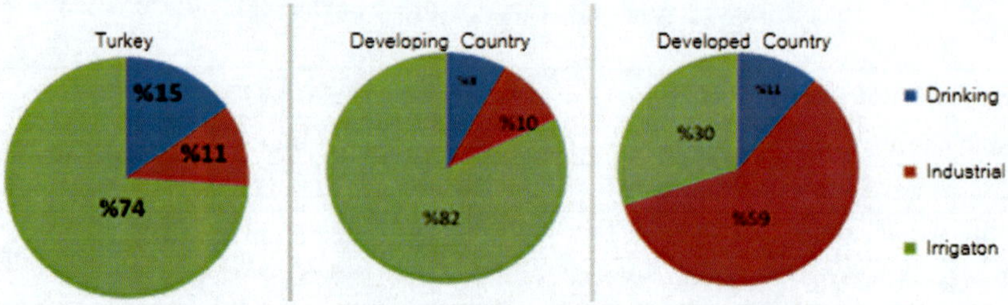

Fig. 2. Sectoral usage rates of water resources [9]

The possible effects of climate change on industry can be listed as a decrease in the efficiency of production processes, an increase in the costs of the maintenance activities of operations, a decrease in labor productivity, a decrease in infrastructure, a decrease in utility services, decreases in raw materials and service supply, changes in customer demand for certain products and services, interruptions in energy supply because of water scarcity, interruptions of operations from flooding and similar disasters, and mandatory changes in investments and business plans. These negative effects on industry will hinder investments and sustainable development.

3 Climate Change Effects on Sakarya Province

Sakarya Province has some of the most fertile agricultural lands in northwest Turkey. It has a mild climate because of it being within the Mediterranean and Black Sea climate transition zone. Although industrialization has been developing rapidly within the region, agriculture is still the most important sector. An increase of ~2 °C in temperature, an average of a 2% decrease in precipitation, and an increase in heavy precipitation are predicted because of climate changes within the region [6] (Fig. 3).

The most important risks posed by climate change in Sakarya are drought, overflows, and floods caused by heavy precipitation; severe weather events; and increases in sea water temperatures and sea levels. In particular, drought has a very adverse effect on fertile soils and the agricultural sector in general (Fig. 4). Recently, meteorological and hydrological drought and its effects have been experienced more frequently [10].

Water consumption rates by sectors in Sakarya differs slightly from Turkey's overall rate. In Sakarya, 80% of the water is used for irrigation, 12% for domestic use, and 8% for industrial use. According to the results of the climate change sectoral affectability analysis, the sectors that will be most affected are agriculture and livestock, water management, ecosystem services, public health, energy, industry, and urban infrastructure, with agriculture and livestock (46%) being the most affected sector, and water resources (42%) being the second most effected sector (Fig. 5) [9].

Industry in Sakarya is also developing. In addition to several huge industrial facilities within the province, there are eight organized industrial zones and eight small industrial sites [11]. Generally, groundwater is used in the industrial facilities. The decrease in industrial production resulting from water scarcity that may occur from climate changes may affect production at a national level. Within this context, it is important to determine

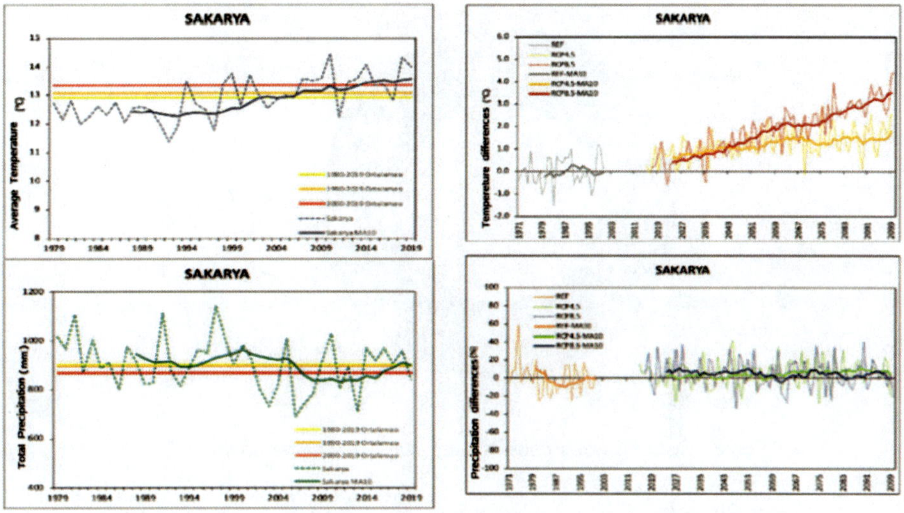

Fig. 3. Temperature and precipitation changes in Sakarya province [3]

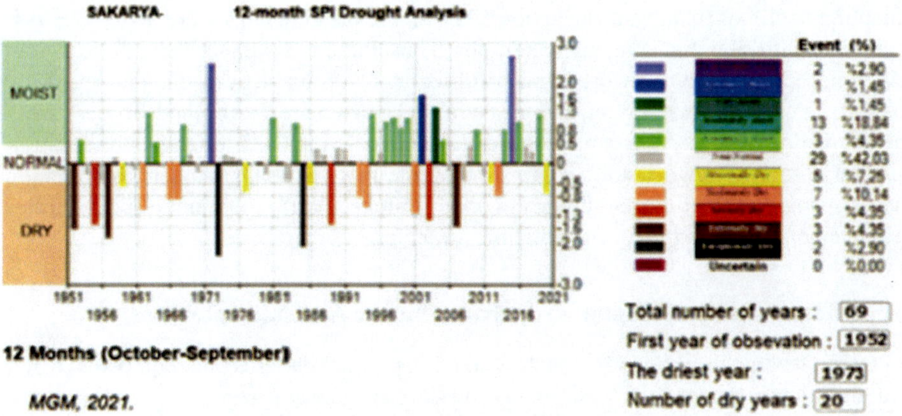

Fig. 4. Drought analysis of Sakarya [3]

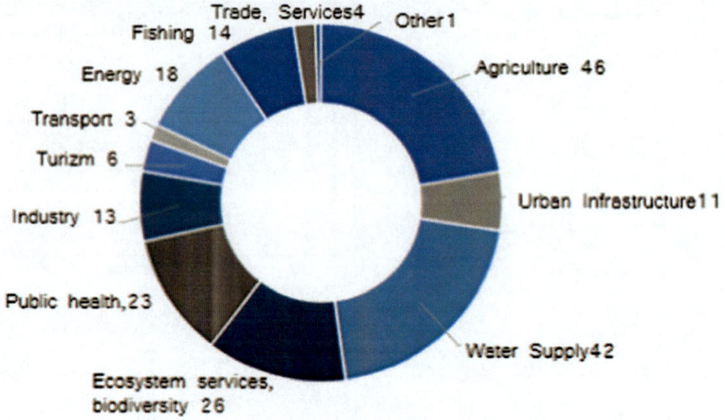

Fig. 5. Sectoral water consumption rates in Sakarya province [9]

the affectability and risks within the industry and implement activities that would help the sectors adapt to the changes.

According to the results of the needs analysis for adaptation to climate change in Sakarya, green conservation, protecting water resources, raising awareness, providing education, using sectoral measures, and helping institutions cooperate emerge as the priorities. For adaptation, first, water should be managed with the perspective of adapting to climate change and water resources should be managed in a holistic way.

4 Adaptation to Climate Change

Adapting to climate change is the action taken and realized to help societies and ecosystems cope with changing climatic conditions. Thus, there is assurance that public awareness is raised for adaptation works, and priorities are determined; everything will be ready for climate change in the region; and measures to be taken are determined. To assist with the adaptation process, adequate resource allocation is required to raise public awareness on climate change issues, compile available information, establish coordination mechanisms, explore different financing opportunities, clarify roles and responsibilities in implementation, and realize adaptation strategies.

4.1 Climate Change Adaptation Activities for the Agricultural Sector

Primarily, technical applications, such as creating product patterns that require less water and are more resistant to disease; using low-loss irrigation systems, such as drip irrigation or sprinkler irrigation methods; building closed system for irrigation channels; installing rain-harvesting systems; using purified and biologically safe water in irrigation; changing planting; and changing the harvest time, should be implemented. It has also been supported by preliminary studies that such technical adaptation activities can yield very positive results. For example, although 87% of the water is used for irrigation in the closed basin of Konya, plants that require quite a bit of water, such as corn, have been cultivated over the last 15 years. The water levels in the wells have dropped by up to 2 m per year. Some of the aquifers in the basin are at the point of depletion, which is why studies are conducted on alternative product production that requires less water. The results of the studies have indicated that there will be a 10% decrease in the irrigation water requirement per hectare when grain cultivation is increased by an average of 10% within the current crop pattern; therefore, even by adapting the product patterns to the climate, there can be a significant reduction in water use [12].

In addition to the technical applications, it important that nontechnical practices, such as promoting new production methods, reducing poverty and providing alternative jobs, conducting disaster risk management, analyzing the economic viability of agricultural policy under different climate scenarios, creating incentive programs for alternative crop cultivation, and disseminating trainings for farmers, are implemented immediately [13].

4.2 Climate Change Adaptation Activities for Drinking and Using Water Resources

Planning the actions for adaptation in the water sector is very detailed and intersects with other sectors. For example, a change in water resources affects sectors such as agriculture, health, energy, and infrastructure; therefore, it is very important that all the concerned institutions and organizations are in accordance with each other in managing water resources.

For adaptation, it is first possible to minimize the use of clean water resources by making savings in domestic uses, conscientious, and controlled consumption of the water allocated for business and raising awareness of society. The necessary technical and nontechnical adaptation activities are listed below.

- Reducing losses in water conveyance and distribution networks.
- Using gray water for domestic and/or industrial purposes.
- Harvesting rainwater.
- Using treated wastewater for agricultural purposes.
- Promoting technical equipment that saves water.
- Protecting water supply structures.
- Retarding salinization caused by the rise in sea level.
- Mapping regions where water resources exist.
- Managing sustainable groundwater resources.
- Taking precautions to reduce the effects of droughts and floods.

4.3 Climate Change Adaptation Activities for the Industrial Sector

To adapt to climate change in industrial production processes, it is very important and urgent that technical practices, such as implementing methods that will save water, developing rainwater collection and use systems, installing water connections for the cooling systems of power plants during dry periods, reusing wastewater by treating, and using gray water or treated seawater, should be applied. Adaptation practices should be supported with nontechnical works, such as ensuring the protection of facilities against heavy precipitation and floods, determining the additional energy required by climate change on a regional and sectoral basis, managing in-plant water, and preferring clean production technologies.

5 Conclusion and Suggestions

The priority activities are the risk analyses that enable the determination of adaptation activities to reduce the effects of climate change and enter a positive cycle, identifying the sectors and groups at risk that could be most affected, taking legal measures, increasing the capacity of institutions, and producing scientific data.

Because of its geographic position, Turkey is one of the countries that will be adversely affected by climate change. These effects are already being felt in Turkey, and this situation is expected to become more evident in the coming decades. Turkey will experience water stress; therefore, determining the sectoral effects of climate change and adaptation practices are very important. According to the results of the affectability analysis study on sectors, it is necessary to determine the adaptation activities, prioritize them, plan them on a sectoral basis, and complete the cost/benefit analyses of the activities; studies on this have already begun. It is necessary to increase the adaptation capacity and endurance at the national level, achieve the national targets, and determine the institutional capacity potential for the development of adaptation applications.

It is essential to raise public awareness about the risks with regard to conservation and the correct use of water resources within the Sakarya region, to use gray water in the urban areas instead of supplying new drinking water resources, install rainwater-harvesting systems urgently, decrease the cultivation areas of products with high water needs in agriculture, and encourage irrigation systems that consume less water. Raising public awareness and interinstitutional cooperation are required for planning and implementing these activities, which are the priority adaptation activities within the region.

Sustainable use of water resources is becoming a priority within the framework of the recent experiences with climate change and the predictions made. Under the effect of a changing climate, reducing consumption by sustainable water management should be a country's first priority, and it should be applied at the regional and even at the basin level. Reducing the impact of climate change on water resources is absolutely necessary to increase the durability of economies and ecosystems. Adaptation studies assist countries in improving water security and thus adapt to climate change, reducing disaster risks, and progressing toward current and future development goals by developing holistic approaches to water resources management.

References

1. IPCC Fifth Assessment Report (2013)
2. Gümrükçüoğlu Yiğit, M.: Urbanization and climate change. In: International Symposium of Urbanization and Environmental Problems, ISUEP 2018, Eskişehir, Turkey, pp. 73–80 (2018).
3. Turkish State Meteorological Service (MGM) (2021)
4. Ministry of Environment and Urbanization (2011)
5. Turan, E., Mollamahmutoğlu, A., Aydoğan, A.: Turkey's drought situation due to climate change. J. Nat. Disaster Environ. **4**(1), 63–69 (2018)

6. Kılıç, S.: Water management in the global climate change, Istanbul University. J. Polit. Sci. Fac. **39**, 161–186 (2011)
7. Turkish Statistical Institute (TUİK) (2021)
8. General Directorate of State Hydraulic Works (DSİ). Technical report (2021)
9. Turkey Ministry of Environment and Urbanization. Project of Climate Change Adaptation Action Report (2021)
10. Determination of Water Quality and Sediment Transport in Sapanca Lake Basin. Sakarya University, Interdiciplinary Research Project Report, Project No: 2012-01-04-016 (2014)
11. Sakarya Chamber of Commerce and Industry. Technical report (SATSO) (2020)
12. Arık, F.: Water Use and Sustainability in Konya Basin, Panel of BlackSea Environment Platform (2020)
13. Çakmak, B., Yıldırım, M., Aküzüm, T..: Water Policy Congress II, Agricultural Irrigation Management in Turkey, Problems and Solution Suggestiıons, Proceeding Book, Ankara, Turkey, pp. 215–223 (2005)

Project Transferring Water from Turkey to Cyprus Island

Necati Agiralioglu(✉)

Antalya Bilim University, Antalya, Turkey
necati.agiralioglu@antalya.edu.tr

Abstract. The island of Cyprus, like other Mediterranean islands, is located in a semi-arid climate zone. Since the annual rainfall in the region is low and it only rains in winter, drought and water shortages have been experienced throughout the history on the island and are still being experienced now. In this study, firstly, the climate and water resources of the island of Cyprus were evaluated. Then, some water resources projects carried out in Southern Cyprus are summarized. In addition, some investigations made to meet the increasing water demand were evaluated. On the other hand, in 2015, a water transfer project was carried out from Turkey to Northern Cyprus with pipes laid under the sea. After evaluating this water project according to today's conditions, the current water situation of Northern Cyprus was examined. Finally, the situation of meeting the agricultural water needed in the region and the climate change and the studies that need to be done for adaptation is emphasized.

Keywords: Water transfer · Water resources · Water development works · Climate · Cyprus · Turkey

1 Introduction

Some studies have been carried out for many years on the surface and underground water resources of the island of Cyprus [1–4]. In this study, some of the applied studies on this subject were evaluated and some suggestions were made regarding the existing water resources problems on the island.

The island of Cyprus is located in a semi-arid climate zone. Since the annual rainfall in the region is low and it only rains in winter, drought and water shortages have been experienced throughout the history on the island and are still being experienced now. In this study, firstly, the climate and water resources of the island of Cyprus were evaluated. Then, some water resources projects carried out in Southern Cyprus are summarized. In addition, some investigations made to meet the increasing water demand were evaluated. On the other hand, in 2015, a water transfer project was carried out from Turkey to Northern Cyprus with pipes laid under the sea. After evaluating this water project according to today's conditions, the current water situation of Northern Cyprus was examined.

The island of Cyprus is located in the Eastern Mediterranean. It is the third largest island of the Mediterranean with 35° north latitudes and 33° east longitudes, with a total

H. Gökçekuş and Y. Kassem (Eds.): NRSEM 2021, EESCI, pp. 28–40, 2022.
https://doi.org/10.1007/978-3-031-04375-8_4

area of 9251 km^2. Cyprus, the largest Mediterranean island after Sicily and Sardinia, is located in the most northeastern corner of the Mediterranean. The island is 71 km south of Turkey, 105 km west of Syria and 370 km north of Egypt. Its average width is between 56 km and 72 km.

2 Climate and Water Resources in Cyprus Island

2.1 Climate in Cyprus Island

Cyprus has an intense Mediterranean climate with a typical seasonal rhythm of temperature and precipitation. It has hot and dry summers from mid-May to mid-September and a rainy climate from November to mid-March.

Cyprus can be divided in four main topo-climatic regions as seen in Table 1.

1. The high altitude areas (500 to 1950 m amsl) of the Troodos mountain range that dominates the central part of the island.
2. The slopes of the Troodos Mountain range at altitudes of 200 to 500 m amsl.
3. The Mesaoria Plain (Mesarya Ovası) dominating the central eastern part of the island.
4. The coastal areas at 0 to 200 m elevation amsl, including also the Pentadactylos Mountain.

Table 1. Topographic regions in Cyprus.

No	Region	Elevation (m)	Area (%)	Annual precipitation (mm)	Annual evaporation (mm)	Aridity index	Classification
1	The high Altitude of Troodos Mountain	500–1950	18	400–700 at 500 m	1400–1700 at 500 m	0.54	Dry sub-humid
				1100 at the top	1000 at the top		
2	The slope of Troodos	200–500	27	300–500 at lower elevations	1600–1900 at lower elevations	0.30	Semi-arid
				400–700 at higher elevations	1400–1700 at higher elevations		
3	Mesaoria Plain	0–200	20	290–350	1650–1850	0.18	Arid
4	Coastal Areas	0–200	35	350–400 south-eastern and southern	1700–2000	0.23	Semi-arid
				450–500 in western and northern			

The overall average aridity index is 0.295, classifying the entire island as Semi-arid. Aridity index can be defined as the ratio between mean annual precipitation (P) and mean annual evapotranspiration- (ETP) calculated with the Penman formula. The aridity index defined as the ratio P/ETP has been proposed by the UNEP identifies areas prone to using five classes.

Nicosia is influenced by the local steppe climate. During the year there is little rainfall. This climate is considered to be BSh according to the Köppen-Geiger climate classification. In Nicosia, the average annual temperature is 20.0 °C. Precipitation here is about 364 mm per year. The Köppen Climate Classification divides the Earth's climate into five main climate groups: A (tropical), B (dry), C (temperate), D (continental), E (polar). These are subdivided by seasonal precipitation and heat.

2.2 Water Resources in Cyprus Island

According to the definitions of the Water Framework Directive, Cyprus has been identified as one River Basin District. Hydrographically the island is subdivided into 9 hydrological regions, including 70 watersheds and 387 sub watersheds.

Average annual precipitation on the island of Cyprus is about 500 mm, ranging from 300 mm in the central plain and southeastern parts of the island to 1100 mm at the top of the Troodos Mountains. The variation in precipitation is not only regional but also interannual, with two or even three years of consecutive droughts often observed. Evapotranspiration is high and accounts for about 80% of the annual average precipitation. In general, full irrigation is required from late spring to late autumn to maintain the production of water-demanding crops during this period.

Most of the streams originate from the Troodos mountains. The seasonal distribution of runoff shows minimum values in summer and maximum values in winter. As a result of the eastern Mediterranean climate's long, hot summers and low average annual precipitation, there are no streams with continuous flow all year round. Most streams flow 3 to 4 months a year and dry up the rest of the year. Some streams in the Troodos regions have a continuous flow only in their upstream parts.

There are 5 natural lakes on the island, which are only brackish or salty. The rest of the surface water bodies are supplied from the dams and ponds constructed.

For Cyprus Island, using the values of Table 2, the average runoff depth can be calculated as $508 \times 10^6/9249 \times 10^6 \times 1000 = 54.9$ mm. The average runoff coefficient can be found as 54.9 mm/478 mm = 0.115.

Table 2. Surface runoff for each hydrologic region.

Region		Catchment area (km^2)	Average rainfall (mm)	Surface runoff (hm^3)
No	Name			
1	Baf	1188	627	125
2	Erenköy	745	585	59
3	Güzelyurt	1585	429	96
4	Girne	455	490	16
5	Karpaz	685	463	22
6	Mesarya	1840	381	53
7	G.D. Mesarya	546	341	4
8	Larnaka	1050	439	39
9	Limasol	1155	555	96
Total		9249		508
Island-wide average			478	

As can be seen from Table 3, the aquifers of the island are fed by precipitation in the order of 273 hm^3. An additional feed of 141 hm^3 comes from runoff seeping into riverbed aquifers and coastal alluvial fans. Some of this additional replenishment is extracted from wells and boreholes, and the rest goes to the sea. Runoff of 8% or 40 hm^3 is diverted for irrigation in late winter or early spring and especially during the rainy

Table 3. Groundwater replenishment (surface runoff and directly from rainfall) and use.

Region		Ground water recharge (hm^3)		
No	Name	Streams	Direct	Total
1	Baf	20	46	66
2	Erenköy	7	23	30
3	Güzelyurt	42	30	72
4	Girne	9	19	28
5	Karpaz	3	26	29
6	Mesarya	41	47	88
7	G.D. Mesarya	0	11	11
8	Larnaka	10	34	44
9	Limasol	9	37	46
Total		141	273	414

season. Sea intrusions have occurred in some important coastal aquifers as a result of long-term overpumping.

3 Water Development Works in Southern Cyprus

3.1 General Remarks

It is widely accepted that in recent years there has been a revolution in the island's water supply industry. Since the last 70 years, the number of dams in Southern Cyprus has increased from 21 to 106. In addition, from 1962 to 2006 the number of large dams increased from 10 to 36. The total area corresponding to irrigated agriculture has increased from 1.600 ha to almost 21.000 ha.

60 years ago, most village water relied on communal sources. Currently, approximately 60 villages are served by centralized domestic water supply systems, either by utilizing local water sources or affiliated with the Cyprus Greater Hydraulic Works (Fig. 1).

Recently, the water economy has begun to shift towards the development of unconventional water resources. Desalination plants were established to eliminate the dependence of large urban and touristic centers on precipitation for drinking water supply. The government's water policy is not limited to desalination plants, but also focuses on the use of other unconventional water sources such as recycled water for irrigation.

Fig. 1. Major water works of Southern Cyprus.

3.2 Effects of Climate Change on Water Resources in Cyprus

Considering the current cultivation and irrigation practices, a decrease in groundwater level close to 1 m and further inland seawater intrusion in Larnaka aquifers are expected, while in Baf' aquifers, the predicted water table fluctuations are not significant. Additionally, SPEI values at the Asprokemos and Kouris dams are correlated with water storage measurements, showing that a SPEI downward trend observed in these reservoirs could set off an alarm to the water authorities with respect to water availability as more severe drought events are expected in the future. The expected pressure on surface waters imposes the need for an improved water management plan that will not depend on the further exploitation of groundwater [5].

A study provides a systematic literature review of smart agriculture technologies towards climate-smart agriculture in Cyprus, including robotics, Internet of Things, and remote sensing. This study starts with a review of climate-smart agriculture, presenting its importance in terms of optimizing agricultural production processes in order to address the interlinked challenges of food security and climate change. An extensive literature review of works published in the areas of robotics, Internet of Things, and remote sensing is undertaken, with particular attention paid to works carried out in relation to agriculture in Cyprus. This study analyzes aspects of the climate-smart agriculture research situation in Cypriot agriculture, identifies gaps, and concludes with new directions [6].

4 Transferring Water from Turkey to Cyprus Island

4.1 Introduction

Cyprus Island has 6 provinces as Lefkoşa, Magosa, Girne, Baf, Limasol, and Larnaka. Northern Cyprus has an area of 3355 km^2, which amounts to around a third of the island.

Northern Cyprus, like most other islands, has problems of fresh water shortage. Because of irregular rain distribution, rain pour down to the Mediterranean Sea directly without infiltration and due to climate change and high temperature, evaporation is increasing in water storage structures. To face the water shortage, groundwater is over-exploited causing a decline of the water table below the sea level. This made island water salty and so Northern Cyprus has one of the lowest domestic water qualities in the world.

4.2 Climate Condition and Natural Resources in Northern Cyprus

Climate Condition: Cyprus has an intense Mediterranean climate. It has hot dry summers from mid-May to mid-September and rainy winters from November to mid-March; separated by short autumn and spring seasons. Climograph of monthly averages climate data - temperature and precipitation in Northern Cyprus are shown in Fig. 2 during 1928–2017.

Months with the largest precipitation are December, January, February with 360 mm precipitation. Most precipitation occurs in December with an average precipitation 129 mm. As seen from Table 4, the annual amount of precipitation in Northern Cyprus is 734 mm. The average annual temperature is 24 °C in Northern Cyprus. The warmest month of the year is August, with an average temperature: 32 °C. Usually January is the coldest month in Northern Cyprus, with average temperature 15 °C. The difference between the hottest month: August and the coldest month: January is: 17 °C. The difference between the highest precipitation (December) and the lowest precipitation (August) is 119 mm.

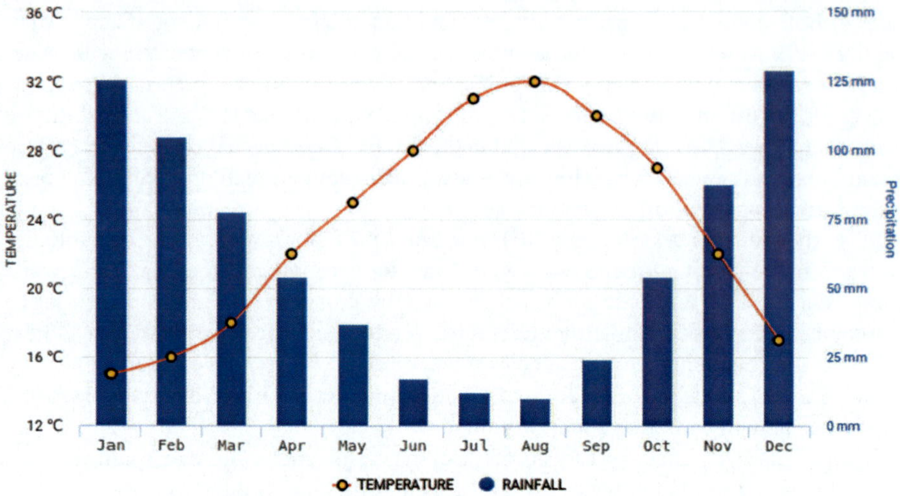

Fig. 2. Climate data graph for Northern Cyprus.

Table 4. Monthly climatic data for Northern Cyprus.

Variable	Jan	Feb	Mar	Apr	May	June	July	Aug	Sep	Oct	Nov	Dec	Annual average
Temperature (°C)	15	16	18	22	25	28	31	32	30	27	22	17	24
Precipitation (mm)	126	105	78	54	37	17	12	10	24	54	88	129	7 34

Climate conditions on the island vary geographically. Winter in the Northern part of Cyprus is cool and rainy. The short spring has unstable weather, occasional heavy storms and the "meltem", or westerly wind. Summer is hot and dry enough to turn low-lying lands on the island brown. Summer is followed by a short, turbulent autumn.

The narrow ridge of the Kyrenia range produces a relatively small rain of about 550 mm along its ridge. Autumn and winter rainfall, on which agriculture and water supply generally depend, is somewhat variable. The average rainfall for the year as a whole is about 480 mm. Rainfall in the warmer months contributes little or nothing to water resources and agriculture. The small amounts which fall are rapidly absorbed by the dry soil and soon evaporated due to high temperatures and low humidity. Statistical analysis of rainfall in Cyprus reveals a decreasing trend over the last 30 years. Snow occurs rarely in the lowlands and on the Kyrenia range. Rainfall is almost negligible but isolated thunderstorms sometimes occurs amounting to less than 5% of the total rain in an average year. In winter, the average rainfall from December to February is about 60% of the annual total. Climate change is another important factor which affect on water resources [7].

Land and Water Resources: Some studies were carried out on water resources potential in Northern Cyprus [8]. Norhern Cyprus has an area of 3,355 km^2, which amounts to around a third of the Cyprus Island. Northern Cyprus suffers from a chronic shortage of water. When 56.7% of the land is agricultural land, 45% of it is irrigable and 20% of it is irrigated. The country relies heavily on rain to provide household water, but in the past 30 years average yearly precipitation has decreased.

Drinking water is normally supplied to homes only every few days but everyone has water tanks to collect it in Northern Cyprus. Water delivery services drain wells in an unregulated way. Water quality is derogating since landfill areas are close to water resources and potable water blends into underground waters. So the quality of water in Northern Cyprus is constantly deteriorating and the water potential which is already limited is being reduced every day.

4.3 Preliminary Investigations for Project

It was considered the installation of a fresh water pipeline between the Turkish mainland and the northern coast of Cyprus. The pipeline diameter will be 1600 mm, and the length of the pipeline will be approximately 80 km. The pipeline material will be High Density Polyethylene (HDPE) as commonly applied for water transportation systems.

The preliminary engineering will develop the design to a sufficient level of detail to allow a decision to be taken regarding the overall project feasibility [9]. Detailed design will commence once feasibility has been confirmed. Preliminary investigations for geological, geotechnical, meteorological and geo-physical properties were conducted.

The proposed route for the pipeline is divided into five blocks as seen from Table 5. These are: 1. Landfall Turkey, 2. Nearshore Turkey, 3. Offshore 4. Nearshore Cyprus, and 5. Landfall Cyprus. The schematic profile of this Project is shown in Fig. 3.

Table 5. Design properties and values for five blocks.

Properties	Landfall Turkey	Nearshore Turkey	Offshore	Nearshore Cyprus	Landfall Cyprus
Water depth	0–20 m	20–250 m	250 m	20–250 m	0–20 m
Line length	1800 m	3600 m	70000 km	3000 m	600 m
Seawater density (kg/m^3)	1025–1029	1025–1029	1028.8–1029.2	1025–1029	1025–1029
Seawater salinity (%)	38.7–39.7	38.8–39.4	38.8–39.35	38.8–39.4	38.7–39.7
Seawater temperature (°C)	16–28	16–28	15–18	16–28	16–28
Biofouling	Trenched pipe	Up to 200 m wd 50 mm	None assumed below 200 m	Up to 200 m wd 50 mm	Trenched pipe

4.4 Northern Cyprus Water Supply Project

General Remarks: Previously, a lot of projects were developed such as convey water with balloons. But it is noticed that the most appropriate way to offer a long term solution is water transfer from Turkey to Northern Cyprus.

This project started to prepare on 27 May 1998. The Project was officially in March 2011 with the construction of Alaköprü Dam in Anamur and structures of the project are planned to finish in 2014, but it was completed in 2015 [2].

This giant project has structures of 3 main parts as Turkey Part Land Structures, Sea Crossing and Northern Cyprus Part Land Structures [10]. Within the scope of this project, water being supplied from Alaköprü Dam constructed in Turkey was passed through the sea and conveyed to Geçitköy Dam constructed in Northern Cyprus. The most critical points of the Project are the pipes carrying the water through the sea.

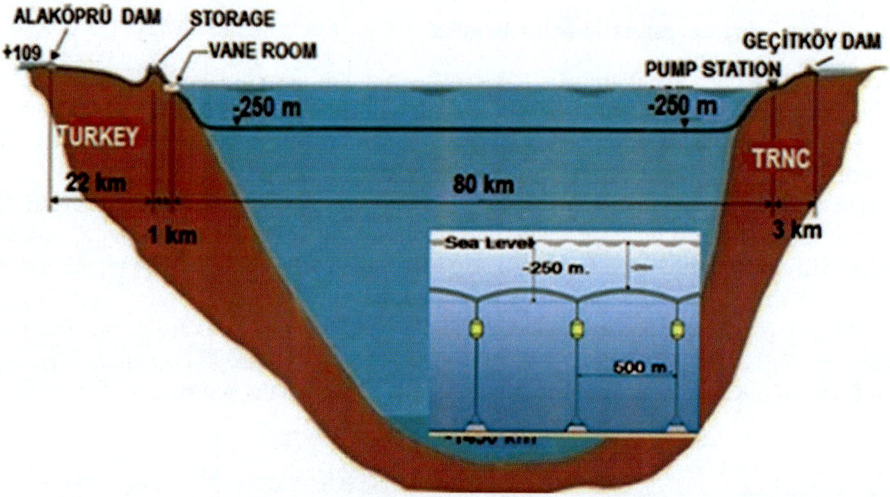

Fig. 3. Schematic profile of the project.

Turkey Part Land Structures of the Project: Turkey part of the project contains: 1. A storage: structure as Alaköprü Dam with water storage capacity is 130.50 million cubic meter. 2. Transmission line: 23 km length, 1500 mm pipe diameter, ductile cast pipe type, and pipe discharge 2.8 m^3/s. Anamuryum balancing room with water storage capacity is 10.000 cubic meter.

Two Battery limits are identified. 1. Physical battery limits 2. Battery limits for hydraulic analysis. Physical battery limit is assumed to be the wet section of the pipeline. This wet section will be between the two tie-in flanges, at Turkish landfall and Cypriot landfall, respectively.

Hydraulic battery limits include the whole system from balancing tank in Turkey to receiving basin in Cyprus.

In order to collect the transferred water, the Alaköprü dam was built on the Turkish side. The source of the water coming to the dam is the Dragon (Anamur) River and

Fig. 4. Water distribution systems in Northern Cyprus.

has an annual capacity of 750 million cubic meters. Approximately 1/10 or 75 million cubic meters of water is transferred to the island of Cyprus. The water coming from Alaköprü Dam is carried to Anamuryum pumping station by pipeline and enters the HDPE pressurized line going to the Mediterranean from there. Alaköprü Dam will also be used for a 26 MW hydroelectric power generation.

Sea Crossing Structures of the Project: Sea crossing of the project contains [11]: 1. High density polyethylene (HDPE) pipe. 2. Yoks (Y Part): Components used to connect pipes and anchorage system to the sea floor. 3. Flotation Tank: Permanent buoyancy providing component to improve the lateral stability of the system. 4. Rope: Cables used to connect the pipes to anchorage [12]. 5. Anchorage: Device provides a link between the suspension ropes and the sea bed.

A basic technical training project for the Cyprus Water Supply is the offshore crossing up to a depth of 1430 m. For the open sea passage, a suspended pipe system passing through the sea has been realized. In this arrangement, HDPE pipe is placed approximately 280 m from the sea surface, each approximately 500 m long [13]. The most important point of the water supply project is the "sea crossing". Producing this project is a world first for such a long-distance transition with a suspended pipe.

Northern Cyprus Part Land Structures of the Project: Northern Cyprus part of the Project contains: 1. Güzelyalı pumping station: 6.25 MVA installation power. 2. Transmission line: 3 km with Pipe diameter as 1400 mm pipe diameter, Ductile Cast pipe type, Pump discharge: 2.8 m^3/s. 3. Storage: Geçitköy Dam with 26.5 million cubic meter storage capacity. 4. Pumping sataion: Geçitköy pumping station with 20.63 MVA installation power. 5. Drinking Water Distribution lines in Northern Cyprus as 477 km (Fig. 4).

Geçitköy dam is storing the water required. Water is transported by transmission line from Güzelyalı Pumping Station to the Geçitköy dam at Northern Cyprus side. Then from Geçitköy Pumping Station water is separated to 4 main distribution lines for irrigation and domestic purposes. Total drinking water distribution line length is 477 km with 80- 1500 mm ductile cast, 3 reservoirs in 500, 1000 and 10.000 m^3 water storage capacity and 19 pumping station for 5 provinces, 20 towns, 141 villages. In project scope, drinking water treatment plant with the capacity of 200.000 cubic meters per day is also constructed, but lines for irrigation and drainage purposes.

5 Water Resources Potential in Norhern Cyprus

Cyprus, located in the Eastern Mediterranean, is divided into two as Northern and Southern Cyprus. Under average meteorological conditions, water is not sufficient in Northern Cyprus. For this purpose, water was transferred from Turkey to Cyprus by underwater pipe in 2015. With this water project, the domestic water needs of the region are met. A significant amount of water is needed for agricultural irrigation, especially in the Mesarya Plain. About the water on the island 75% is required for irrigation and 25% is required for home use [4]. Other critical issue in water management includes the overuse of groundwater, which accounts for about 85% of the water used in agriculture. The limited use of unconventional water resources, such as the reuse of wastewater, is another important problem to be addressed. Some areas that can be irrigated in agriculture may show low adaptation potential to climate variability. There is an urgent need to develop appropriate water governance frameworks that encourage the development of integrated water management plans. These, may require investment in research and innovation.

Any study for water resources management in Northern Cyprus may include: 1. Uneven water resources in Northern Cyprus; 2. Overuse of ground water; 3. Water losses in the region are very high; 4. Reuse of some wastewaters; 5. Use alternative water production, such as by desalination; 6. Inability to provide necessary and sufficient energy in the management of water; 7. Central and local governments should be interested with such water projects.

6 Conclusions and Recommendations

6.1 Conclusions

Climate change, repeated droughts in Cyprus and overuse of aquifers have reduced the amount of water available in the region. In addition, due to excessive water withdrawal from underground in the island, salt water intrusion occurs on the coasts. In this regard, the amount of usable water is limited in the region. As a conclusion, water resources are insufficient in the whole island of Cyprus.

In order to meet the water need in Cyprus, especially the drinking water need, water was transferred from Turkey to Cyprus by laying pipes under Mediterranean Sea. This project is a unique and complex project worldwide. This project is advertised as the Project of the Century. Turkey named this water as Water of Peace. If the Greek Cypriots want, Turkey will use this water and share Turkey's water with this project. With this project, only drinking water needs were met. New studies are needed for other agricultural irrigation needs.

In addition to water resources, energy and transportation needs also grow over time [14]. The island of Cyprus should be considered as a single basin and the solution of these problems should be evaluated as a whole.

Water was transferred to Cyprus from Turkey, which is the closest land to the island. This transferred water is used for drinking and utility water purposes in Northern Cyprus. However, there is an important and great water shortage on the island for agricultural irrigation. More water can be transferred from Turkey to Cyprus to contribute to the use of the island's water needs. Some of this water will be able to be distributed to Southern Cyprus. Electricity can also be transferred with the infrastructure system to be created in water transfer. Meanwhile, developing electrical energy such as solar and wind energies should also be planned. In addition, studies are needed to improve the transportation from Turkey to the island, especially the sea transportation.

It is clear that there is a need for more research, examination and investment for all these infrastructure works.

6.2 Recommendations

It is recommended to carry out the following studies in order to determine the existing underground and surface resources in Northern Cyprus and to estimate the drinking water and agricultural water needs: 1. Estimation of future population within 50 years. 2. Estimation of climate change within 50 years. 3. Estimation of annual average hydrological cycle components of normal, dry and wet years. 4. Estimation of drainage basin coverage of Northern Cyprus by different land use as agricultural and natural and artificial lands. 5. Determine the general hydrological data (area, precipitation, volume of precipitation, evapotranspiration, percolation, surface runoff, and theatrical water potential as annual average values of a normal year). 6. Prediction of withdrawals from surface and underground waters and overall water use (Agricultural, domestic and livestock). 7. Determination of total and irrigated agricultural areas. 8. Determine dams and their capacities in Northern Cyprus. 9. Water potential of major hydrogeological units of Northern Cyprus. 10. Current status of wastewater treatment in Northern Cyprus (for return use).

References

1. Zachariadis, T.: Residential water scarcity in Cyprus: impact of climate change and policy options. Water **2**(4), 788–814 (2020)
2. Ağıralioğlu, N.: Kıbrıs Su Projesi ve Bölgeye Etkileri, Kalyon Yayınevi, İstanbul, 478 p. (2016). (in Turkish)
3. Koloz, M.: An analysis of the water supply project within the context of water politics and neoliberalism: the case of the Northern Cyprus, Northern Cyprus. Degree of Master of Science in the Department of Political Science and International Relations (2019)
4. Tzanakakis, V.A., Angelakis, A.N., Paranychianakis, N.V., Dialynas, Y.G., Tchobanoglous, G.: Challenges and opportunities for sustainable management of water resources in the island of Crete, Greece. Water **12**(6), 1538 (2020)
5. Papadopoulou, M.P., et al.: Agricultural water vulnerability under climate change in Cyprus. Atmosphere **11**(6), 648 (2020)
6. Adamides, G.: A review of climate-smart agriculture applications in Cyprus. Atmosphere **11**, 898 (2020)
7. Angın, M., Çubukçuoğlu, B., Gökçekuş, H.: Case studies on the impacts of climate change on historical buildings in Northern Cyprus. Int. J. Built Environ. Sustain. **7**(1), 57–65 (2020)
8. Numan, T., Ağıralioğlu, N.: Long Term Prediction of Water Demand and Suggestions for Solution in Northern Cyprus. II. Technical Congress on the Developments in Civil Engineering, pp. 287–296. Bosphoros University, Istanbul (1995). (in Turkish)
9. INTEC Engineering BV, Cyprus Water Supply Project, Design Basis-Initial, 31016001-CWP-INT-Z-DBD-4110-Rev 0 4-2, June 2006
10. Ağıralioğlu, N., Mehr, A.D., Akdeğirmen, Ö., Taş, E.: Cyprus water supply project: features and outcomes. In: 13th International Conference on Advanced Civil Engineering, İzmir, Turkey, p. 7, 12–14 September 2018
11. Güngör, A.P.: International water transfer project: Nortern Cyprus Turkish Republic water supply project. World Irrigation Forum, Chiang Mai, Thailand W.2.4.04, 6–8 November 2016
12. Özturk, I., Ağıralioğlu, N., Ozdemir, Ö., Akinci, N.: Water supply from Turkey to Cyprus Island with suspended marine pipeline. In: International Conference on Civil Infrastructure and Construction (CIC 2020), Doha-Qatar, pp. 818–827, February 2020
13. Tas, E., Agiralioglu, N., Danandeh Mehr, A., Tur, R.: Energy loss investigation in submarine pipelines: case study of Cyprus water supply project. Adv. Res. Civ. Eng. **2**(2), 31–44 (2020)
14. Öner, H.: Economic feasibility assessment of solar powered seawater desaliation plants: unconventional freshwater supply for Güzelyurt, Northern Cyprus. Degree of Master of Science in the Sustainable Environment and Energy Systems Program (2019)

Opportunities and Challenges Facing the Future Development of International Environmental Law

Abbas Poorhashemi(✉)

Canadian Institute for International Law Expertise (CIFILE), Toronto, Canada
info@cifile.org

Abstract. This article aims to discuss the opportunities and challenges for developing international environmental law. It briefly describes "international environmental law" as a new branch of public international law. It also attempts to provide some knowledge on the sources and principles of international environmental law. The contemporary world's ecological problems are the most urgent and require an immediate collective response from the international community, including states. However, despite developing international environmental law, there have been different severe obstacles and challenges. One of the significant challenges is the States' unwillingness to delegate or limit their sovereignty to the benefit of environmental organizations. Furthermore, since environmental protection is a transboundary issue, a single country's effort is insufficient, and all countries must collaborate for this purpose. As a result, international cooperation is primordial to developing and implementing international environmental law.

Keywords: Environmental protection · International environmental law · Climate change · Biodiversity · Human rights

1 Introduction

This research aims to provide some essential elements of the formation and development of international environmental law and present some challenges facing the international community. As a new branch and discipline of public international law, International Environmental Law emerged basically in the 1970 decade. Since then, global environmental protection has become a significant concern and preoccupation of the international community. Climate change, air pollution, drinking water, wastes and hazardous Substances, soil protection, loss of biodiversity, noise and light pollution are the contemporary preoccupations of humanity. As a result, the international community, including States, has recognized that environmental protection must be addressed holistically and globally. In this context, the purpose of environmental law is to prevent or reduce environmental harm by imposing rules and regulations.

A. Poorhashemi—President.

H. Gökçekuş and Y. Kassem (Eds.): NRSEM 2021, EESCI, pp. 41–47, 2022.
https://doi.org/10.1007/978-3-031-04375-8_5

2 Emergence of International Environmental Law

For understanding the emergence and development of international environmental law, it is crucial to describe its three generations:

The first-generation Starts before the formation of international environmental law until the 1972 Stockholm Conference. This generation began mainly from the 18th century with the bilateral agreements in the field of fisheries, high seas and marine life resources. The majority of treaties were bilateral based on the common interest of states. For this reason, the governing principle was reciprocity of obligations, which is called *Sic Omnes* obligations.

Following industrial development and economic revolution in the western countries, environmental activists and scientists warned that if industrial progress continued the same way, the world would be destroyed. In this perspective, at the end of the 1960 decade, the states were forced to hold the Stockholm Conference of 1972 [1].

The second generation starts from the Stockholm Conference until the Rio Conference. Based on the recommendation of the United Nations General Assembly, the International Conference on the Protection of the Environment was held in June 1972 in Stockholm.

The Stockholm Declaration recognized the right of development to be closely linked to the environment by identifying the right to a healthy environment as a fundamental human right.

According to this document, environmental issues should be at the forefront of international cooperation [2].

This declaration marked the start of a dialogue between developed and developing countries on the link between economic growth, air pollution, water and oceans pollution, and people's well-being worldwide.

This generation of environmental law has recognized a fundamental right to the environment. According to the Principle First of the Stockholm Declaration:

"Man has the fundamental right to freedom, equality and adequate conditions of life, in an environment of a quality that permits a life of dignity and well-being, and he bears, a solemn responsibility to protect and improve the environment, for present and future generations" [3].

Another important outcome of the Stockholm Conference was establishing an international body called the United Nations Environment Program (UNEP) [4]. The UNEP is a United Nations-affiliated environmental protection program that has been one of the essential pillars of ecological protection globally.

States have negotiated and adopted several international treaties in this period, including conventions, protocols, and agreements.

These multilateral treaties are concluded by the active participation of non-state actors and non-governmental organizations. Most of the treaties were based on "treaty-laws," distinguished from "treaty-contracts" on the first generation.

Finally, the third generation of International Environmental Law begins with the 1992 Rio Conference on Environment and Development [5]. The United Nations Convention on Climate Change (UNFCC) and the Convention on Biological Diversity, drafted before the Rio Conference, were opened for states' signatures. The Rio conference established some regulations to achieve sustainable development and marked a new turning point in developing international law. In addition, the Rio Conference concluded with the adoption of three non-binding instruments, including the Conference Final Statement Agenda 21 and the Principles of Forest Conservation [6].

The Rio Declaration affirms that States should always be concerned with the environmental preoccupations in their economic development and industrial growth. Furthermore, States should consider sustainable development as an approach in their national plans.

Also, they should consider the "principle of sustainable development" in the national laws and international agreements. Rio Declaration also recognized the role of social groups and non-governmental organizations in protecting the national, regional, and global environment.

Another remarkable event in this period is the Agenda 2030 for Sustainable Development. This Agenda aims at a plan of action for sustainable development, strengthening universal peace, and eliminating poverty.

In this generation, the right to the environment and the right to development are recognized by international law. Therefore, the third period of the development process of international environmental law could be called the period of realism, universalism, and reforms.

3 Sources of International Environmental Law

International environmental law is based on the two types of non-binding and binding sources. Non-binding Sources or soft law is not obligatory and has no specific warranties. This source includes statements, resolutions, agendas, action plans, and the rules that guide governments. The primary purpose of soft law is to express the non-obligatory regulations and principles that guide governments, to protect the environment in a flexible manner [7].

Although these resources are not binding in themselves, they have a significant impact on the development of international environmental law. The most important examples of these sources are the Stockholm Declaration, the Universal Charter of Nature of 1982, the Rio Declaration and Agenda 2030. Again, the drafting and implementation of these action plans or statements allow governmental and non-governmental actors to cooperate together.

The second source of international environmental law is binding Law or Hard Law which imposes obligatory rules and regulations on governments to take global responsibility in environmental issues. It is essential to consider that the sources of international environmental law go beyond public international law. Although Article 38 (thirty-eight) of the Statute of the International Court of Justice (ICC) [8] mentioned the long-established sources of public international law, the other sources also apply to environmental law, such as *Jus cogens* rules, *Erga omnes*, and the obligatory resolution of international organizations [9].

4 Fundamental Principles of International Environmental Law

One of the crucial approaches to analysis international environmental law is the examination of its basic principles and concepts. This model can facilitate understanding many international environmental treaties [10].

Today, almost all international environmental instruments, binding or non-binding, contain or refer to these principles and concepts. State Sovereignty is one of the basic and fundamental principles of international law. However, the new development of international law clearly shows that the concept of sovereignty is not absolute. It is considered general duty not to cause environmental damage to t As stated in the Rio Declaration:

"*States have, under the Charter of the United Nations and the principles of international law, the sovereign right to exploit their resources according to their own environmental and developmental policies, and the responsibility to ensure that activities within their jurisdiction or control do not cause damage to the environment of other states or areas beyond the limits of national jurisdiction*" [5].

The principle of International cooperation is another principle, and it is an integral part of the United Nations Charter. This principle is also one of the features of contemporary international law, which is mandatory and takes the status of customary global commitment. Due to the universality and transboundary nature of the environment, states are obliged to cooperate in all circumstances and situations in good faith to protect the environment. Therefore, it is evident that the principle of cooperation is based on universal obligations and is rooted in customary international law.

Another principle of international law is the "Precautionary Principle." According to Article 3 of the UNFCC, the concept of this principle is where there are threats of severe damage lack of complete scientific certainty, which should not be used as a reason for postponing such measures. Under this rule, a state may be obligated to apply precautionary measures to prevent damage within its own or territory beyond its jurisdiction [11].

The principle of notification is another fundamental principle that can be traced back to the International Court of Justice, from the 1949 Corfu Strait case, and other international sources such as environmental treaties and agreements.

The principle of Sustainable development is another fundamental principle of international law. The 1987 Brundtland Commission presented the most well-known definition of sustainable development "*as meeting current needs without forgetting the needs of the next generation*" (Commission, 1987) [12]. The collective efforts of the international community to formulate and achieve sustainable development goals have continued since the 1990s. One of the essential documents for developing the Sustainable Development Goals was the Rio declaration of 1992. Then, the Millennium Development Goals (MDGs), adopted by the UN General Assembly in 2000, [13] is for the purpose to ensure environmental sustainability by encouraging states to address extreme poverty, achieving universal primary education, fight to hunger, promoting gender equality and empowering women, reducing child mortality, improving maternal health, diseases, and expanding global partnerships for development. However, the implementation of sustainable development confronted some challenges in its forms and contents. In terms of content, it lacks a comprehensive approach for including indigenous people, local cultures, good governance, consuming resources, producing goods and services, freedom of expression, employment, identifying the roots of poverty and gender discrimination.

Moreover, the Sustainable Development Goals (SDGs) are considered a universal call to action to protect the planet, end poverty, and improve lives and prospects for human beings worldwide. The majority of States members of the United Nations in 2015, as part of the 2030 Agenda for Sustainable Development, have adopted the 17 goals by which sets out a 15-year plan to achieve the goals [14].

There are also some other principles and concepts, such as the common heritage of humanity in international law and the rights of future generations.

5 Challenges Facing the Future Development of International Environmental Law

Despite all global efforts to develop binding and non-binding legal instruments, environmental concerns and threats such as global warming, climate change, desertification, and deforestation remain and increase in many regions and countries. Today's world's ecological problems and threats go beyond the predictions of the experts and scientists present at Stockholm's first environmental Conference in 1972.

Regardless of the concept of the sovereignty of states in international law, one of the main obstacles for the development of international environmental law is the non-acceptance of governments to delegate or limit their sovereignty in favour of environmental organizations.

Another fundamental principle of Public International Law is the principle of sovereignty. However, the concept of sovereignty is not absolute, and it is subject to a general duty not to cause environmental damage [6].

Another conflict also exists between developed and developing countries in enforcing regulations. For instance, according to the principle of Common but Differentiated Responsibilities, developed countries should take additional actions such as transferring technologies or contributing finance to developing countries, but this principle is not respected correctly. This principle is clearly set out in many international environmental treaties such as UNFCC. Based on this obligation, the developed countries are responsible for providing financial resources to assist developing countries in implementing climate change's international commitments.

Another substantive challenge to developing international law is the diversity of binding and non-binding sources, which is caused confusion and vagueness in the implementation of international law.

The Insufficient guarantee of implementation of international environmental law is another fundamental challenge of international environmental law. Many international rules and regulations, including multilateral treaties, are confronted with a deficiency of compliance mechanisms to protect the environment.

Institutional Challenges are also another gap in international environmental law. With the growing importance of global environmental protection since the 1972s, international governmental organizations have tried to play their role in ecological protection. As active subjects of international law, international organizations have a serious task to achieve. As is mentioned above, one of the essential reflections holding the Stockholm Conference of 1972 was establishing the United Nations Environment Program as the global and executive arm for international protection of the environment. However, the UNEP has limited capacity to protect the global environment. As a result, is it time to evaluate possibilities for creating a new organization for environmental protection like World Environmental Organization?

Furthermore, there are also new challenges on the planet, such as environmental immigrants, ecological purposes refugees, environmental confrontations between states such as water war, international arms conflicts, ecocide, and ecoterrorism, which have to discuss on another occasion.

6 Conclusion

The progressive development of international environmental law is based on qualitative and quantitative rules and principles, international regulations and standards, national legal systems, the practices of States and the promotion of soft law. However, the increased risks, threats, and significant environmental damage have raised concerns generalized for the law international environment's ineffectiveness in facing these challenges. In this context, the codification of multilateral treaties, the attitude of judges of the International Court of Justice, the reform of UNEP and the recognition of the role of non-governmental actors could be some solutions for the legal gaps and ineffectiveness of international environmental law.

The "internationalization of environmental protection" for the formulation and implementation of international rules and regulations is one of the future perspectives of international environmental law. In this approach, a global codification in this domain must be developed and adopted to define a relationship between sustainable development, human rights and the right to the environment. In addition, sovereign states must consider this interrelationship in all projects, programs, policies and political, economic, social and cultural activities in their territories. Through elsewhere, can we imagine that creating a possible international organization of the environment could remove the obstacles associated with the development of international law?

Finally, the critical approach to international environmental law can analyze the challenges and gaps. Reconstruction and modernization of international environmental law are based on forming a legal framework favorable to sustainable development according to the geographic, economic and cultural and political situation of States.

It is also crucial to recognize the right of public participation in the environmental decision-making process and implementation. Therefore, the "right to environment" should be considered as a human right and inserted in the concept of sustainable development.

References

1. Khalatbari, Y., Poorhashemi, A.: Environmental damages: challenges and opportunities in international environmental law. CIFILE J. Int. Law **1**, 21–28 (2019). https://doi.org/10.30489/CIFJ.2019.93906

2. Poorhashemi, A., Khoshmaneshzadeh, B., Soltanieh, M., et al.: Analyzing the individual and social rights condition of climate refugees from the international environmental law perspective. Int. J. Environ. Sci. Technol. **9**, 57–67 (2012). https://doi.org/10.1007/s13762-011-0017-3
3. Declaration of the United Nations Conference on the Human Environment, Stockholm, 16 June 1972. https://legal.un.org/avl/pdf/ha/dunche/dunche_ph_e.pdf
4. United Nations Environment Program (UNEP), United Nations, Nairobi, Kenya. https://www.unep.org/
5. Declaration on Environment and Development, Rio de Janeiro, 14 June 1992. https://legal.un.org/avl/pdf/ha/dunche/rio_ph_e.pdf
6. Poorhashemi, A.: Emergence of "International Environmental Law": as a new branch of International Public Law". CIFILE J. Int. Law **1**(2), 33–39 (2020). https://doi.org/10.30489/CIFJ.2020.218985.1013
7. Alexandre, K., Dinah, Sh.: Guide to International Environmental Law. Martinus Nijhoff Publishers, Boston (2007)
8. Statute of the International Court of Justice (1945). https://www.icj-cij.org/en/statute
9. Sand, P.H.: Accountability for the commons: reconsiderations. In: Westra, L., Gray, J., Gottwald, F.-T. (eds.) The Role of Integrity in the Governance of the Commons, pp. 3–21. Springer, Cham (2017). https://doi.org/10.1007/978-3-319-54392-5_1
10. Tayebi, S., Moosavi, F., Poorhashemi, A.: Interaction and cooperation for achievement of global peace. J. Polit. Law **9**(9) (2016). https://doi.org/10.5539/jpl.v9n9p150
11. United Nations Framework Convention on Climate Change, New York, United States. https://unfccc.int/
12. Jarvie, M.E.: Brundtland Report. Encyclopedia Britannica, 20 May 2016. https://www.britannica.com/topic/Brundtland-Report
13. United Nations Millennium Development Goals, Sustainable Development Goals (SDGs). https://www.un.org/millenniumgoals/
14. The 2030 Agenda for Sustainable Development, United Nations. https://sdgs.un.org/goals

Effects of Undervaluation of Ecosystem Services: Highlighting Cost of Water

Ömer Eker(✉)

Faculty of Forestry, Forest Engineering Department, Kahramanmaraş Sütçü İmam University, Avşar Campus, Onikişubat, Kahramanmaraş, Turkey
omereker@ksu.edu.tr

Abstract. Water scarcity has been increasing worldwide with the accelerated population growth rate. In the 20th Century, water use has risen up twice than the rate of population increase globally and caused water shortages with the effect of intensified climate change. By the end of the last century various researches underlined closing threats to the environment and specifically on ecosystems. Ecosystem services are essential for human beings in terms of providing clean water, aesthetic value, ecotourism activities, timber, non-wood forest products, pollination, flood and erosion control, carbon sequestration, climate regulation, recreation, and also cultural services. Among these provisioning, regulating, cultural, and supporting services water-related ones such as the creation of water cycle and wetland habitats, purification of water, production of drinking water, supporting biodiversity and wildlife habitats are vital for the continuity of the world. While estimating the cost of water, indirect use, option and non-use values are often neglected which causes the improper calculation of total economic value. Inaccurate estimation of water value also creates externalities that are not reflected in the total economic value, leading to excessive use of sources. Sustainable use of water requires rational use of natural resources without wasting, polluting, and unsustainably managing them. This study highlights the need to better understand the sustainable use of water resources, internalising water-related externalities, accurate valuation of water costs and also water management related policy issues. Furthermore, it draws attention to the drawbacks on the issue of considering water as a "free good".

Keywords: Water cost · Ecosystem services · Forests · Total economic value · Economics

1 Introduction

In recent decades, water scarcity has been one of the major problems due to the degradation of forests and other aquatic ecosystems. Around 1.2 billion people live in areas of physical scarcity, and 500 million people are approaching the same situation. Another 1.6 billion people, face an economic water shortage [1]. Water stress can change from one place to another, in some cases causing immense damage, comprising public health, economic growth, and global trade. It can also mobilize huge migrations and trigger conflict. Currently, pressure is rising on countries to implement more

H. Gökçekuş and Y. Kassem (Eds.): NRSEM 2021, EESCI, pp. 48–55, 2022.
https://doi.org/10.1007/978-3-031-04375-8_6

sustainable and innovative practices and to support international collaboration on water management [2].

Global problems arising in the axes of population growth, degradation of agricultural areas, shrinking of forest areas, the gradual extinction of plant and animal species, decline in water reserves, and the increase in temperature associated with the greenhouse gasses have reached an alarming level that put the future of humanity at serious risk. Moreover, these problems, which are embodied based on the deterioration, pollution, and destruction of natural resources, are more dangerous than expected, as they are not sufficiently understood and noticed. One of the most decisive dimensions of the negative process that is tried to be defined is the problems experienced in the direction of the protection of water resources and their proper use. In this context, the hydrological function of forests and their role in watershed management need to be well-understood regarding supplying sustainable water services for society and other public and private sectors.

Overlooking the functions of forests that are related to water production and improvement of water and air quality destroys the environment and life quality on earth. The fact that the social benefits of the goods and services produced by forest resources with the above functions are seen as "non-material benefits" is the belief of a common understanding that the benefits of these goods and services cannot be measured. However, many non-wood outputs that emerge with these functions of forest resources, such as water regulation and water quality, have physical measurable benefits.

This study underlines understanding the sustainable use of water resources, internalising forest-water related externalities, accurate valuation of water costs, and also water management related policy issues. Furthermore, it includes the drawbacks on the issue of considering water as a "free good".

2 Importance of Water Costs in Watersheds and Total Economic Value

Successful management of watersheds is possible by effectively analyzing the water function of the watersheds in economic, social, and environmental terms. The cost of water is mainly related to the institutions that make management and investment expenditures for the basin. Determining the cost of water and the amount of unit sales price applied to consumers will play a critical role in the provision of water production, transmission and distribution services to the private sector. Calculating the cost of water is a guide in obtaining compensation or price (contribution share) from those who use the water. This means taking a step towards the internalization of the deliberate externalities that apply to water. Internalization means that the production of water, which is a public good, is financed to a certain extent by its users. This price will also guarantee a fair distribution between those who directly benefit from water and those who indirectly benefit from it. To achieve this goal, the basin must be accepted as a water production system, and the whole expenditures made for this system must be considered.

As has been stated by Rizal et al. [3], the total economic value (TEV) of a resource can be categorized into two groups which are use value (UV) and nonuse value (NUV). Use value comprises direct use value (DUV), indirect use value (IUV), and option value

(OV). Nonuse value has shown to be both difficult to describe and measure. Nonuse value includes existence value (EV) which evaluates willingness to pay for some moral, altruistic, or other reason, and unconnected to use or option value, and bequest value (BV) which measure an individual's willingness to pay to ensure that his/her heirs will use the resource in the future (Fig. 1) [3]. TEV can be expressed by:

$$TEV = UV + NUV = (DUV + IUV + OV) + (EV + BV) \tag{1}$$

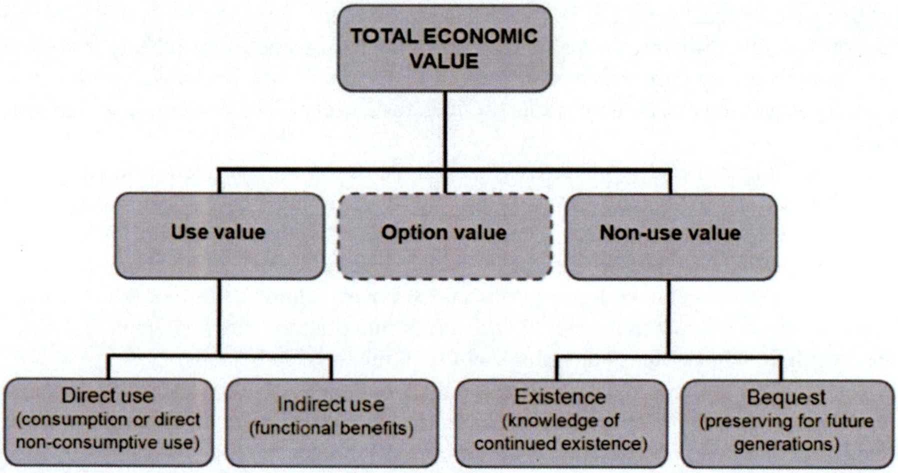

Fig. 1. The total economic value (TEV) framework for valuation of ecosystem services [4].

The value of water differs greatly according to factors such as the socio-economic characteristics of users, its presence in space and time, and the quality and reliability of supply [5].

3 Forest-Watershed Linkages

Watersheds with forests are highly stable hydrological systems compared to the other land uses [6]. Healthy forests:

- highly affect both the quantity and quality of water;
- discharge lower storm flow peaks and volumes for a given input of rainfall;
- weaken variation in stream flow during the year;
- stabilize soil and hinder gully and surface erosion;
- export the minimum level of sediment downstream
- reduce the treatment costs (Fig. 2).

Ecological services provided by forests and wetlands have been increasingly recognized in recent years. These services are necessary for economic development. However, their economic value is usually disregarded. This induces a threat on many areas

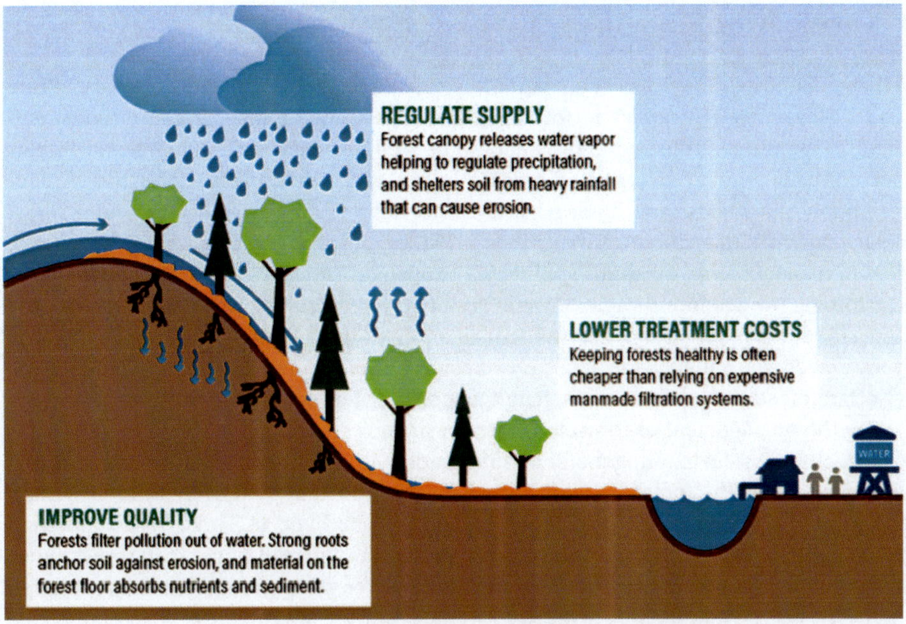

Fig. 2. Forest ecosystem services and water linkages [7].

as a result of poor management and unsustainable use, causing environmental degradation. When ecosystems are threatened, their ecological services are declined. This often creates high costs for the local communities who heavily depend on products in their surrounding environment. As a result of the growing awareness of both human and ecosystem connections between water and forests, various watershed management programs have targeted to protect watersheds by the protection of forests. This ecosystem approach is extremely beneficial to conventional infrastructure approaches, which often have negative effects on forest and wetland environments. One of the most significant watershed management programs was created in New York City in 1997; similar programs have been carried out all over the world. Tokyo and Sydney are the other two cities that have given priority to preserving the water quality by protecting their forests [8]. Therefore, understanding the wide range of values on water-related forest services is important in decision-making. However, more scientific studies are needed to support this. Monetary values are often better understood by decision-makers. That may help to better identify the importance of services, as well as reduce uncontrolled exploitation. A wide economic framework should ideally consider all the values systematically. Currently, existing economic valuation methods such as derived demand functions, market analysis, travel cost method, hedonic price method, contingent valuation method, and others, are useful, but many remain controversial because of the limitations of economic tools to accurately capture all the values and associated challenges [9].

4 Considering Water as a "Free Good"

In many countries, water has been extremely subsidized and often considered as a "free good". However, its scarcity is forcing authorities to determine its value more realistically. In contrast, the economic benefits of well-administrated or protected forests have not been counted in terms of avoided losses from soil erosion, debris flows, sedimentation and floods, and other hazardous negative effects [10]. Disregarding forest ecosystem services also causes inaccurate valuation of forest related outputs. In general, forestry management operations applied to the upstream are not taken into account when determining the economic value of water. To provide water from forest ecosystems continuously, some interventions such as control of forest cover density with the silvicultural operation, changes in tree species, the extension of the rotation length and management period, reforestation of open areas, maintenance of forest cover waterways are required. During the management of forests in water basins for water production, there are some costs such as reforestation, general administration, and research and development that were incurred by the forestry sector. However, repayment system of hydrological costs incurred by the forestry sector has not been established yet, in Turkey. This situation creates undervaluation of forest ecosystem services and loss of benefits in forest enterprises whose forest management plans are allocated for water production and regulation functions. Facing and dealing with water-related administration costs weaken the sustainable management of forest enterprises and financially non-support them in terms of making investments to improve the forest ecosystem services.

In Turkey, the valuation of forest ecosystem services within the framework of water production has been carried at by two studies in 2008 and 2020 so far [11, 12]. In the second study, the WACUM software was also coded and developed for the accurate calculation of water value in the forest watersheds by using the PYTHON programming language. Both studies used the cost method to calculate the expenses for each 1 m^3 of water incurred by the forestry sector. Both findings show that the forestry sector has an important contribution in terms of providing clean water to society and other sectors.

5 Sustainable Forests Management for Water and PES Approach

Currently, water security in the world is becoming one of the major problems due to the expansion of urban settlement areas, rapid population growth, land degradation, and climate change. Forest management can assure a nature-based solution [13]. As forests are the greatest terrestrial habitats for fauna and flora, forest-protected areas are extremely important in various conservation strategies. Watershed protection can also play a key role in helping biodiversity conservation strategies [14]. According to the projections, climate change and extreme weather events will considerably affect hydrology and water resources in the future. They will also cause catastrophic events such as floods and droughts and landslides. Experiences have proven that restoration and maintenance of degraded forest ecosystems can play a major protective role and minimize the effects of climate change. However, ecosystem services of forests have always been disregarded when developing water management plans and policies. In many countries, awareness of interrelated and interdependent linkages between forest and watershed management is still insufficient. Therefore, cooperation between the water and forest sectors is needed to address this issue in a consistent way [15].

Today, the arrangement between forest and water plans and policies are being established under the framework of PES (Payment for Ecosystem Services), in many countries. PES is a useful mechanism in terms of financing various sustainable development processes as well as sustainable forest and water management. [16] (Fig. 3). PES is also an environmentally effective, economically efficient, and socially equitable useful tool for accomplishing integrated water resources management (IWRM). The Kyoto Protocol and REDD+ have brought PES to the international agendas. Several mechanisms are available for PES. Some mechanisms are well established, such as where urban users pay for watershed protection in upland areas [17]. Kommet Program in Sweden is an important forest and habitat conservation PES approach that was started by Swedish Government in 2010. The objective of the program is to inspire forest landowners to protect their lands without converting to other land uses and inform them of which options exist for habitat protection. Agreements may last for between 1 to 50 years, depending on the site's significance. Landowners receive payments to compensate for limitations placed on their management in the interests of nature conservation. For habitat protection sites and nature reserves, owners receive full compensation plus an additional 25%. A payment for drinking water from forested watersheds (Switzerland) is another example that also highlights connections between water and forest management. 12% of the Canton of Basel-Stadtis covered with forests. 429 hectares of the area are dominated by broadleaved trees, of which 90 hectares are the property of 330 private forest owners. Nearly half of the drinking water for the canton is supplied from the Langen Erlen catchment area. In this area, water from the Rhine River is purified naturally and sustainably by forest stands. Among other good practices, this also required changes in species composition, such as replacing hybrid poplars, which have damaged the soil,

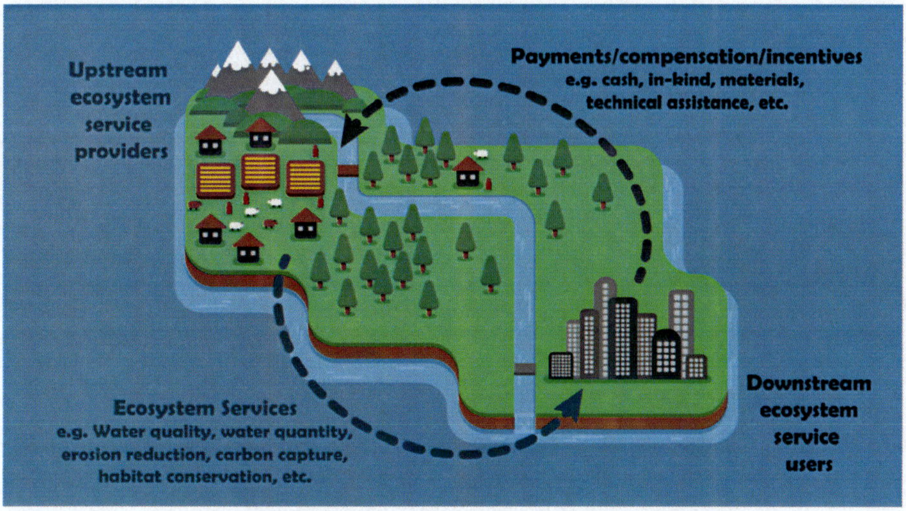

Fig. 3. Application of PES in upstream and downstream

with willows and *Prunus avium* (wild cherry tree). In addition, water users pay an extra charge in their water bill for the sustainable management of forests belonging to the city of Basel [18].

In the Catskill example, the New York City Department for Environmental Protection funds a Watershed Protection Program to supply high-quality drinking water for approximately nine million water consumers. Landowners in the Catskills example are paid to take measures that reduce diffuse pollution [19]. In short, PES schemes create markets for ecosystem services. Such markets must be understood not simply as simple economic exchanges between interested groups but rather as exchanges embedded in a particular socio-political framework to support the sustainable use of forest resources and also local livelihoods [20].

6 Conclusions

Forests make an important contribution to the sustainable management of water ecosystems and resources, while water is crucial for the sustainable management of forest ecosystems. Policy-makers and planners should be aware of the various interactions between forests and water. Undervaluation of forest ecosystem services creates various bottlenecks related to the use-protection balance of the source which causes degradation due to deficit financial support. Disregarding forest ecosystem services not only hinders rehabilitation and protection of the source, but it also declines the efficient management process. Considering water as a free good implies that forest ecosystems in water production basins are excluded from the water production process. This situation will lead to the fact that the nonuse value of forests within the scope of ecosystem services will not be reflected in the economy. Finally, ignoring forest ecosystem services will undermine the forestry sector's share in gross national product.

References

1. UNDP, Human Development Report 2006: Coping with water scarcity. Challenge of the twenty-first century. UN-Water, FAO (2007)
2. CFR, Council on Foreign Relations: https://www.cfr.org/backgrounder/water-stress-global-problem-thats-getting-worse
3. Rizal, A., Sahidin, A., Herawati, H.: Economic value estimation of mangrove ecosystems in Indonesia. Biodiversity Int J. **2**(1), 98–100 (2018). https://doi.org/10.15406/bij.2018.02.00051
4. Grant, S.M., Hill, S.L., Trathan, P.N., Murphy, E.J.: Ecosystem services of the Southern Ocean: trade-offs in decision making. Antarct. Sci. **25**(5), 603–617 (2013). https://doi.org/10.1017/S0954102013000308
5. Briscoe, J.: Water as an economic good: the idea and what it means in practice. Paper presented at the World Congress on Irrigation and Drainage, Cairo, September (1996)
6. FAO: Sassari Declaration on Integrated Watershed Management: Water for the Future (2003). www.fao.org/forestry/site/36420
7. WEF (World Economic Forum): https://www.weforum.org/agenda/

8. Blumenfeld, S., Lu, C., Christophersen, T., Coates, D.: Water, Wetlands and Forests. A Review of Ecological, Economic and Policy Linkages. Secretariat of the Convention on Biological Diversity and Secretariat of the Ramsar Convention on Wetlands, Montreal and Gland. CBD Technical Series No. 47 (2009)
9. UNECE: Forest and water: valuation and payments for forest ecosystem services (2018) ISBN: 978-92-1-117175-4
10. FAO: Sustainable use and management of freshwater resources: the role of forests. https://www.fao.org/3/y7581e/y7581e09.htm#TopOfPage
11. Eker, Ö.: Ormanların Su Üretim İşlevinin Ekonomik Analizi. KSÜ J. Sci. Eng. **11**(1), 74–81p (2008)
12. Görücü, Ö., Eker, Ö. ve Yalçınkaya, S. C.: Valuation of water as an ecosystem service and a case study from Turkey: Yamula Water Dam. In: 1st International Forestry and Nature Tourism Congress "New Approaches and Trends in Forestry", Kastamonu Turkey, pp. 46–50, November 25–27 2020
13. FAO, IUFRO and USDA: A Guide to Forest–Water Management. FAO Forestry Paper No. 185. Rome (2021). https://doi.org/10.4060/cb6473en
14. The World Bank: The Importance of Forest Protected Areas for Drinking Water, Running Poor (2003) ISBN 2-88085-262-5
15. MCPFE: Sustainable forest management and influences on water resources – coordinating policies on forests and water. In: Workshop on Forest and Water, Antalya, Turkey, 12–14 May 2009
16. Calder, I., Hofer, T., Vermont, S., Warren, P.: Towards a New Understanding of Forests and Water. Unasylva, vol. 58, no. 229. Rome, FAO (2007) http://www.fao.org/docrep/010/a1598e/a1598e02.htm
17. Fripp E.: Payments for Ecosystem Services (PES): A Practical Guide to Assessing the Feasibility of PES Projects. CIFOR, Bogor (2014)
18. Viszlai, I., Barredo, J.I., San-Miguel-Ayanz, J.: Payments for Forest ecosystem services – SWOT analysis and possibilities for implementation. EUR 28128 EN (2016). https://doi.org/10.2788/957929
19. DEFRA: Payments for ecosystem services: a best practice guide. Department for Environment, Food and Rural Affairs, 28 p (2013)
20. To, P.X., Dressler, W.H., Mahanty, S., et al.: The prospects for payment for ecosystem services (PES) in Vietnam: a look at three payment schemes. Hum Ecol **40**, 237–249 (2012). https://doi.org/10.1007/s10745-012-9480-9

Hydropower Outlook of Turkey in 2021

İbrahim Gürer(✉)

Engineering Faculty, Civil Engineering Department, Başkent University, Bağlıca, Ankara, Turkey

igurer@baskent.edu.tr

Abstract. Turkey is located partly on Europe and also on Asia, and has a surface area of 780 580 km^2 and a population of 84.6 million and electric energy is essential for the everyday life. The annual per capita consumption of electricity is mainly used to measure the prosperity of a nation.

The energy use in a country increases not only with population increase but also by improving the living standards of people and industrial development. Turkey being a non-oil-producing country at present, therefore needs to import fossil fuels (like natural gas and fuel oil) as the main source of electric energy. The utilization of fossil fuels for energy production by thermal power plants is one of the main sources of air pollution. The consumption of imported energy sources should be reduced for the environmental, economic, and political reasons. As the electricity need of the country is increasing continuously, more energy from the renewables and national resources must be produced.

Hydropower plants (HPPs) provide clean, fast, flexible electricity generation. The amount of electricity that a HPP can produce basically depends on the available flow rate (Q) and the head (H). Very briefly the greater the flow rate and the net head, the more electricity can be produced in a HPP.

The first hydropower station of Turkey had been opened at Tarsus town in 1929. Number of hydropower installations and total installed capacity has increased especially after the 1960s.

In 2020, the Turkish electric production values were 32.9% from HPPs, 27.2% from thermal plants working with natural gas, 21.3% from thermal plants working with coal, 8.8% from wind power plants, 6.8% from the sun, 1.6% from geothermal plants, and about 1.4% from other types of sources.

Technically and economically feasible total hydroelectric energy potential of Turkey is 180 billion kWh/year, of which 160 billion kWh/year has been developed and completed. With 714 completed facilities and 31,391 MW installed power and 108.0 billion kWh/year of it has been put into operation. With the projects to be established until 2023, a total installed power capacity of 40,000 MW and a generation potential of 135 billion kWh/year will be reached.

Because of the very fast increase in energy consumption, and only 22% of total energy is produced from national sources, and 125 billion dollars total investment is needed. Due to the climate change, and the change in the oil prices, the renewable sources such as solar, wind, geothermal, biogas, biomass, hydropower etc. should be developed and put in service as quickly and efficiently as possible.

In this paper, the current level of hydropower production and energy saving, and power consumption by different sectors, and the subject matter of the most recent energy figures are given and the future projections are also presented.

H. Gökçekuş and Y. Kassem (Eds.): NRSEM 2021, EESCI, pp. 56–65, 2022.
https://doi.org/10.1007/978-3-031-04375-8_7

Keywords: Turkey · Renewable energy sources · Hydropower · Energy demand · Energy saving

1 Introduction

Electricity is used as the common energy type in the modern life. The need of Turkey for electricity is growing at a rapid pace, therefore the planning, design and construction of new power plants and the energy transmission facilities are needed for the future. The main sources of electric energy are thermal, geothermal, nuclear, solar, hydrogen, wind and hydropower energies. The hydropower (being a renewable one) accounts for about 19% of global generating capacity (in 2020s), and about 20% of the energy produced each year.

"Water energy" is originated from the sun through the processes in hydrologic cycle. When the creeks, streams and rivers are formed on the earth's surface under the action of gravity, along their path, between the source (spring) and the end point (e.g., lake or sea level), streams have certain potential energy being proportional to the product of the "discharge" (Q) and the "elevation difference" (ΔZ) or "head" (H). If this energy is not harnessed, it is mainly dissipated in the nature because of the friction between flowing water and riverbed, also as formation of eddies and sediment transport in streams.

Hydropower plants (HPPs) can operate more rapidly and economically and therefore basically used to meet "peak loads" which occur daily, weekly, and seasonally. Large hydropower plants with a relatively high electrical production capacity due to big water volumes stored and high heads are very effective to meet peak demands. The large hydropower plants may also be used as the "base-load plants" when large fossil fuel and nuclear power plants are not present in the region to supply the base load.

The basic advantages of HPPs can be listed as relatively simple technology, high efficiency, long useful life, no thermal pollution of water and do not consume water. However, there are some disadvantages such as: the number of favorable sites is limited, cavitation and water hammer problem, high initial cost (especially for low head plants), loss of arable lands because of the floods of reservoir water and resettlement of local people, sedimentation in the reservoir entrance and bank erosion in the downstream fields.

2 The Sources of Energy in Turkey

The first hydropower station of Turkey had been opened at Tarsus town in 1929. DSI constructed its first HPP at Çukurova, Adana in 1956, and opened its number of hydropower installations and total installed capacity has increased especially after 1960's. In 2016, the Turkish electric production was as 33.9% from thermal plants working with coal; 32.1% from thermal plants working with natural gas; 24.8% (about 68 billion kWh) from HPPs; 5.7% from wind power plants; 1.8% from geothermal plants; and about 1.7% from other types of plants. At the end of 2016, the installed capacity of Turkey was 78497 MW with an annual energy production of 274 billion kWh. At the beginning of 2017, the "installed capacity" of Turkey was shared as: 35.4% HPPs; 29% thermal plants of natural gas; 22.1% thermal plants of coal; 6.1% wind power plants; 0.9% geothermal plants; and 7.4% other types of plants [1, 2]. In 2020, the electric production values were 32.9% from HPPs, 27.2% from thermal plants working with natural gas, 21.3% from thermal plants working with coal, 8.8% from wind power plants, 6.8% from sun, 1.6% from geothermal plants, and about 1.4% from other types of thermal sources.

Technically and economically feasible total hydroelectric energy potential of Turkey is 180 billion kWh/year, of which 160 billion kWh/year has been developed and completed. With 714 completed facilities and 31,391 MW installed power and 108.0 billion kWh/year of it has been put into operation. With the projects to be established until 2023, a total installed power capacity of 40,000 MW and a generation potential of 135 billion kWh/year is planned to will be reached [3].

The change in figures during the last five years, show that the use of non-renewable energy sources decreased about 17.5% and it was replaced by renewable energy sources, this helped to decrease the air pollution. The growth in energy markets slowed down during 2019, due to economic growth decrease at the same period. Renewable energy consumption share increased to 3/2 EJ globally increased. Hydropower share 'n Turkey raised to 0/3 EJ. For longer periods such as 1990 2019 energy demand increased 175.2% and during 2002–2019 period 87.3%. Every year the Energy Outlook's renewed by state and also by nongovernmental organization such as Mechanical Engineers Chamber (MMO) and Electrical Engineers Chamber (EMO) of Turkish Engineers and Architects' Chambers (TMMOB).

Hydrological inputs are essential for the preparation of a hydropower project. Hydrological data collection, estimation/assessment and data analysis are performed to establish a reliable flow quantity with its time variability and also the peak flood discharge at the project site. The longer the length of data more is the confidence on the reliability of the analysis. Generally hydrologic data set of at least 30 year-duration is considered adequate for a statistical analysis. Especially for the run-off-river plants hourly discharge data are very valuable. Unfortunately, for many Turkish streams there are either limited or no recorded data.

Since the "flow duration curve" (FDC) represents the variability of flow at a proposed hydropower site in a graphical way, it is useful for evaluation of available water for hydropower design and planning studies. For the feasibility study of a HPP at a particular site, the designer needs at least the average flow and available head information. Once the flow has been ascertained and the head measured, the available power can be estimated. Maximum capacity of the hydropower plant is decided on the basis of minimum quantity of available water with a storage reservoir constructed for the maximum quantity of water so that it can store the water even in the peak periods.

As the source of hydroelectric energy there are 25 river basins in Turkey (Table 1). The basin name, surface drainage area, volume of the water stored in the basin, estimated installed capacity, and average power production are given with 2021 figures. The cumulative values of Installed capacity and average energy production values for Turkey are also given in Table 1. Euphrates and Tigris rivers system is given separately, but they are interpreted as a single transboundary basin system.

Table 1. River Basin based distribution of installed power and energy production at operational stage by DSI [4].

Basin number	River basin name	Land area	Stored water	Installed capacity	Average power generation
		(km^2)	(hm^3)	(MW)	(GWh)
1	Meriç Ergene	14.486	1.657	0	0
2	Marmara	23.074	7.442	12	48
3	Susurluk	24.319	4.963	199	735
4	Kuzey Ege	9.861	1.985	3	8
5	Gediz	17.137	1.776	69	192
6	K. Menderes	6.963	624	0	0
7	B. Menderes	25.960	3.047	305	741
8	B. Akdeniz	20.956	6.499	431	1.439
9	Antalya	20.248	12.944	926	2.510
10	Burdur	6.294	234	0	0
11	Akarçay	7.995	375	0	0
12	Sakarya	63.303	6.487	651	1.834
13	B. Karadeniz	28.855	10.797	406	1.309
14	Yeşilırmak	39.595	7.046	1.798	6.495
15	Kızılırmak	82.181	7.004	2.205	6.797
16	Konya	49.930	2.407	2	3
17	D. Akdeniz	21.150	7.560	1.166	4.281
18	Seyhan	22.035	6.183	719	1.963
19	Asi	7.886	1.782	11	35
20	Ceyhan	21.391	7.734	2.091	6.580
21	Euphrates Sb	121.448	31.133	8.562	31.801
	Tigris Subbasin	54.695	25.183	2.497	5.959
22	East Karadeniz	22.846	16.427	2.447	8.203
23	Çoruh	20.248	6.981	2.373	7.505
24	Aras	27.775	4.480	340	1.135
25	Van	17.861	2.602	24	74
	TOTAL	778.492*	185.352	27.237	89.647

3 Hydropower Terminology

Water "discharge" is the rate of flow as m^3/sec through the hydropower plant. When turbine gates and valves are fully open it is called "Full gate discharge".

"Hydraulic head" is defined as the difference of elevations the water falls as passing through the hydropower plant (HPP. "Gross head" of a HPP is the difference between "headwater elevation" and "tailwater elevation".

A typical installation of HPP, where various losses in energy and various types of heads as "gross head", "net head", "total head", "velocity head", and the useful energy developed by the turbine are indicated in Fig. 1.

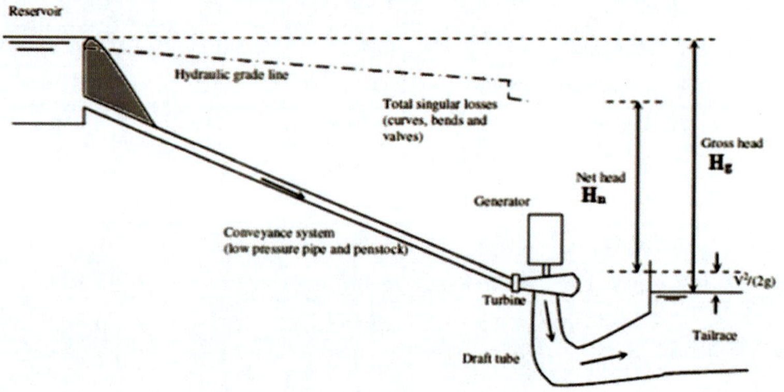

Fig. 1. Typical hydroelectric scheme [5].

As given in Fig. 1, the different types of heads are listed below:

Gross head, H_g, is the elevation difference between levels of head water and tail water, when no water is flowing.

Net head, H_n, is the head available for doing work on the turbine, that is, the difference between the total heads at inlet and outlet of a turbine ($H_n = H_g - h_l$).

Total head, H_t, at a given section, is the summation of velocity head, $V^2/2g$, pressure head, P/γ, and potential head, Z over a reference level.

Maximum head is the maximum steady net head acting on the turbine axis.

Minimum head is the minimum steady net head acting on the turbine axis.

"Net head" is equal to the gross head minus all the energy losses before entrance to the turbine and outlet losses (Fig. 2.) "Effective head" is the head at which the turbine In designing a turbine for the best speed and efficiency; "Rated head" is the lowest head at which the full-gate discharge of the turbine will produce the rated capacity of the generator. The turbine manufacturers give guarantee to costumers for "the rated net head" (effective head). The "critical head " is used at "turbine setting" and also cavitation.

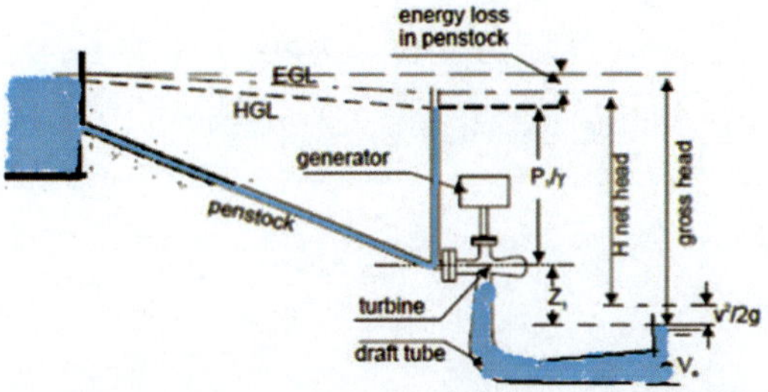

Fig. 2. Terminologies used in hydropower [6].

If there is a barrage (dam) in the stream (Fig. 3), then the velocity of water at section B is VB → 0 and friction loss from A to B is $h_f \rightarrow 0$, the "theoretically maximum utilizable power"; that is regained from the energy dissipated in overcoming frictional losses from section A to section B of the riverbed. This value becomes impossible when there is allowable flow from the reservoir at a diminished velocity VB (Fig. 2). Thus, h_f is directly proportional to V, n and L_o. There are three methods to reduce the bed resistance (h_f). Reducing V by constructing a dam; decreasing n by lining the canal; and reducing Lo along which between two sections, h_f represents the frictional head loss.

In big HPPs, the necessary slope of the flow is reduced by erecting dams which increase the flow depth and reduce the flow velocity. Figure 3 gives the idea about the reduction of bed resistance by this method.

If a certain section of the river is utilized by a chain of dams as in Coruh River Basin, the degree of utilization can rise to 50 or 60% [7].

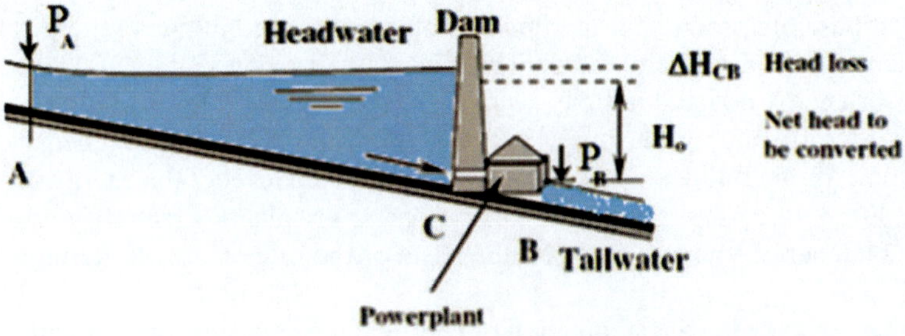

Fig. 3. Damming a stream and forming a reservoir [8].

In run-off-river type power plants, the frictional head loss "h_f" necessary to maintain the flow can also be reduced by weirs and diverting part or whole discharge of the river into an artificial channel (Fig. 4) with favorable flow conditions (for example, with smaller roughness coefficient; n and small bottom slope, S_0). The water level difference between artificial channel and the main river can be used for power production, as in the most of the run-off-river power plants located at Eastern Black Sea region.

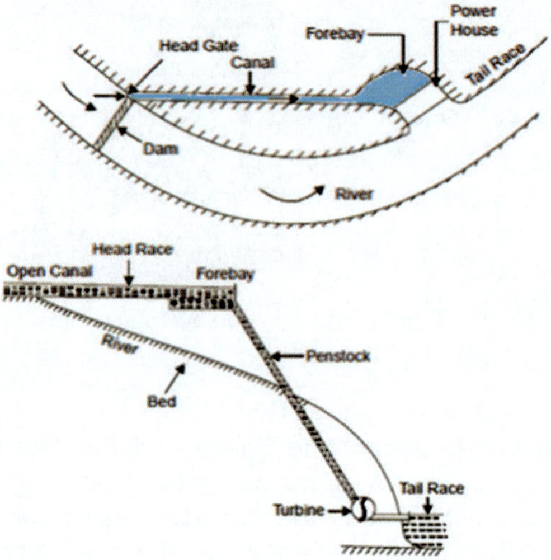

Fig. 4. Diversion of a part discharge of the river into an artificial channel [9].

4 Energy Demand

The annual energy demand of a country increases in time parallel to economic development and population growth. The demand of energy is not constant in time. It has hourly, daily, monthly, and yearly variations. It has peak values at an instant in a day, in a week, etc. (Figs. 5a and b). Energy use in time is closely related to the power requirement, called the load. The amount of total production of energy in "kilowatt-hours" divided by the total possible production in kilowatt-hours (expressed in percent) are called the annual plant factor or capacity factor. This factor is always less than 100% due to the variations of energy demand in time.

In planning of the power supply systems engineers should make energy-demand extrapolation for the future electric uses. After getting the level of the energy demand or future load they study the installation capacity of power plants which may use thermal or renewable energy sources. curve on the basis of the rate of growth during previous years.

Instantaneous load is the summation of energy consumptions for residential as well as industrial and commercial areas at a fixed time. The maximum load occurs at certain time when all energy requirements or certain of them make peaks. Figure 5a represents the daily load for a summer day in an average community. Figure 5b indicates the December peak–load curve for a community where domestic and lighting use of electricity is predominating. The "energy demand for the peak day in a year" determines the required generating capacity of the energy production system. For example, the energy requirements of the peak week or peak month dictate the amount of pondage and storage of water for HPPs.

The load factor is expressed as the ratio of the average load to the peak load of the given period of time. The area below the load curve represents the number of the "kilowatt-hours" required during a period of time.

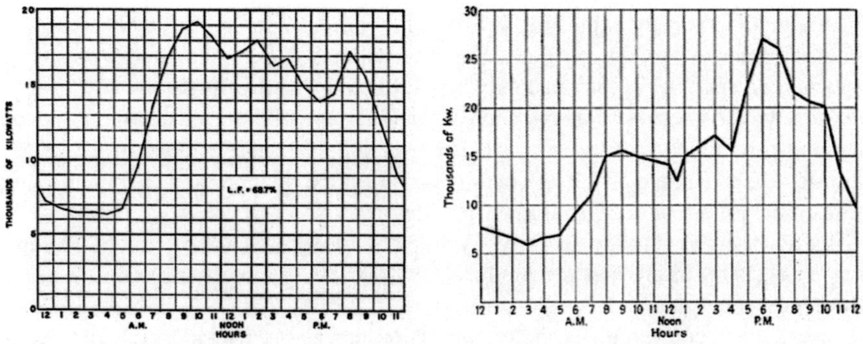

(a) (b)

Fig. 5. (**a**) Daily load for a summer day in an average community [10]. (**b**) Daily load for winter day in an average community [11]

4.1 Load Prediction

Demand for electricity is not constant and fluctuation in load generally occurs. The power required from a power system at any instant fluctuates depending upon whether the system is single or interconnected one. The capacity of hydropower plants is selected differently for single and interconnected systems.

Load Prediction for a Single System. In some special cases the demand may be greatly affected or even caused entirely by one particular industry (e.g., aluminum industry in Seydisehir, Konya) or isolated location (e.g., Bozcaada). The production would be such that the load (demand) is always to be supplied. The ideal case is the production of power at a rate exactly what is needed.

Load Prediction for Interconnected systems. All the power productions are connected together (by high voltage, high-capacity transmission lines) near the load center (e.g. Adapazarı in Turkey). Steam plants (thermal or nuclear power plants) run steadily (as base load units) while hydropower plants run and stop when it is required. For such national system there is only one market of which the requirements can be estimated easily with a high precision by considering the statistics of the previous years. In interconnected systems, the shape of the daily load curve does not change for many years.

5 Energy Outlook of Turkey In 2021

Especially in recent years, energy conservation and also decrease in energy loss of energy distribution network could help to improve the system, especially negative environmental impacts could be minimized and shift to renewable energy sources and effective transmission and distribution of energy were positive steps [12].

Next decade environmental considerations in energy sector will be more important, for Turkey because "Green Deal, is not only a climate change policy but also for EU and USA, a new industry policy, commerce policy and a new foreign policy frame. Since

EU is the most important export market of Turkey, in near future Turkey has to review its industry, energy, agriculture and commerce policies from green energy conversion perspective and prepare a new road map synchronized with EU follow it [13].

Recently “the security of energy supply” and increase of domestic oil and gas exploration and production from various oil and gas supply sources, more investment on infrastructure, and manage with less energy consumption by increased energy efficiency have been the pillars of the Turkish energy strategy.

To strengthen the energy security, Turkey has moved toward the liberalization of energy markets by improving the predictability and being more transparent in pricing [14].

In particular, renewable electricity generation has nearly tripled in the last decade, and its share in total power generation reached 44% in 2019 (including notable growth in distributed solar generation). Turkey’s share of renewable energy in electricity generation was 44% in 2018, higher than the median in the IEA; for hydropower, Turkey ranked the seventh-highest [15].

Under the Renewable Energy Support Mechanism (YEKDEM), Turkey offers feed-in tariffs for renewable power plants, including wind, solar, biomass, hydro and geothermal. The feed-in tariffs are currently set at USD 0.133 per kWh for solar and biomass, USD 0.105. From energy consumption point of view, in 2018, about 70% of total energy consumption in Turkish households was used for space heating and water heating [16].

Turkey’s total final electricity consumption was 258 TWh in 2018. The industry sector dominated with 44% of total demand, followed by services at 33% and residential consumption at 21%. The rest were minor shares used in energy industries and transport. Electricity consumption has steadily increased since 2000, and in the decade from 2008 to 2018, electricity demand grew by 60%. The most rapid growth has been in the services sector with an increase of 82%, industry with 58%, and residential use with 38% [17].

Turkey has a well-established transmission and distribution infrastructure, grid system by TEİAŞ. Also, through cross-border interconnections with Bulgaria and Greece, Turkey can import 600 MW of electricity and export 500 MW to Europe.

6 Conclusions

HPPs may have some negative effects on the environment, especially from ecological point of views. The environmental impacts of HP projects are closely related to how it affects river’s ecosystem and living habitat.

When there are some ecosystem changes at the project site as in the South Eastern Anatolia, a new pattern of biological activities and dynamic equilibrium emerge. Then, to fit new equilibrium, the plants, fish, and wildlife existing in the project area may change in quality and quantity.

The energy outlook of Turkey during 2021, hints the next decade will bring more energy expense to Turkish people. Especially the foreseen Climate change may duplicate even triplicate the cost that Turkey will pay, for the energy sources imported from abroad. Increase in internal tax may cause rather cold winters and too hot summers in coming years.

References

1. DSI: SU DÜNYASI, sayı 162, Ocak 2017 (2017)
2. Günyaktı, A., Gürer, I.: Applied Hydropower Engineering. Nobel Yayınevi, Ankara (2021)
3. DSI: Personal Communication with Dr. Murat Hatipoğlu, Deputy Head of Reconnaissance and Planning Department in DSI (2021)
4. DSI: Havza Plan Özetleri DSI Ankara (in Turkish) (2021a)
5. ESHA: Guide on How to Develop a Small Hydropower Plant. European Small Hydropower Association – ESHA, Belgium (2004)
6. Warnick, C.C.: Hydropower Engineering. Prentice Hall, Englewood cliffs, NJ (1984)
7. Garbrecht, G.: Hydropower. Course notes, METU, Civil Engineering Department, Hydraulics Lab. Public (1969)
8. Ramos, H.: Guideline for Design of Small Hydropower Plants. Western Regional Energy Agency & Network and Department of Economic Development, Belfast, North Ireland (2000)
9. Raja, A.K., Srivastava, A.P., Dwivedi, M.: Power Plant Engineering. New Age International, New Delhi, India (2006)
10. Doland, J.J.: Hydro Power Engineering. The Roland Press Company, New York
11. Creager, W.P., Justin, J.D.: Hydroelectric Handbook. Wiley, New York (1958)
12. TMMOB-MMO: Türkiye Enerji Görünümü (2021). https://enerji.mmo.org.tr
13. IKV: SABANCI IPM (IPC), TOBB TEPAV 9.3.2021
14. IEA. https://www.iea.org/reports/turkey-2021
15. IEA: IEA World Energy Statistics and Balances (database) (2020a). www.iea.org/statistics
16. IEA: Energy Efficiency Indicators 2020 (database) (2020b). www.iea.org/statistics
17. IEA: Energy Prices and Taxes (database) (2020c). www.iea.org/statistics

Economic Feasibility of Large-Scale Renewable Energy Projects in Mountain Location, Northern Cyprus

Youssef Kassem[1,2](✉), Hüseyin Gökçekuş[2], and Rifat Gökçekuş[2]

[1] Department of Mechanical Engineering, Engineering Faculty, Near East University, 99138 Nicosia (via Mersin 10, Turkey), Cyprus
yousseuf.kassem@neu.edu.tr, youssef.kassem1986@hotmail.com
[2] Department of Civil Engineering, Civil and Environmental Engineering Faculty, Near East University, 99138 Nicosia (via Mersin 10, Turkey), Cyprus
huseyin.gokcekus@neu.edu.tr, rifatgokcekus@hotmail.com

Abstract. Selvili-Tepe location has average wind speed of 5.2 m/s and solar horizontal irradiation of 5.46 kWh/m^2/day. Therefore, this paper investigated the economic validity of 5 MW grid-connected wind and solar energy with various models and sun-tracking system, respectively at Selvili-Tepe location in Northern Cyprus. For the economic validity of installing wind and PV systems, RETScreen Expert software was used. It is found that the cost of the production energy is within the range of 0.059–0.066 $/kWh and 0.087–0.121 $/kWh for wind and solar farms, respectively. Moreover, the results indicate that the developed wind farms can considered as an economic option for generating clean energy in the selected location. Consequently, using renewable energy systems will help reduce the dependency on fossil fuels, the effect of global warming, and enhance the country's sustainable technological development.

Keywords: Selvili-Tepe · RETScreen · Economic validity · Grid-connected · Northern Cyprus

1 Introduction

The high electricity consumption has motivated the researchers to find an alternative energy source to reduce energy and environmental impacts due to the use of fossil fuels [1, 2]. Several researchers concluded that using renewable energy like solar and wind energy as power sources helped to reduce greenhouse gas (GHG) emissions [3]. Moreover, several scientific researchers investigated the potential and the economic feasibility of renewable energy in various countries around the world [4–9]. For instance, Chauhan and Saini [4] investigated an integrated renewable energy system (biogas, biomass, wind, and solar) in village hamlets of Chamoli district of Uttarakhand state, India. It was found that the most suitable option for the selected location was a micro-hydropower-Biogas-Biomass-Wind-PV-array-Battery based configuration. Himri et al. [5] evaluated wind energy's potential and economic viability in the South-West region, Algeria. The

H. Gökçekuş and Y. Kassem (Eds.): NRSEM 2021, EESCI, pp. 66–71, 2022.
https://doi.org/10.1007/978-3-031-04375-8_8

results indicated that the project was economically viable with good energy production and capacity factor. Ahmed [6] investigated the techno-economic feasibility of wind farms in Sidi Barrani province, Egypt. The author concluded that a developed wind farm could produce approximately 988 GWh/year with an electricity production cost of 1.7 USC/kWh. Kassem et al. [7] evaluated solar and wind energy's economic viability and potential to meet the electricity demand for a small household in Northern Cyprus. It was found that the most suitable option for the selected location was a solar PV system. Enongene et al. [8] investigated the economic viability of small-scale PV systems in Nigeria. They found that the developed system could reduce GHG emissions (mainly CO_2) in the country.

Thus, the study aims to investigate the economic viability and environmental sustainability of grid-connected wind and PV systems in Selvili-Tepe location in Northern Cyprus. Numerous economic indices such as net present value, payback period, annual life cycle savings, internal rate of return, and the Levelized Cost of energy were estimated using RETScreen software.

2 Material and Methods

2.1 Study Area

Cyprus is the third largest island in the Mediterranean after Sicily and Sardinia, with an area of 9251 km^2, of which 1733 are covered with forests. It is located at the end of the southwestern part of the Mediterranean Sea, about 380 km from northern Egypt, 105 km from western Syria and 75 km from southern Turkey. The climate of Cyprus is a strong Mediterranean climate with full seasons and a large difference in temperature in the seasons, as well as with regard to rain and the weather in general. Selvili-Tepe, 35.320° N and 33.161° E, is located at Kyrenia region as shown in Fig. 1.

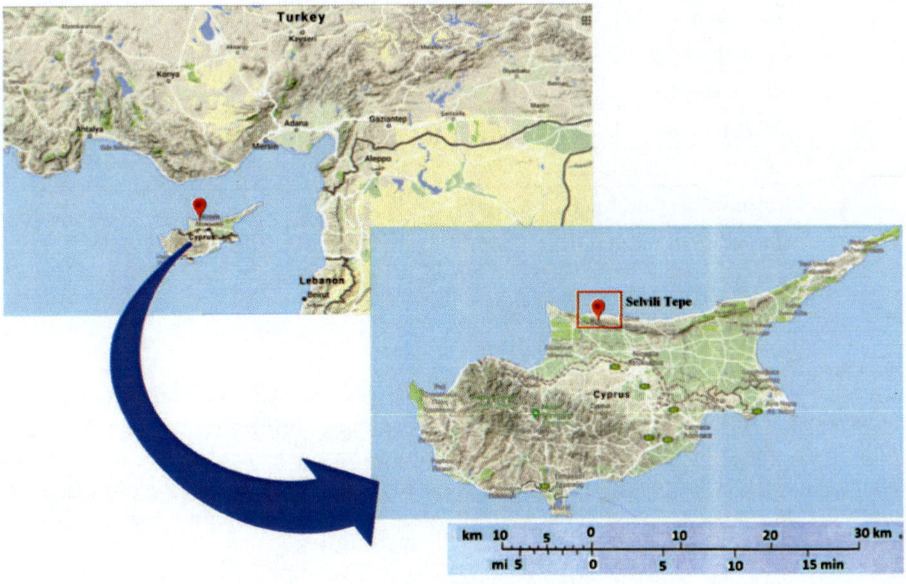

Fig. 1. Map of Cyprus.

2.2 Estimation of the Wind Turbine Output

For designing the wind farm, the average wind power density (WPD) is calculated to assess wind power potential. Eq. (1) is utilized to determine the WPD [10].

$$\frac{\overline{P}}{A} = \frac{1}{2}\rho\overline{v}^3 \tag{1}$$

$\overline{P}$ is mean wind power density in W, A is swept area in m^2, ρ is the air density in kg/m^3, $f(v)$ is the probability density function (PDF), and $\overline{v}$ is the mean wind speed in m/s.

Moreover, the power-law model is used to estimate the wind speed at various hub heights, and it can be expressed as Eq. (2) [10].

$$\frac{v}{v_{10}} = \left(\frac{z}{z_{10}}\right)^{\alpha} \tag{2}$$

where v is the wind speed at the wind turbine hub height z, v_{10} is the wind speed at the original height z_{10}, and α is the surface roughness coefficient (Eq. (3)) [10].

$$\alpha = \frac{0.37 - 0.088 ln(v_{10})}{1 - 0.088 ln(z_{10}/10)} \tag{3}$$

In this study, two horizontal axis wind turbines with various characteristics are selected in this study. Table 1 shows the specification of the selected wind turbines.

Table 1. Specification of selected wind turbine.

Characteristics	EWT DW61-900	Vestas V47
Hub height [m]	69	55
Rated power [kW]	900	660
Rotor diameter [m]	61	47
Sweep area [m^2]	2922.5	1735
Design life [years]	20	20
Cut-in wind speed [m/s]	2.5	4
Rated wind speed [m/s]	11.5	15
Cut-off wind speed [m/s]	25	25

2.3 Design a Solar Plant

In this work, grid-connected PV systems are developed for electricity generation for buildings in the selected location. In the literature, power generating factors, energy demand, solar PV energy required, PV module sizing, and inverter sizing are essential factors for designing a PV system [11]. The description of designing a solar plant is discussed in Ref. [12].

2.4 RETScreen Software

RETScreen has been widely utilized to investigate the feasibility of PV power plants. Natural Resources Canada (NRC) developed RETScreen software. It helps to estimate the energy production, capacity factor, emissions reduction and financial viability [12]. The software uses the long-term NASA monthly average meteorological data. Numerous economic indices, including net present value (NPV), Simple payback (SP), annual life cycle savings (ALCS), internal rate of return (IRR), Equity payback (EP) and the Levelized Cost of energy (LCOE) were estimated to measure the performance indicator for the proposed projects.

3 Results and Discussion

3.1 Wind Farm

In this study, Table 4 lists the financial parameters, which assumed based on the previous studies in the literature. Also, Table 4 summarized the value of the capacity factor (CF), energy production and economic indices. It is found that EWT DW54 has the highest generating electricity and CF compared to Vestas V47 due the cut-in wind speed (see Table 1). Moreover, it is observed that EWT DW54 has better economic indices compared to other turbine (Tables 2 and 3).

Table 2. Values and assumptions for the economics of wind energy.

Particular	Value
Wind turbine capacity [kW]	660/900
Number of turbines	8/9
Turbine hub height [m]	69/55
Wind farm capacity [MW]	5
A lifetime of the project [year]	20
Array losses [%]	3
Airfoil losses [%]	1
Miscellaneous losses [%]	2
Inflation rate [%]	7.2
Discount rate [%]	12
Electricity export escalation rate [%]	5
The initial cost of wind turbine [$/kW]	1300
Wind turbine cost share [%]	70
Grid connection cost share [%]	12
Civil work cost share [%]	9
Other capital cost share [%]	8
O & M cost [%]	1.5

Table 3. Performance of 5 MW wind farm projects.

Wind turbine model	CF [%]	Generated electricity [MWh/year]			Electricity export revenue [$]		
EWT DW54	28.5	13472			1347232		
Vestas V47	25.4	11769			1176911		
Economic performance							
Wind turbine model	*NPV* [$]	*EP* [year]	*SP* [year]	*LCOE* [$/kWh]	*B-C*	*ALCS* [$/year]	*IRR* [%]
EWT DW54	8763716	2.3	5.2	0.059	5.2	1173276	48.7
Vestas V47	7042439	2.7	5.9	0.066	4.4	942833	42

3.2 Solar Plant

In the present study, AS-M60-310W PV module and Sunny Central 850CP XT inverter have been selected for the proposed PV plant. The specification of the selected module and PV module can be found in Refs. [11, 12]. Financial metrics such as inflation rate (7.2%), discount rate (12), reinvestment rate (9%), debt ratio (70%), debt interest rate (15). electricity export escalation rate (5%) are used as input variables in this study.

Table 4. Performance of 5 MW PV power plants.

Tracking mode	CF [%]	Generated electricity [MWh/year]			Electricity export revenue [$]		
Fixed	19.0	8328			832796		
One-axis	25.3	11060			1106030		
Two-axis	26.5	11604			1160410		
Inclined axis	24.3	10623			1062279		
Economic performance							
Tracking mode	NPV [$]	EPB [year]	SPB [year]	LCOE [$/kWh]	BCR	ALCS [$/year]	IRR [%]
Fixed	2100912	8.7	11.5	0.121	1.7	267866	17.1
One-axis	5383008	5.1	8.6	0.091	2.9	686333	25.3
Two-axis	6036219	4.7	8.2	0.087	3.1	769618	27.0
Inclined axis	4857469	5.5	9.0	0.095	2.7	619327	24.0

Table 4 summarizes the major findings of the designed system's economic performance of a 5 kW grid-connected rooftop PV system. It is found that Two-axis system has the highest energy production, NPV, ALCS, and IRR and lowest value of EP, SP and LCOE.

4 Conclusions

This paper investigated the validity of grid-connected wind and PV systems in Selvili-Tepe, Northern Cyprus. The feasibility of large-scale wind and PV system was evaluated using RETScreen software. The results demonstrated that the amount of energy output from the proposed systems would contribute significantly to reducing the effect of global warming as well as enhancing the sustainable technological development of the island. The result of the present study will help investors in the energy and building sectors and accelerate an informed transition towards a more sustainable future.

References

1. Riahi, K., et al.: The shared socioeconomic pathways and their energy, land use, and greenhouse gas emissions implications: an overview. Glob. Environ. Change **42**, 153–168 (2017)
2. Arreyndip, N.A., Joseph, E.: Small 500 kW onshore wind farm project in Kribi, Cameroon: Sizing and checkers layout optimization model. Energy Rep. **4**, 528–535 (2018)
3. Shahsavari, A., Akbari, M.: Potential of solar energy in developing countries for reducing energy-related emissions. Renew. Sustain. Energy Rev. **90**, 275–291 (2018)
4. Chauhan, A., Saini, R.P.: Techno-economic feasibility study on integrated renewable energy system for an isolated community of India. Renew. Sustain. Energy Rev. **59**, 388–405 (2016)
5. Himri, Y., Merzouk, M., Merzouk, N.K., Himri, S.: Potential and economic feasibility of wind energy in south West region of Algeria. Sustainable Energy Technol. Assess. **38**, 100643 (2020)
6. Ahmed, A.S.: Technical and economic feasibility of the first wind farm on the coast of Mediterranean Sea. Ain Shams Eng. J. **12**(2), 2145–2151 (2021)
7. Kassem, Y., Azoubi, R., Gökçekuş, H.: The possibility of generating electricity using small-scale wind turbines and solar photovoltaic systems for households in Northern Cyprus: a comparative study. Environments **6**(4), 47 (2019)
8. Enongene, K.E., Abanda, F.H., Otene, I.J.J., Obi, S.I., Okafor, C.: The potential of solar photovoltaic systems for residential homes in Lagos city of Nigeria. J. Environ. Manage. **244**, 247–256 (2019)
9. Solyali, D., Altunç, M., Tolun, S., Aslan, Z.: Wind resource assessment of Northern Cyprus. Renew. Sustain. Energy Rev. **55**, 180–187 (2016)
10. Alayat, M.M., Kassem, Y., Çamur, H.: Assessment of wind energy potential as a power generation source: a case study of eight selected locations in Northern Cyprus. Energies **11**(10), 2697 (2018)
11. Kassem, Y., Çamur, H., Alhuoti, S.M.A.: Solar energy technology for Northern Cyprus: assessment, statistical analysis, and feasibility study. Energies **13**(4), 940 (2020)
12. Kassem, Y., Gökçekuş, H., Güvensoy, A.: Techno-economic feasibility of grid-connected solar PV system at Near East University Hospital. Northern Cyprus. Energies **14**(22), 7627 (2021)

A Climate Action with Developing Five Minutes Walking Inside and Near Public Centers Software

Vahid Nourani[1,2], Hüseyin Gökçekuş[1], Farhad Bolouri[1(✉)], and Ali Sheikhbabaei[3]

[1] Civil Engineering Department, Faculty of Civil and Environmental Engineering, Near East University, Nicosia, Turkish Republic of Northern Cyprus
farhad.bolouri@neu.edu.tr

[2] Center of Excellence in Hydroinformatics, Faculty of Civil Engineering, University of Tabriz, Tabriz, Iran

[3] Civil Engineering Department, Faculty of Civil Engineering, University of Tabriz, Tabriz, Iran

Abstract. Today, clean air as the major requirement in improving the quality of lives is highly in demand which is overshadowed by climate change as well as air pollution. In this regard, policy-making and planning to reverse climate change and air pollution are of great importance. Also, with advances in technology, mobile applications can play a key role in environmental awareness and taking suitable environmental action. Hence, by having benefited from this aspect the idea of "5-min walking inside and near public software" has been proposed. An environmental application which can develop in accordance with the user enthusiasm and reveals the 5 up to 10 walking passways to the nearest and famous places and also, by considering the suggested discount presented by various stores. The over mentioned project was successfully implemented in Boston, USA and also, inside the university of Tabriz as the 5-min walking map to reduce air pollution. The idea managed to win first place in Climate Projection Project by United Nations Development Program (UNDP). The project follows lots of advantages encompasses using fewer cars which results in less pollution, individual health promotion, clean air improvement, environmental respect, Encourage exercise and etc.

Keywords: Clean air · Climate change · Environmental application · Ideation · Sustainable development goals

1 Introduction

Department of Development Plans of the Ministry of Science, Research and Technology; Defines green management as: Management of energy consumption, materials and environmental protection in the organization by effectively and efficiently using all material and human resources, organizing and planning to guide them to achieve environmental goals [1]. The most important reason for the need for green management is that our

H. Gökçekuş and Y. Kassem (Eds.): NRSEM 2021, EESCI, pp. 72–79, 2022.
https://doi.org/10.1007/978-3-031-04375-8_9

lives and our economy depend on the services we receive from nature, and it is nature that should support our lives and work. On the other hand, the way we manage and the consequences of our actions affect the continuity and sustainability of nature in providing the services we need. Therefore, it is our important responsibility to make a committed effort to protect and improve the functioning of natural systems that meet our needs, because our survival depends on their survival, and their survival is directly and indirectly affected by our management performance [2].

How the plan responds to the problems raised is; i) Influencing people's decisions; As the previous information obtained from people's opinions is used by artificial intelligence and influences their decisions effectively. ii) Use of advertising tricks and joint programs and cooperation; For example, to take advantage of discounts on online taxis. iii) Reducing the use of private cars, which is known as the most important cause of urban air pollution in many parts of the world. Because with this software, walking and using bicycles instead of personal cars will be more. iv) Encouraging people to use shared bikes and walking is the main purpose of this program.

The cost is just installing the app and it is the concern of the people, but creating concern and encouragement by the software team is very important and earning money through the advertisements of entertainment and food centers in the software.

For example, by developing a campus map and encouraging students and academics to walk around the university instead of using existing transportation systems such as telephone taxis, buses, and vans; Multiple goals are available seamlessly. First, we will see a reduction in air pollution. Secondly, noise pollution is reduced and thirdly, the benefits of walking for the human body and brain are undeniable, some of which are recounted below [4]; Walking helps in weight loss, improves blood circulation from the benefits of walking, strengthens muscles with walking, benefits of walking in improving digestion and digestion, helps absorb vitamin D through walking, benefits and effects of walking in reducing stress, Increase productivity from the benefits of walking and increase creativity from the benefits of walking.

This article will address this issue in detail and map the 5-min walkways near public centers in detail. This type of map, which was first prepared and used in Boston, is now used at the University of Tabriz [3].

In the following, first, a number of studies related to clean transportation are reviewed, then the desired area is introduced, and after describing the research method, the prepared roadmap is explained, and finally, the conclusion and Suggestions are made.

2 Literature Review

In this section, articles related to clean transportation near public centers such as universities are reviewed.

Parsa et al. (2015) in an article entitled "Study of the effect of travel purpose on the choice of transportation mode, a case study: Amirkabir University of Technology" acknowledged that; The results of this study can be used directly or indirectly to improve the transportation issues of students of this university and by understanding the group tendency of the community in choosing the mode of transportation in different situations can be used to improve Transportation conditions and the provision of efficient and useful transportation systems in different corridors, depending on the uses covered by these modes, took a major step [4].

Kadkhodai et al. (2017) in their article entitled "Evaluation of intra-university transportation systems with a sustainable development approach: A case study: Ferdowsi University of Mashhad" stated that; In this research, intra-university transportation options have been evaluated and prioritized using Analytic Hierarchy Process (AHP). Based on the results of this study, the most important criteria and indicators affecting the prioritization and evaluation of intra-university transportation modes with a sustainable approach are travel cost, travel time, waiting time. Intra-university modes of transportation have also been preferred by shared bicycles, buses, walks, and vans, respectively [5].

Mohammadi et al. (2013) in an article entitled "Transportation planning with a sustainable development approach (Case study: Amirkabir University of Technology)" stated that the purpose of this article is to evaluate the position of public transport fleet in sustainable transport and present the optimal travel model in this regard is the proposed model, based on the principles of linear planning and taking into account the main criteria of sustainable transport, i.e. the criteria is economic, social, and environmental. Finally, according to the results obtained from the model, the optimal travel pattern is presented. According to the presented results, traveling from areas that have good access to the public transport system. Costs are up to 50% and NOx emissions are up to 100% lower than in areas that do not have adequate access to public transportation. In this regard, a survey conducted at the Amirkabir University of Technology has been used [6].

Khajeh Nouri and Arabshahi (2017) in their article entitled "The role of cable transportation in reducing the cost of air pollution in the Islamic Azad University of Tehran, Science and Research Branch" concluded that; According to the plan to build urban cable transportation in the Islamic Azad University of Tehran Campus, Science, and Research Branch, there is a need to study the current transportation of the university and compare it with cable transportation to estimate the benefits of reducing pollution for decision-makers. In this research, statistics and field studies have been used. The results of this study show that the construction of cable transport reduces the cost of current transportation pollution [7].

Elahi (2014) in his article entitled "Study of the urban collective space with a sidewalk approach: a case study: the campus of the University of Cincinnati Ohio" discussed walking in the university environment and announced that; The best function of the sidewalk is to use the experience of movement to understand both the body and the memory of the city. If crossing the pedestrian axis can be an effective factor in limiting civic life and cultural promotion of society. The economic role of sidewalks as a stimulus for tissue development and a place for urban walks and leisure for residents is not negligible. The innovation of this research is the study and analysis of the role of urban pedestrian axes as a collective space, if further by examining and analyzing the case study of the collective space of the Cincinnati Ohio campus using exploratory research, strategies, and executive suggestions to improve and update the quality and upgrade roles. Sidewalks are provided [8].

Therefore, as it is clear, clean transportation and reduction of pollution caused by improper transportation in public centers such as university environments is one of the main concerns of university green management and this article tries to continue the way of previous researches and present It has a solution without imposing costs and with many values.

3 Materials and Method

This plan will be software and using applications such as Google Map. But because it is done on paper for the University of Tabriz, for example, the University of Tabriz is explained.

In this research, library studies have been used to collect information and field observations to prepare a map, then the results have been analyzed. In order to collect information about green management, clean transportation and the benefits of using measures, etc., articles, books and reports, etc. have been studied as a library and then from field observations is used for proper navigation to prepare a 5-min map.

Google maps and other related maps were also used to prepare the 5-min walking map, and finally it was drawn in AutoCAD software with a centimeter accuracy and then using the information obtained. Comes from databases about people's interests, sent to them in government centers (such as their free time at lunch-time at the university) by the message software, which is close to and within 5 to 10 min' walk, where there is a favorite. Cycling delivery centers and other eco-friendly items are also shown.

4 Results and Discussion

The paper sample of this project has been implemented for Tabriz University, which will be explained first. Using satellite maps and related software such as Google Earth and GIS, a detailed map of Tabriz University was prepared by observing the neighborhood boundaries, and then using related software such as Google map, the time of arrival from a point to Another point in the campus was precisely identified. Then, in the AutoCAD map of the university, by observing the scale and with centimeter accuracy, the origin and destination of the obtained neighborhoods for a 5-min walking distance, which is equivalent to 330 m on the ground with a zero degree slope, were determined. Thus, the campus on the map was associated with a number of nodes and lines that connect the nodes. The nodes represent the origins and destinations of the 5-min walk, and their lines represent the walking route for students and academics. The next part of this map is displayed.

4.1 Map of 5 min' Walk

In this section, in Fig. 1, a 5-min walking map of Tabriz University is shown.

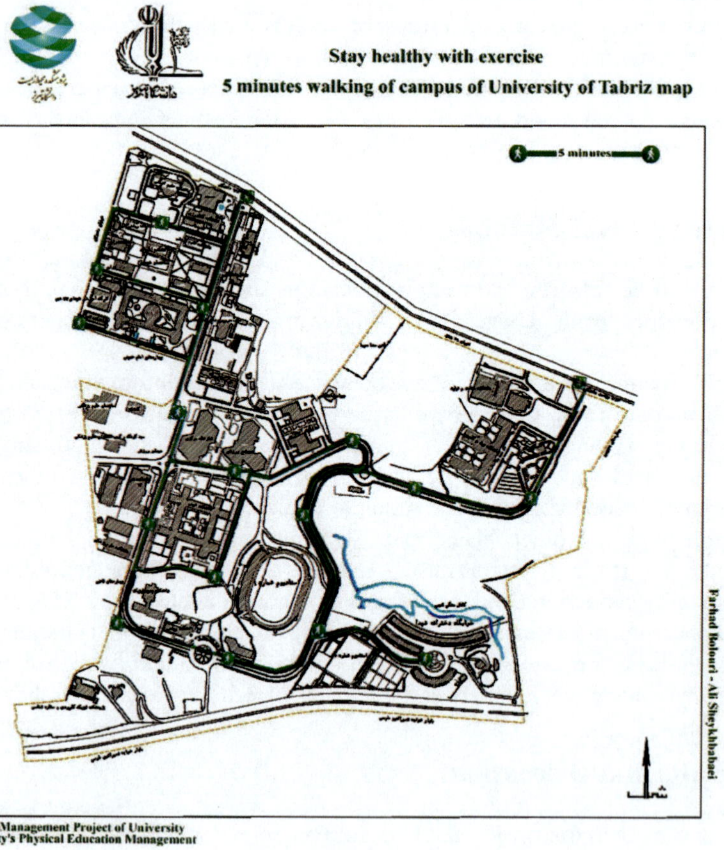

Fig. 1. 5-min walking map of University of Tabriz [9]

4.2 The Horizon Ahead

The horizon for this research is as follows; i) The ability to develop the idea and continue the way is reserved for this project and this project is in progress. This idea has won first place in the "Climate Promise Project" Startup Ideation Event hold by UNDP (March, 2021), the certificate of which is given in the appendix. ii) augmented reality glasses; It is also one of the concepts that can be done in this project and in its future. iii) Logical software and audio receiver for closer proximity to the user; Can be embedded and used in this software. iv) It is possible to work with other applications for advertising and cooperation for this project.

5 Conclusion

In the case of university, the following suggestions are also provided in this regard:

- Informing academics The program of opening university doors, which has a very high role in managing the time and movement path of academics (there have been cases where professors have referred to the derby without notice, which was closed a few minutes ago).
- Managing student visits and providing public transportation such as minibuses and buses for missions and academic visits
- Promote the culture of cycling for use on long distances
- Accurate forecasting of university stations for public transportation
- Correct timing of inter-university transportation services
- Accommodation of students inside the university and elimination of student service (centralization of dormitory complexes)
- Opening of a university cycling site in student dormitories (study plan and follow-up to set up a cycling system at the university level)
- Removing worn-out vehicles (buses, minibuses, cars, etc.) from the university transport fleet
- Creating a special place for e-bicycle park on the university site
- Implement strict policies to restrict the entry of cars into the university

Appendix

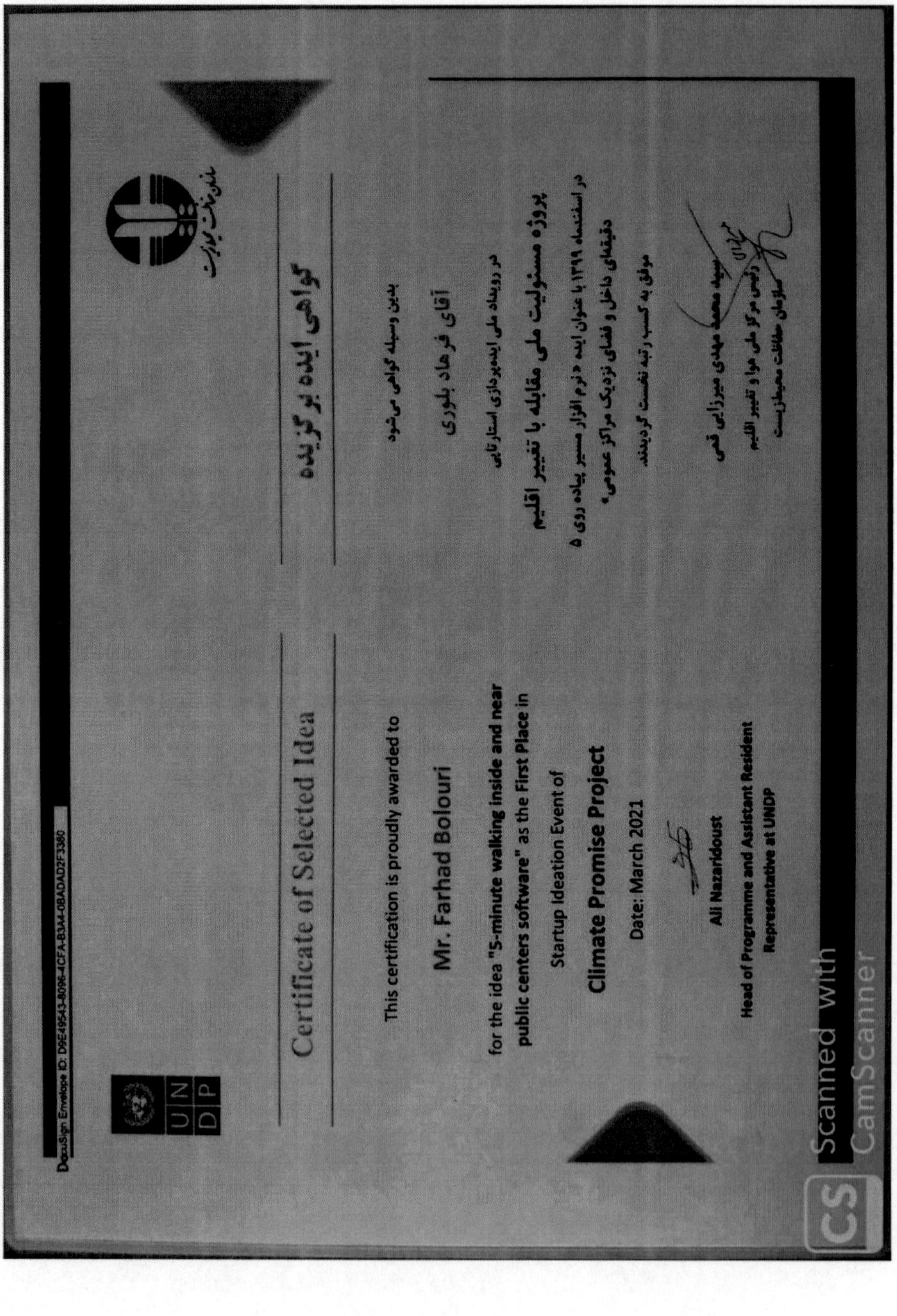

UNDP

Certificate of Selected Idea

گواهی ایده برگزیده

This certification is proudly awarded to

Mr. Farhad Bolouri

for the idea "5-minute walking inside and near public centers software" as the First Place in

Startup Ideation Event of

Climate Promise Project

Date: March 2021

Ali Nazaridoust
Head of Programme and Assistant Resident Representative at UNDP

بدین وسیله گواهی می‌شود

آقای فرهاد بلوری

در رویداد ملی ایده‌پردازی استارتاپی

پروژه مسئولیت ملی مقابله با تغییر اقلیم

در اسفندماه ۱۳۹۹ با عنوان ایده «نرم افزار مسیر پیاده روی ۵ دقیقه‌ای داخل و فضای نزدیک مراکز عمومی»

موفق به کسب رتبه نخست گردیدند.

سید محمد مهدی میرزایی قمی
رئیس مرکز ملی هوا و تغییر اقلیم
سازمان حفاظت محیط‌زیست

References

1. https://omrani.msrt.ir/fa/page/833/
2. Zahedi, S.S.: Green Management. Tehran, University Jihad Publishing Organization (2013) (In Persian)
3. https://walkboston.org/
4. Parsa, A.B., Mahmoudzadeh, A.R., Naderi, A., Golro, A.: Investigating the effect of the purpose of the trip on the choice of transportation mode, a case study of Amirkabir University of Technology. In: International Conference on Civil Engineering, Architecture and Urban Infrastructure, Tabriz, Permanent Secretariat of the Conference (2015) (In Persian)
5. Kadkhodai, M., Shad, R., Kadkhodai, M., Zadeh, S.E.: Evaluation of intra-university transportation systems with sustainable development approach Case study: Ferdowsi University of Mashhad. In: 17th International Conference on Transportation and Traffic Engineering, Tehran, Vice Chancellor and Traffic Transportation Organization (2017) (In Persian)
6. Mohammadi, R., Aghashahi, N., Dibaj, S., Golro, A.: Transportation planning with a sustainable development approach (Case study: Amirkabir University of Technology). In: 13th International Conference on Transportation and Traffic Engineering, Tehran, Deputy and Transport and Traffic Organization (2013) (In Persian)
7. Nouri, A.H.K., Arabshahi, N.: The role of cable transportation in reducing the cost of air pollution in the Islamic Azad University of Tehran Campus, Science and Research Branch. In: Sixth National Conference on Air and Noise Pollution Management, Tehran, Scientific Association of Clean Air (2017) (In Persian)
8. Elahi, S.: A study of urban collective space with a sidewalk approach case study: Ohio University of Cincinnati Campus. In: First National Conference on Urban Planning, Urban Management, and Sustainable Development, Tehran, Iranian Institute, Iranian Architectural Association (2014) (In Persian)
9. http://tabrizu.ac.ir/

Non-stationary Temperature Duration Frequency Curves for the North-West Homogeneous Region of India

Meera G. Mohan(✉) and S. Adarsh

TKM College of Engineering, APJ Abdul Kalam Technological University, Kollam, Kerala 691 005, India
meeragmohan90@gmail.com

Abstract. The long-term rise of global temperatures due to climate change has resulted in a larger increase in the probability of occurrence of extreme temperature events. The warming trend with an increase in the intensity, frequency and duration of heat waves is observed mainly in the North western regions of India. The current development of Temperature Intensity Duration Frequency (TDF) curves relies on the assumption of stationarity which does not hold valid due to the recent evidences in global warming. In this study, India Meteorological Department (IMD) $1^\circ \times 1^\circ$ gridded temperature dataset is used to examine the frequency of occurrence of extreme temperatures over the North Western homogeneous temperature region of India covering 50 grid points during the period of 1951–2019. The maximum daily temperatures for six different durations of 1-day, 2-day, 4-day, 6-day, 8-day and 10-day were extracted. This study proposes a non-stationary approach to the development of TDF curves keeping time as a covariate. Five Non-Stationary TDF models were developed varying location and scale parameters linearly, exponentially and their combinations with respect to time. The goodness-of-fit is improved when using a non-stationary approach with time covariates. Model, NSGEV-1 for which location parameter was linearly varied with respect to time turned out to be the best fitting model for more than 84% of the grid points for different durations. A comparison of Stationary TDF curve and Non-Stationary TDF curves were also carried to illustrate the impacts of considering non-stationarity in extreme temperature analysis.

Keywords: Climate change · Non-stationarity · Trend · Extreme temperature · Covariate · TDF curves

1 Introduction

Urbanization and industrialization, the effect of natural climatic variability, and increased greenhouse gasses in the atmosphere have been suggested as the leading causes of the increase in duration, magnitude, and frequency of temperature extremes and heatwaves in the context of the current climate evolution [1]. Study of temperature extremes is important for understanding climate variability which can vary spatially and temporally at different local, regional and global scales. Rising global temperatures are causing increases

H. Gökçekuş and Y. Kassem (Eds.): NRSEM 2021, EESCI, pp. 80–89, 2022.
https://doi.org/10.1007/978-3-031-04375-8_10

in the frequency and severity of extreme climatic events, such as floods, droughts, and heat waves [2]. Mean temperatures across India have risen by more than 0.5 °C over the period of 1960–2009, with statistically significant increases in heat waves.

The development of Temperature-Duration-Frequency (TDF) curves, which relate the intensity of heat events of different durations to their frequencies, can be useful tools for the analysis of heat extremes [3]. The TDF curves are developed based on historical maximum temperature time series data by fitting a theoretical probability distribution of annual maximum extreme temperature series. TDF curves are developed similar to Rainfall Intensity-Duration-Frequency (IDF) curves. Conventionally, frequency analysis is based on the concept of stationary extreme value theory i.e., the probability distribution parameters are invariant with time. This assumption of stationarity does not hold valid due to the recent evidences in global warming. Various studies analysing the patterns of change in temperature extremes in terms of heatwaves and cold waves [4–8], summer and winter temperature extremes [9] and trend analysis [10, 11] were carried out in the past years. Khaliq et al. [12] developed the concept of TDF curves assuming that the characteristics of the probability distribution of extreme heat events are invariant through time. This study was conducted for four stations in southern Quebec region, Canada. This work was further extended to the non-stationary concept of TDF curves by Ouarda [3] for six stations in the Province of Quebec, Canada. As far as India is concerned, the climate is interesting and complex due to diverse topographical features and large geographical area, hence in case of a local variability in climate change, it is important to know which area could be more affected by these changes. These results can aid in verification of climate model studies, especially for the future. High resolution daily gridded temperature data (1969–2005) was developed by Srivastava et al. [13] which has supported many researchers in their temperature analysis. Even though stationary and non-stationary frequency analysis in terms of rainfall intensity, flood, drought severity has been explored by various researchers in the past, the notion of establishing TDF curves for India still remains to be unexplored. The semi-arid north-western part of India will likely warm more rapidly than the remaining parts making it more prone to extreme temperatures.

This study aims to conduct the frequency analysis of temperature extremes with time-varying distribution functions to characterize temperature extremes in a Non-Stationary (NS) context for the North-West region of India. The TDF analysis is carried out to explore the following objectives: (i) to develop Stationary (S) TDF model assuming location, scale and shape parameter to be a constant and five NS-TDF models by varying location and scale parameter linearly and exponentially with respect to time and the shape parameter as a constant; (ii) to identify the best fit model among the S and NS models for each duration extreme temperature series; (iii) to build and compare S-TDF curve and NS-TDF curves based on the best fit NS model for quantifying the percentage variation; and (iv) to compute the temperature return levels for various return periods with reference to the NS-TDF curves.

2 Study Area and Data

India is a vast country stretching between latitudes 8°4′N and 37°6′N and longitudes 68°7′E and 97°25′E. It is fractionated into seven homogeneous temperature regions namely, North-West (NW), Western Himalayas (WH), West Coast (WC), East Coast (EC), North-East (NE), North Central (NC) and Interior Peninsula (IP) by the Indian Institute of Tropical Meteorology, Pune (online at www.tropmet.res.in). In this paper, the NW homogenous region which comprises of the Indian desert is considered as the study area due to its prominent temperature. The daily maximum temperature gridded data (1°latitude × 1°longitude) observed by the India Meteorological Department (IMD) is procured for NW region for a period of 1951–2019 (69 years). A total of 50 grid points spread out over Gujarat, Rajasthan, Haryana, Punjab, Uttar Pradesh and Himachal Pradesh belongs to the NW region (see Fig. 1).

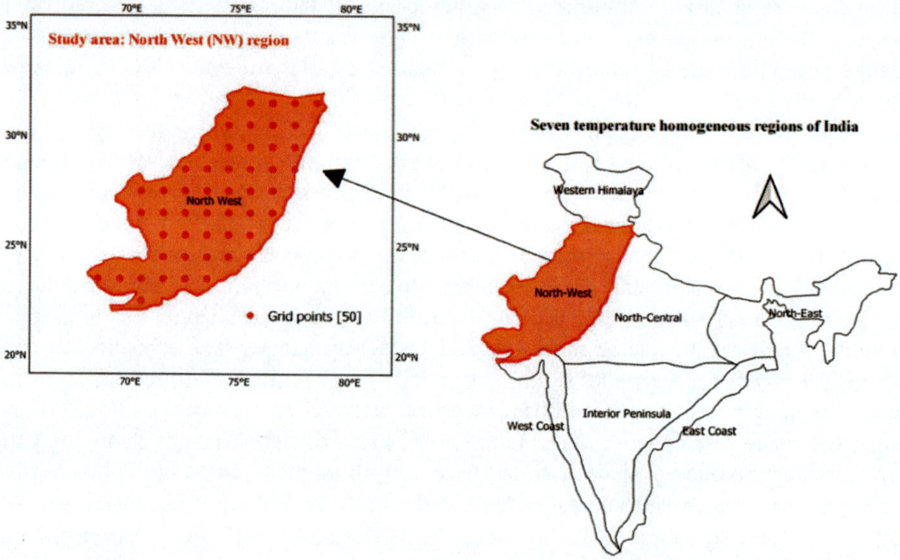

Fig. 1. Study area – North West temperature homogeneous region of India

3 Methodology

3.1 Significance of Trend

Firstly, Mann Kendell (MK) Trend analysis test [14, 15] was carried out to ascertain the presence of statistically significant trend. For a time series, X(t), the Mann–Kendall statistic is defined as:

$$S = \sum_{t=1}^{N-1} \sum_{i=t+1}^{N} sign\big(X(i) - X(t)\big) \tag{1}$$

where N is number of data and sign (•) represents a signum function given by:

$$sign(a) = \begin{cases} \frac{a}{|a|}, if\ x \neq 0 \\ 0, if\ x = 0 \end{cases} = \begin{cases} 1, if\ a > 0 \\ 0, if\ a = 0 \\ -1, if\ a < 0 \end{cases} \tag{2}$$

Sign and value of the S statistic show the direction and intensity of the trend. The increasing or decreasing trends are indicated by positive or negative values of S respectively, while zero represents no trend. A normalized test statistic z can be used to statistically quantify the significance of the trend.

$$z = \begin{cases} \frac{S-1}{\sqrt{var(S)}} & if\ S > 0 \\ 0 & if\ S = 0 \\ \frac{S+1}{\sqrt{var(S)}} & if\ S < 0 \end{cases} \tag{3}$$

The positive values of z represent increasing trend while negative values indicate decreasing trend. At 95% confidence interval, if the value of z is more than 1.96 it represents significant non-stationary increasing trend. Similarly, if z is less than −1.96, it represents significant non-stationary decreasing trend.

3.2 Stationary and Non-stationary Temperature Models

Extreme Value Theory (EVT) is a framework used for analysing the risks associated with climate extremes and their return levels [1]. The Generalized Extreme Value (GEV) distribution is a continuous probability distribution that combines Gumbel, Fréchet and Weibull extreme value distributions. The frequency model using this family of distributions is often referred to as the block maxima approach [1]. The GEV distribution depends on three parameters: location (μ), scale (σ), and shape (ξ) with a cumulative distribution function expressed as:

$$F(x, \mu, \sigma, \xi) = \begin{cases} exp\left\{-\left[1 + \xi\left(\frac{x-\mu}{\sigma}\right)\right]^{-1/\xi}\right\}, \sigma > 0, 1 + \xi\left(\frac{x-\mu}{\sigma}\right) > 0, \xi \neq 0 \\ exp\left\{-exp\left[-\frac{(x-\mu)}{\sigma}\right]\right\}, \sigma > 0, \xi = 0 \end{cases} \tag{4}$$

In this study Stationary (S) model and five Non-Stationary (NS) models were developed as shown in Table 1. The NS setting for the location and scale parameters of GEV distribution was introduced as a function of time. The shape parameter was assumed to be a constant for all models.

Table 1. Development of stationary and non-stationary temperature models

Model name	Model parameters		
	Location	Scale	Shape
SGEV	$\mu_t = \mu$	$\sigma_t = \sigma$	$\xi_t = \xi$
NSGEV-1	$\mu_t = \mu_0 + \mu_1 Time$	$\sigma_t = \sigma$	$\xi_t = \xi$
NSGEV-2	$\mu_t = \mu$	$\sigma_t = \sigma_0 + \sigma_1 Time$	$\xi_t = \xi$
NSGEV-3	$\mu_t = \mu_0 + \mu_1 Time$	$\sigma_t = \sigma_0 + \sigma_1 Time$	$\xi_t = \xi$
NSGEV-4	$\mu_t = \mu$	$\sigma_t = exp(\sigma_0 + \sigma_1 Time)$	$\xi_t = \xi$
NSGEV-5	$\mu_t = \mu_0 + \mu_1 Time$	$\sigma_t = exp(\sigma_0 + \sigma_1 Time)$	$\xi_t = \xi$

The distribution parameters are estimated using Maximum Likelihood Estimation (MLE) method. MLE involves defining a likelihood function for calculating the conditional probability of observing the data sample given a probability distribution and distribution parameters. The log likelihood is derived from Eqs. (5) and (6). The location and scale parameters in the above equations are replaced with the corresponding equations in Table 1 based on the model application. MLE is done by maximizing the likelihood estimates obtained from Eqs. (5) and (6).

For $\xi \neq 0$,

$$\log L(\mu, \sigma, \xi|X) = -n\log\sigma - \left(1 + \frac{1}{\xi}\right)\sum_{i=1}^{n} log\left[1 + \xi\left(\frac{x_i - \mu}{\sigma}\right)\right] - \sum_{i=1}^{n}\left[1 + \xi\left(\frac{x_i - \mu}{\sigma}\right)\right]^{-1/\xi}, 1 + \xi\left(\frac{x_i - \mu}{\sigma}\right) > 0 \tag{5}$$

For $\xi = 0$,

$$\log L(\mu, \sigma|X) = -n\log\sigma - \sum_{i=1}^{n}\left(\frac{x_i - \mu}{\sigma}\right) - \sum_{i=1}^{n}\exp\left[-\left(\frac{x_i - \mu}{\sigma}\right)\right] \tag{6}$$

The Akaike Information Criterion (AIC), which penalizes the minimized negative log likelihood for the number of parameters estimated, is used to select the best model [16]. AIC is used to estimate the best fit GEV model among the SGEV model and 5 NSGEV models. It is computed using Eq. (7)

$$AIC = 2K - 2\ln(L) \tag{7}$$

where K is the number of independent variables used and L is the log-likelihood estimate.

3.3 Quantification of Temperature Intensities for Various Return Periods

Once the best NS model is identified, the NS temperature intensity is estimated using the best NS model parameters. Unlike S model, the NS models' location and/or scale parameter value are bound to be time dependant. In this study for NS framework models, 95 percentiles of the location and scale parameter values from the historical observation obtained from Eqs. (8) and (9) is substituted into Eq. (10) for estimating the non-stationary temperature intensities for various return periods (T in years).

$$\widehat{\mu_{95}} = Q_{95}(\widehat{\mu_{t1}}, \widehat{\mu_{t2}}, \ldots\ldots, \widehat{\mu_{tn}}) \tag{8}$$

$$\widehat{\sigma_{95}} = Q_{95}(\widehat{\sigma_{t1}}, \widehat{\sigma_{t2}}, \ldots\ldots, \widehat{\sigma_{tn}}) \tag{9}$$

$$I = \begin{cases} \widehat{\mu} + \frac{\widehat{\sigma}}{\widehat{\xi}}\left[\left(-\log\left(1 - \frac{1}{T}\right)\right)^{-\widehat{\xi}} - 1\right], \widehat{\xi} \neq 0 \\ \widehat{\mu} + \widehat{\sigma}\left[-log\left(-\log\left(1 - \frac{1}{T}\right)\right)\right], \widehat{\xi} = 0 \end{cases} \tag{10}$$

4 Results and Discussion

The daily maximum temperature data obtained from IMD for 50 grid points belonging to NW region was used to extract the annual maximum temperature over 6 durations (1-day, 2-day, 4-day, 6-day, 8-day and 10-day). The trend analysis was carried over the NW region using MK test and it was found that number of grid points following non-stationarity trend increases as the duration increases. The same is represented in a bar graph in Fig. 2.

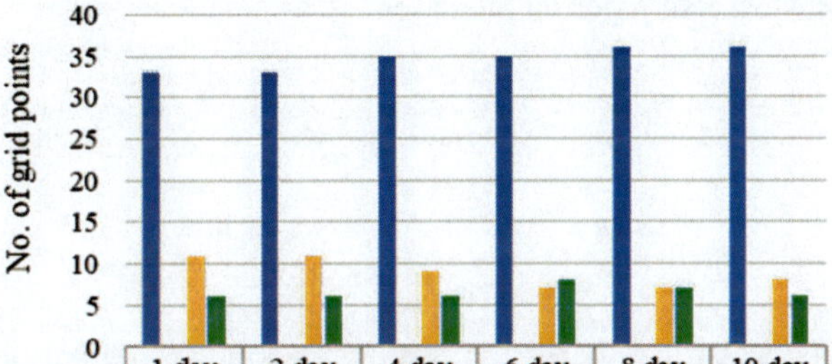

	1 day	2 day	4 day	6 day	8 day	10 day
NS increase trend (z ≥1.96)	33	33	35	35	36	36
NS decrease trend (z < -1.96)	0	0	0	0	0	0
Increase trend (0 ≤ z < 1.96)	11	11	9	7	7	8
Decrease trend (0 > z > -1.96)	6	6	6	8	7	6

Fig. 2. Significance of trend in grid points of NW region

The S TDF and NS TDF analysis was carried out by developing GEV models as mentioned in Sect. 3.2 and the best fit model was identified with the help of AIC values. From Fig. 3 it can be seen that the best fit models vary for the same grid point as the duration changes in some cases while in others the best fit model remains constant throughout for all durations.

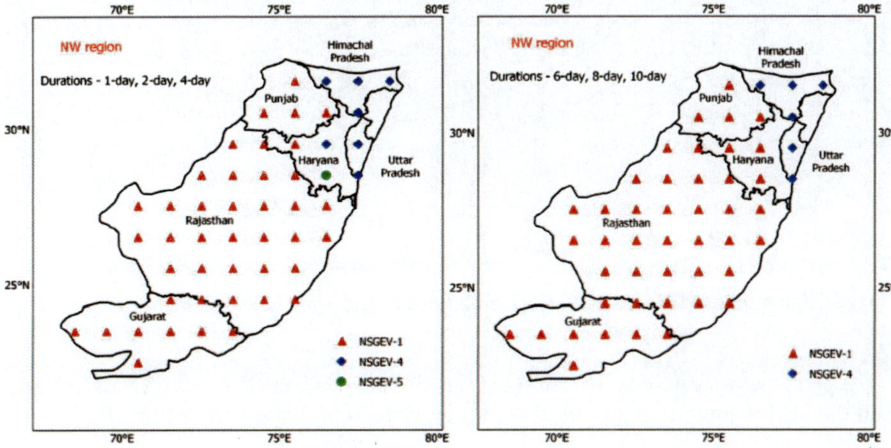

Fig. 3. Best fit models for NW temperature homogenous region for all durations

It can be concluded that more than 84% of the grid points have NSGEV-1 as the best fit model where location parameter is linearly time varying with a constant scale and shape parameter for all durations (see Fig. 4). S TDF curves and best fit NS TDF curves were developed for all the grid points for various return periods (RP) of 2 years, 5 years, 10 years, 25 years, 50 years and 100 years. Due to space constraints, a sample grid point (23.5°N, 68.5°E) of Gujarat is chosen for a detailed representation of S and best fit NS TDF curves for all return periods in Fig. 5. The dashed curves represent S-TDF curves and the thick line represent NS-TDF curves for various return periods. It was found that the temperature intensities decrease with increase in the daily maximum durations and increase when return period increases.

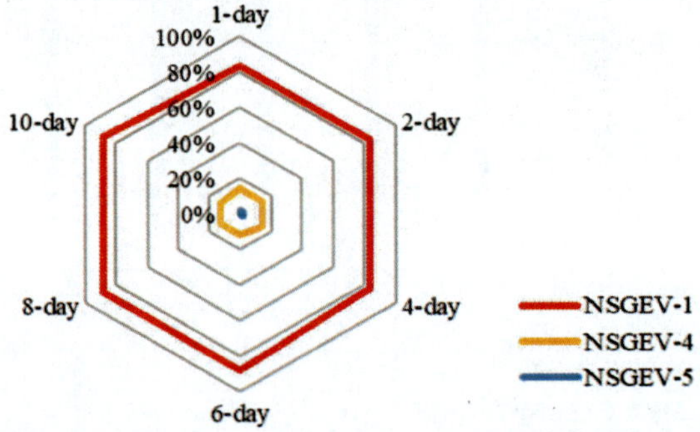

Fig. 4. Percentage of grid points for best fit models in NW region

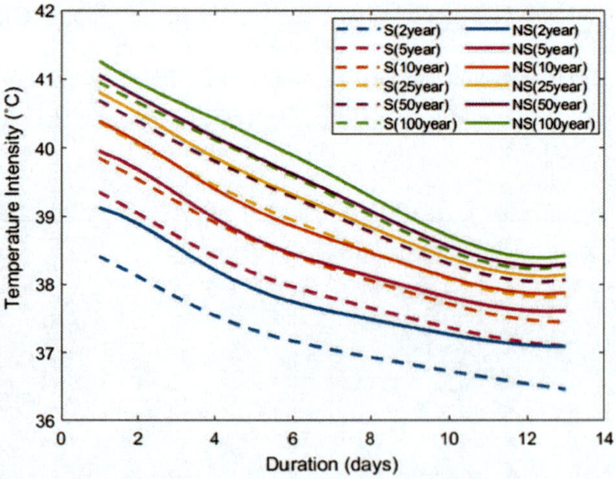

Fig. 5. S and best fit NS (NSGEV-1) TDF curves for grid point (23.5°N, 68.5°E) Gujarat

The percentage variations (PV) of return levels between S & NS cases were calculated for all the grid points. A spatial representation of the temperature intensities for both S and NS case along with the PV for 1 day duration for all return periods for the entire NW region is shown in Fig. 6.

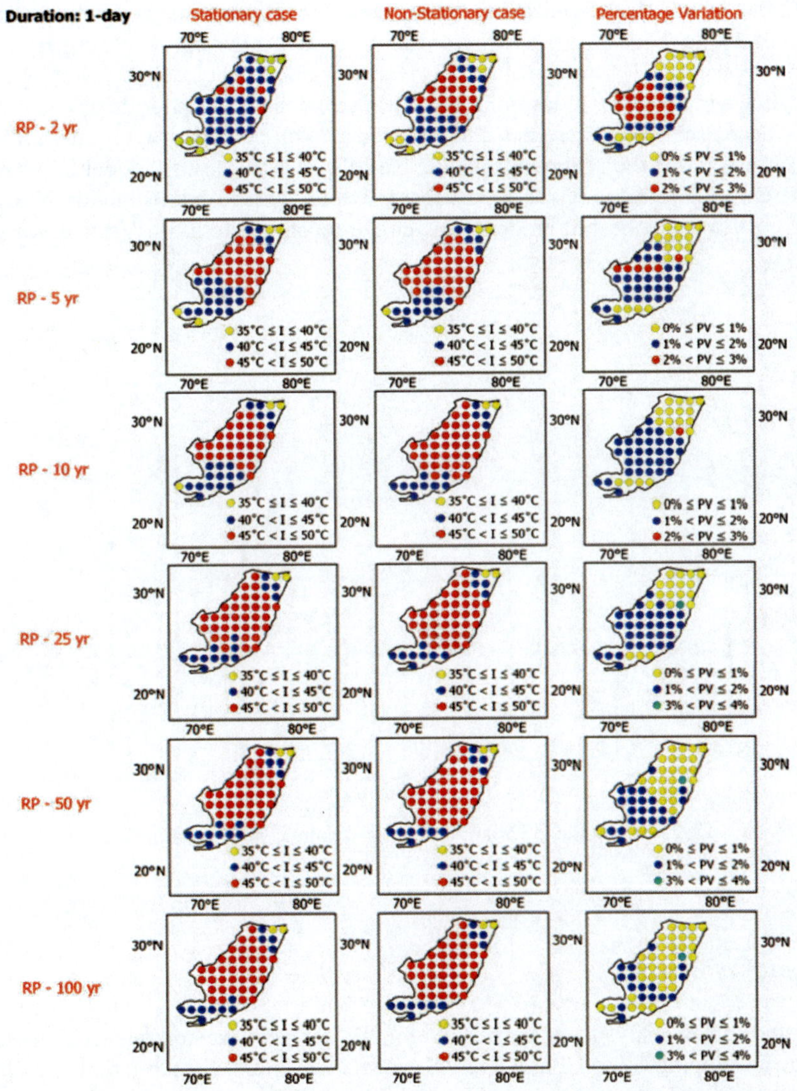

Fig. 6. S and NS temperature intensities and percentage variation for 1-day duration

When the stationary case is compared with non-stationary case it was found that the temperature intensities increased beyond 45 °C for many grid points when non-stationarity was considered. This points out that the stationary temperature intensities are lower than all of the corresponding non-stationary values for all grid points. The same was the case for all the other durations as well. Majority of the percentage variations range between 0% and 2% i.e., about 0 °C to 0.9 °C. In the case of return period of 2 years around 36% of the grid points showed a percentage variation between 2% to 3% which reduced to the category of 0% to 2% as the return period increased. One grid point (28.5°N,76.5°E) showed a consistent increase in the percentage variation reaching a value of 3% to 4% (up to 1.6 °C) from 25 years to 100 years of return period. The pattern of percentage variations of return levels was observed closely by plotting a stem plot as shown in Fig. 7 for 1-day duration. It was seen that the graph follows two patterns i.e., PV decreases as RP increases and vice-versa. While comparing Fig. 6 with Fig. 3 it was observed that the grid points having NSGEV-1 as the best fit model followed the first pattern of PV decreases as RP increases while the grid points having NSGEV-4 and NSGEV-5 as the best fit model followed the second pattern of PV increases as RP increases.

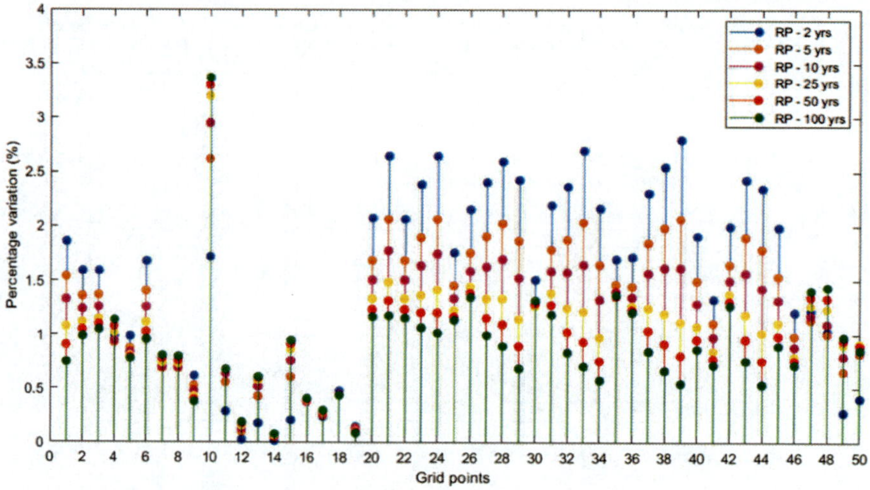

Fig. 7. Percentage variation of all grid points for 1-day duration

5 Conclusions

In this study, TDF analysis was carried out for NW temperature homogeneous region of India by developing S model and five NS models with time as covariate. It was found that more than 84% of the grid points approve NSGEV-1 model with location parameter linearly varying with respect to time as the best fit model with the help of AIC values. The best fit model tends to change for each duration for certain grid points. Temperature intensity versus duration plots were established for both S and NS case with each curve representing return period of 2 years, 5 years, 10 years, 25 years, 50 years and 100 years. A comparison of S-TDF curves and NS-TDF curves were carried for all the 50 grid points and it was found that the NS temperature intensities were more than the S estimates. The temperature intensity increased beyond 45 °C when non-stationarity was considered.

As the return period increases, the temperature intensities tend to increase for S and NS case. A PV of 0% to 2% was observed for majority of the grid points. In general, it can be concluded that for more than 84% of the grid points PV decreases as RP increases while for the remaining 16% PV increases as RP increases based on the best fit model for each grid point.

References

1. Hamdi, Y., Duluc, C.M., Rebour, V.: Temperature extremes estimation of non-stationary return levels and associated uncertainties. Atmosphere **9**(4), 129 (2018). https://doi.org/10.3390/atmos9040129.hal-02871800
2. Mazdiyasni, O., et al.: Increasing probability of mortality during Indian heat waves. Sci. Adv. **3**, e1700066 (2017)
3. Ouarda, T.B.M.J., Charron, C.: Nonstationary temperature-duration-frequency curves. Sci. Rep. **8**(1), 1–8 (2018). https://doi.org/10.1038/s41598-018-33974-y
4. Dash, S.K., Mamgain, A.: Changes in the frequency of different categories of temperature extremes in India. J. Appl. Meteorol. Climatol. **50**(9), 1842–1858 (2011). https://doi.org/10.1175/2011JAMC2687.1
5. Perkins, S.E., Alexander, L.V.: On the measurement of heat waves. J. Clim. **26**(13), 4500–4517 (2013). https://doi.org/10.1175/JCLI-D-12-00383.1
6. Rohini, P., Rajeevan, M., Srivastava, A.K.: On the variability and increasing trends of heat waves over India. Sci. Rep. **6**, 1–9 (2016). https://doi.org/10.1038/srep26153
7. Mishra, V., Mukherjee, S., Kumar, R., Stone, D.A.: Heat wave exposure in India in current, 1.5 °C, and 2.0 °C worlds. Environ. Res. Lett. **12**(12) (2017). https://doi.org/10.1088/1748-9326/aa9388
8. Khan, N., Shahid, S., Ismail, T., Ahmed, K., Nawaz, N.: Trends in heat wave related indices in Pakistan. Stoch. Env. Res. Risk Assess. **33**(1), 287–302 (2018). https://doi.org/10.1007/s00477-018-1605-2
9. Loikith, P.C., Broccoli, A.J.: The influence of recurrent modes of climate variability on the occurrence of winter and summer extreme temperatures over North America. J. Clim. **27**(4), 1600–1618 (2014)
10. Kothawale, D.R., Revadekar, J.V., Kumar, K.R.: Recent trends in pre-monsoon daily temperature extremes over India. J. Earth Syst. Sci. **119**(1), 51–65 (2010). https://doi.org/10.1007/s12040-010-0008-7
11. Ghasemi, A.R.: Changes and trends in maximum, minimum and mean temperature series in Iran. Atmos. Sci. Lett. **16**(3), 366–372 (2015). https://doi.org/10.1002/asl2.569
12. Khaliq, M.N., St-Hilaire, A., Ouarda, T.B.M.J, Bob'ee, B.: Frequency analysis and temporal pattern of occurences of Southern Quebec heatwaves. Int. J. Climatol. **25**, 485–504 (2005). https://doi.org/10.1002/joc.1141
13. Srivastava, A.K., Rajeevan, M., Kshirsagar, S.R.: Development of high resolution daily gridded temperature data set (1969–2005) for the Indian region. Atmos. Sci. Let. (2009). https://doi.org/10.1002/asl.232
14. Mann, H.B.: Non-parametric tests against trend. Econometrica **13**, 163–171 (1945)
15. Kendall, M.G.: Rank Correlation Techniques.Charles Griffen. London ISBN: 195205723 (1975)
16. Katz R.W.: Statistical methods for nonstationary extremes. In: AghaKouchak, A., Easterling, D., Hsu, K., Schubert, S., Sorooshian, S. (eds.) Extremes in a Changing Climate. Water Science and Technology Library, vol 65. Springer, Dordrecht (2013). https://doi.org/10.1007/978-94-007-4479-0_2

A Comprehensive Strategy Against COVID-19 and Further Pandemics

Walter W. Kofler[1,2] and Oleg S. Glazachev[2(✉)]

[1] Medial University of Innsbruck, 6020 Innsbruck, Austria
walter.kofler@ias-he.com
[2] I.M. Sechenov Moscow State Medical University (Sechenov University), Mokhovaya 11, bld 4, 125009 Moscow, Russia
glazachev@mail.ru

Abstract. For almost two years, COVID-19 dominates the world despite the highest financial and scientific efforts. There is no end in sight to the pandemic. The successes in the development of vaccines and their use should not obscure the fact that essential characteristics of SARS-CoV-2 and its health consequences have not yet been sufficiently clarified. Nevertheless, decisions on action and inaction must be made without delay. This can often only be done on the "basis of everyday experience and the laws of reasoning." Omicron recalls that highly infectious pathogens can suddenly appear, evading the antibodies of the recovered and the vaccinated. The long phase until the new vaccine is effective worldwide can only be bridged with the help of non-specific methods, such as antiseptics. Governments are not prepared in this respect.

Keywords:: Omicron · Antiseptic · Unspecific defense · Hypoxia-hyperoxia · Emergency law

1 The Actual Situation

For almost two years, COVID-19 dominates the world. There is no end in sight to the pandemic. Two positions determine the situation here: Some assume that the methods of contact restriction and vaccination of the population used so far would be sufficient if they were only implemented consistently enough. They are opposed by those, who doubt that there are sufficient reasons for this. The naturally existing defensive power would normally be sufficient to ward off the infection. Therefore, the restriction is disproportionate and a threat to democracy. For this reason alone, COVID-19 also determines political events. The economic impact is considered to be the most serious since World War II.

This has not been changed by vaccines developed at a previously unimaginable rate. But more people died from COVID 19 in the U.S. in 2021 in spite of vaccination than in 2020 without vaccination. One of the reasons for this is the emergence of more virulent and more pathogenic mutants, with the risk of increasingly evading the effectiveness of the vaccines. However, Pfizer announced to be able to provide a new vaccine within

H. Gökçekuş and Y. Kassem (Eds.): NRSEM 2021, EESCI, pp. 90–99, 2022.
https://doi.org/10.1007/978-3-031-04375-8_11

100 days [3]. But this vaccine would also have to be produced, distributed and inoculated in sufficient quantities worldwide. However, it was not even possible in the USA to achieve the necessary vaccination rate within one year. Long enough for a new escape mutant to appear? Whether Omicron turns out to be "only" a wake-up call or the predicted worst-case scenario remains to be seen: In any case, it is a reminder: We need a precaution to outlast the phase known from 2020 until vaccination can bring success. During this time, besides lockdowns etc., only non-specific measures are available, e.g. antiseptics such as N-chlorotaurine or NO.

Mutants arise in host cells. Any measure that helps prevent transmission from leading to infection - and therefore any measure that strengthens the non-specific defenses - lowers the risk of mutant formation.

2 Analysis

2.1 The Challenge: Insufficient Knowledge

The Legal Framework

Decisions with limited knowledge must be made not only in the case of an epidemic with a pathogen that has not yet been adequately researched. This is the normal case in operational approval processes. The legislator therefore stipulates that experts be consulted who are restrained to make the interrelationships comprehensible for the questions not covered by the state of knowledge "*by applying the laws of reasoning and the experiences of everyday life*" in such a conclusive way that the decision-maker can come to a judgment [10].

The need for action despite limited knowledge is especially true during a pandemic. Therefore, the legislature has authorized the government or the responsible minister to take action via emergency decrees. He is therefore authorized to make medical products without CE marking or medicines without the usual approval procedure available to the public.

Appropriate Accompanying Empirical Review, Communication and Adjustment of the Strategy

It makes sense to clarify as soon as possible with studies whether logically conclusive but empirically inadequately supported decisions that are particularly influential for health should continue to be upheld. This requires transparency. Therefore, it is not a sign of a lack of responsibility if it is clearly communicated where the current lack of knowledge is seen, where the state of knowledge is better, and how this has currently changed, so that changes in procedure are set when new findings are available.

Progress in Interaction Between Researchers, Industry, Regulators and Government

The urgency to act has once again made it clear that formal requirements can delay procedures disproportionately. A relationship between researchers, industry and government that is not problem-oriented can also contribute to this. After all, authorities are not only control bodies, but also crucial for ensuring that socially significant government goals can be implemented in an appropriate manner.

Expanding Public Interest

Necessary tools in the fight against COVID 19 are also urgently needed in areas other than vaccination and drug development. This applies to all areas of nonspecific coping methods. They are also effective for many other pathogens and - hopefully - for the next pandemic resulting from one of the approximately 800,000 potential candidates according to IPBES [7].

One should think of the possibilities to support the extracorporeal nonspecific defenses, e.g. by antiseptics like N-chlorotaurine (NCT) [5, 15]. However, physiological processes can also be promoted in the body in a non-specific manner. One proven method is based on improving oxygen uptake and availability in the target organ. This generally improves the adaptive capacity. Extensive data are available on [2]. Systematic use of this method could be important for prevention, since oxygen deficiency is critical for the progression of COVID-19 and precautionary training can be expected to be beneficial in this respect [4]. Adequate oxygen should also be useful in rehabilitation and long-COVID.

2.2 Non-specific Defense and/or Specific Defense

The differences in opinion on the importance of the two physiological principles surprise physiologists and epidemiologists. Since the end of the 19th century, thanks to Semmelweis (1847), Lister (1865), Pasteur (1870), Koch (1871, 1881) and many other heroes of medicine, it has been textbook knowledge: Without non-specific defenses, mankind would have died out long ago. Natural defenses very often prevent pathogens from the nasal cavity from reaching the inside of the body, and so many people are kept alive long enough, despite infection, for specific defenses to defeat the disease. Both physiological principles are thus essential. Both can be artificially improved. They are essential in the fight against the era of pandemics.

How Constant Is the Non-Specific Defense?

The view is widely shared, that every healthy person has a sufficiently good natural defense by nature, if only he or she behaves in a healthy manner, and that this level remains constant. Only the virulence, i.e. the infectivity of the pathogens, can be changed. This assumption is wrong. The identical pathogen (A/H1/N1) of the Spanish flu caused "only" about 6500 deaths per million in the USA, where World War 1 only took place outside the country, and about 35 000 in the rest of the world, where war, destruction, hunger, fear of death, etc. affected the defense situation.

How inconspicuous influencing factors on the defense can be, is proven by the following study on healthy, voluntary, randomly distributed students. Under the pretext of testing how strongly the odor and pain thresholds vary in healthy young people to SO_2 "and "SO_3", and if a "good smeller" is good for different substances a compulsory internship was used for a study with olfactometry. The concentration of secretory immunoglobulin A (sIgA) in saliva before and after the exposures was also examined, ostensibly to use only completely healthy students. The students learned the health relevance of SO_2, SO_3, and the method of olfactometry. In olfactometry, odorless synthetic air and the test gas are alternately offered one breath at a time (2.3 s) through a mask up to the concentration until the subject just perceives the test target and terminate the test himself. Therefore, it was clear that a health relevant exposure even with the much more toxic SO_3 was excluded. But the test persons were only told that they were inhaling SO_3. They were only ever given the much more harmless SO_2. All predictions were significantly confirmed: There were responders and nonresponders in terms of odor and pain thresholds as it was to expect according to the toxicopy principle [9]. The level of sIgA also changed markedly: it decreased massively in the third with the highest baseline values and increased slightly in the third with the lowest baseline values (Fig. 1).

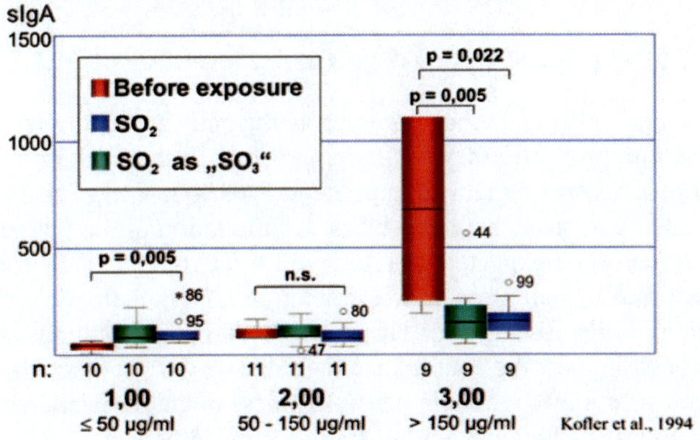

Fig. 1. Secretory immunoglobulin A (sIgA, mg/ml) in healthy students, sorted according to their sIgA concentration before exposure.

2.3 Infectivity and Susceptibility

Changes in defense status also suggest a change in susceptibility. This should be reflected in the length of the incubation period. In the case of SARS-CoV-2, it is essential to consider that germ carriers are already infectious during the incubation period, even though they do not yet have symptoms. This was demonstrated by He et al. in 2020, corrected by Ashcroft et al. [1] and accepted by He et al. [6]. Thus, the risk of contamination by an asymptomatic germ carrier is not only 1–2 days (Fig. 2. A, dashed line), but longer. Both demonstrate the importance of the asymptomatic phase for cumulative infectivity (Fig. 2. B).

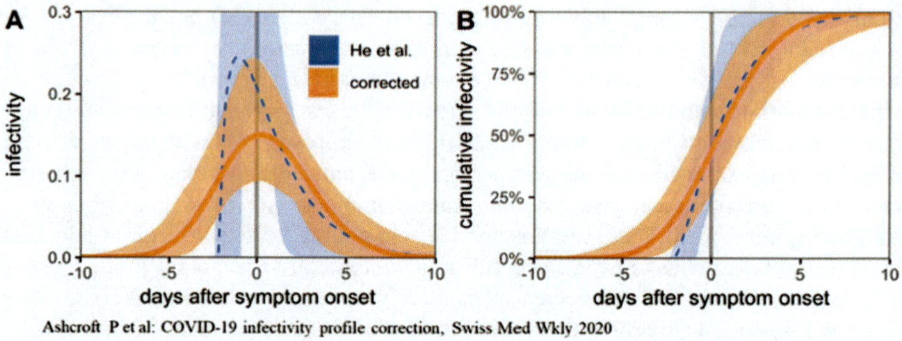

Fig. 2. Infectivity before and after onset of symptoms of the germ carrier.

2.4 Is There a Viral Dark Net in COVID-19?

The goal of vaccination is that pathogens entering the body are neutralized, particularly by the antibodies, thus preventing severe illness and death. It should therefore be expected that vaccinated persons will remain susceptible to SARS-CoV-2, so that after infection - as in unvaccinated persons - there will be an accumulation of the viruses in the nasal cavity. It is therefore to be expected that they will infect others. They themselves will remain symptom-free if their vaccination protection is sufficient. This also applies if the infected person is also vaccinated. Then both remain inconspicuous, as do all other vaccinated persons who become contaminated. In this way, a "viral dark net" is created, which leads to the seemingly incomprehensible illness of an unvaccinated person when he or she has been infected by a symptom-free vaccinated person.

2.5 The Mystery of the Temporary Disappearance of Endemic Occurrences of COVID 19

The transition from epidemic to sporadic occurrence of COVID-19 cases in the summer of 2020 presents a particular challenge in many states. The problem can be demonstrated with data from Carinthia. Carinthia is a relatively small province in Austria that is a particularly popular vacation destination. In summer 2020, therefore, tourists took advantage of the freedom regained after the lockdown to an unprecedented extent on their vacation in Carinthia. Therefore, a very high rate of contacts also with non-locals can be expected. Nevertheless, all findings related to COVID-19 decreased massively: The last death of the 1st wave (Nr 13) was registered in Carinthia on May 3. Nr 15 did not die until Oct 23, 2020. The number of daily PCR-positive new cases can be seen in Fig. 3. Although findings from across the country were considered over 13 days, over long periods of time the effective reproductive rate could not be determined [16].

In any case, these infections and especially the extent of the second wave with 507 deaths cannot be explained by vaccinated carriers. Vaccination was not available in Austria until January 2021. Currently, it is being demonstrated - but only using immuno-suppressants and AIDS as examples - that SARS-CoV-2 can be present in the body for an extended period.

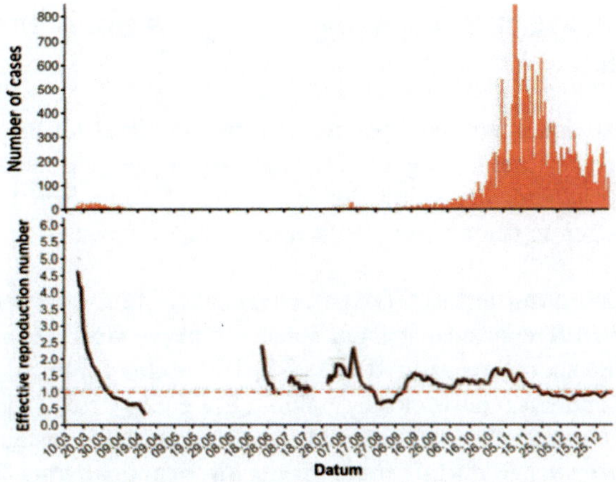

Fig. 3. Daily incidence of laboratory-confirmed new disease, Carinthia and estimated effective reproductive rate related to the previous 13 epidemic days, Carinthia 2020 (AGES).

2.6 Prognosis of the Course of an Epidemic

The forecasts for pandemic progression are based on simulation models created. Their progenitors are Kermack and McKendrick. They also extended the two ways of dealing with infectious epidemics given in 1927 [8]: To completely isolate places from the outside world during the epidemic, as in the Middle Ages, or to immunize people artificially. If it is possible to delay the frequency of contact between germ carriers and infected persons, the number of new cases can be reduced. In order to prove this mathematically, they had to assume that all other influencing variables were constant. This also applies to the infection rate. In practice, however, this rate changes if the susceptibility of the persons or the virulence of the pathogen changes. They were aware that this can happen all the time in practice. They therefore cautioned against overlooking these biological, physiological, and psychosocial influences, stating, "Thus, a small increase in the infectivity rate can cause a very pronounced epidemic in an otherwise epidemic-free population," while even a small decrease would dampen the epidemic. However, simulation models are mostly based only on behavioral surveys. COVID-19, however, is an infectious disease, so it is ultimately the expression of a biological process for which, for example, "behavior" can only be a necessary explanation, but never a sufficient one. As long as the physiologically decisive influences are not built into the models as variables, adverse changes that have occurred or beneficial possibilities cannot be detected. Any deviations of observed values from predicted values can only be attributed to noncompliance with the predicted variable, e.g., the assumed frequency of contact. The existing models do not appear to many to be powerful enough, prompting calls for a new U.S. federal institute. However, it also seems necessary to rethink the entire strategy [14].

3 Approach to a Comprehensive Strategy Against the Era of Pandemics

SARS-CoV-2 is just one of so many potential virus candidates for the next pandemic - as IPBES reports [7]. Therefore, a generally applicable strategy is needed [11, 13]. It is recommended to systematically follow the causal chain leading from the appearance of the unknown pathogen to death from the disease it causes in 6 steps.

#1 **SARS-COV-2 is a mutant of CORONA viruses that has jumped from animals to humans:** IPBES reported that about 800,000 types of viruses can jump to humans in a comparable short time as SARS-CoV-2. The reason for this is the way we live together with animals, constrict their habitat, use energy, etc. Therefore, an end of the pandemics is not to be expected without comprehensive sustainable action.

#2 **The viruses are spread mainly through the air, practically not through contact with solid objects:** Therefore, filtering the air is a priority, e.g., in hotels, cruise ships, department stores, schools… and the mask is helpful; other measures are secondary

#3 **Without contact, no contamination:** Therefore, preventive blocking of access to and leaving of hotspots and segregation of contact persons are useful.

#4 **Without contamination no penetration of viruses through e.g. nasal mucosa;** Therefore no infection: CONTAMINATION - i.e. the entry of SARS-CoV-2 into the nasal, pharyngeal or lung of a person - is NOT identical with infection, but the beginning of the incubation period. Contamination is thus a necessary condition, but not a sufficient explanation for the occurrence of infection.

To avoid misinterpretation, the term "infection" should only be referred to as the end point of the incubation period, i.e. when viruses have been able to penetrate into the organism, but not to mean the "infection DISEASE". The disease is the consequence of the infection. Step 4 is thus the most critical phase for the spread of the epidemic.

The non-specific extracorporeally effective defense becomes decisive: cells of the outer boundary of the body are able to produce, for example, the substance N-chlorotaurine. This has an antiseptic effect also against SARS-CoV-2 without affecting the body's own cells. For more than 20 years, this tolerant substance can be produced synthetically [5]. It could be used as a nasal spray, which has not been done so far by governments.

Another antiseptic is nitric monoxide. This substance was approved as medical product in Israel by the Minister of Health by emergency decree as a nasal spray even for children 12 years of age and older.

#5 **Without infection, no disease and no illness: thus no need to provide ICU beds, no indirect consequences (lockdowns…) etc.:** If an antiseptic were used at the same time as the test, it would decrease the likelihood that the person who is currently in the incubation period would become ill. The antiseptic would reduce the viral load in the nose. This would also reduce the risk of contaminating healthy people on one of the days until the next test. Not giving an antiseptic at the same time is a missed opportunity in the fight against new mutants also. Purely theoretically, an epidemic could be prevented with the help of systematically applied and used antiseptics, as long as all infected persons could be captured in the process and further contacts prevented. This option is no longer available for COVID-19. But it should be central in all considerations to counter future epidemics with unknown pathogens: The extracorporeal natural defense and its support, e.g. by synthetic NCT, is effective against a large number of viral, bacterial pathogens, protozoans and fungi. But this would not make the persons thus kept healthy immune. Antibody formation requires stimulation within the body. This can only be achieved by disease or by vaccination. After the successful use of antiseptics, the risk of the emergence of a new wave would remain.

#6 **Without severe disease - no need for specific medication, no admission to ICU, no death from or with COVID-19, no Long COVID etc:** To date, there is no specific therapy against COVID-19, so one can only promote the non-specific possibilities of the organism to cope with the disease. Particu-larly critical in COVID-19 is the lack of oxygen.

Oxygen is a prerequisite for the function of all eukaryotic cells. Therefore, precautionary measures that increase the uptake and availability of oxygen are expected to enhance the natural healing process. Support of the healing process can also be expected by inhalation of a lung-tolerated antiseptic. This would reduce the reinfection by viruses that are released into the lungs [15].

The lack of specific therapeutic agents is part of the reason for the frequency with which COVID-19 patients must be referred to ICU. This creates competition with others who would need this care. This can and has led to triage situations. Something similar must be considered e.g. in the care of children who need psychiatric care to prevent, for example, suicide.

If one assumes that the lack of available resources leads to the occurrence of symptoms, because no homeostasis can be achieved, then the lack of available oxygen may be one of the reasons why pre-damaged persons suffer more from severe courses of COVID-19 and why COVID-19 diseases can lead to an intensification of the consequences of other, already existing, pre-damages that were at best still sufficiently adaptable. This would make the connection between "died of" or "died with" COVID-19 more insightful. These assumptions also justify the hope of successfully influencing processes of rehabilitation after COVID-19 by using hypoxia-hyperoxia approaches [2, 4].

4 Summary

Omicron should be seen as a wake-up call to also exploit previously overlooked opportunities in the fight against COVID-19. In particular, this relates to opportunities to strengthen nonspecific mechanisms of action, whether through the use and development of compatible antiseptics or techniques to enhance adaptive capacity, e.g., but improving the uptake and delivery of oxygen. Non-specific methods have the advantage of reducing the risk of infection when a pathogen first appears, thereby helping to bridge the time gap until specific tools (vaccines, drugs) have been developed and distributed worldwide.

Strategies are needed that take into account the changing pandemic situation in a problem-oriented manner. In doing so, it will be essential to consistently address, now and after pandemic containment, the reasons identified by IPBES why some 800,000 potential pathogens threaten humanity.

COVID 19 is ominously exacerbating divisions within communities and between North and South. Everyone should strive to counteract this. It could help, for example, to make clear why, in the absence of knowledge but in the presence of a massive threat, measures have to be taken on the basis of evaluative logical considerations, if only in order to act ethically and in conformity with the law. It is in the nature of a pandemic occurring for the first time that much may not be known, but action and inaction must be justified equally well.

The pandemic is not over until it is over everywhere.

References

1. Ashcroft, P., et al.: COVID-19 infectivity profile correction. arXiv preprint http://arxiv.org/abs/2007.06602 (2020)
2. Afina, A.B., et al.: The effects of intermittent hypoxic-hyperoxic exposures on lipid profile and inflammation in patients with metabolic syndrome. Front. Cardiovasc. Med. **8**, 700826 (2021)
3. Bourla, A.: Corona-Impfstoffe könnten bald unwirksam sein (2021). https://www.businessinsider.de/wissenschaft/corona-impfstoffe-koennten-bald-unwirksam-werden-pfizer-chef-will-impfstoff-entwicklung-innerhalb-von-100-tagen-d/42
4. Glazachev, O.S., Kryzhanovskaya, S.Y.: Prospects for adaptive medicine techniques in the era of the new coronavirus pandemic. Herald Int. Acad. Sci. Russ. Sect. **1**, 58-63 (2021)
5. Gottardi, W., Nagl, M.: N-chlorotaurine, a natural antiseptic with outstanding tolerability. J. Antimicrob. Chemother **65**, 399–409 (2010)
6. He, X., et al.: author correction: temporal dynamics in viral shedding and transmissibility of COVID-19. Nat Med. **26**(9), 1491–1493 (2020). https://doi.org/10.1038/s41591-020-1016-z
7. Daszak, P.: IPBES: Workshop Report on Biodiversity and Pandemics of the Intergovernmental Platform on Biodiversity and Ecosystem Services. IPBES secretariat, Bonn, Germany (2020). https://doi.org/10.5281/zenodo.4147317
8. Kermack, W.O., McKendrick, A.G.: A contribution to the mathematical theory of epidemics. Proc. R. Soc. Math. Phys. Eng. Sci. **115**(772), 700–721 (1927). https://doi.org/10.1098/rspa.1927.0118
9. Kofler, W., et al.: Toxicopy and the model of complex coping. In: Proceedings of the 7th IUAPPA Regional Conference on Air Pollution and Waste Issues, 2 – 4 November, Taipei, vol. I, pp. 37–46 (1994)
10. Kofler, W.: Environmental medical expertise in administrative proceedings In: Janauer, G.A., Kerschner, F., Oberleitner, F. (eds.). The Expert Witness in Environmental Proceedings. Vienna, Manz, pp.152–215 (1999). (in German)
11. Kofler, W., Glazachev, O.: Vaccination of the population is necessary but not sufficient against the era of pandemics, e-Letter to Plotkin SA: What have we learned from the COVID-19 plague? Sci. Trans. Med. **13**, 611 (2021). https://doi.org/10.1126/scitranslmed.abl9098
12. Kofler, W., Glazachev, O.S., Lyshol, H., Tellnes, G.: Is fighting against COVID 19 enough? Scand. J. Pub. Health **49**(1), 9–13 (2020). https://doi.org/10.1177/14043948209069539
13. Kofler, W., Glazachev, O.S.: A Guide Through The COVID-19 Jungle. Herald of the International Academy of Science, Russian Section, Special Issue 1 (2021). http://www.heraldrsias.org/journals/2021/special/2/

14. Kofler, W.W., Nagl, M.: The whole strategy has to be extended now – beyond modeling, eLetter to Press WH, RC Levin: modeling, post COVID-19. Science **370**, 101 (2021). https://doi.org/10.1126/science.abf7914
15. Lackner, M., et al.: N-chlorotaurine, a novel inhaled virucidal antiseptic is highly active against respiratory viruses including SARS-CoV-2 (COVID-19). Res. Sq (2020). https://www.researchsquare.com/article/rs-118665/v1
16. Richter, L., Schmidt, D., Chakeri, A.S., Maritschnik, S., Pfeifer, S., Stadlober, E.: Epidemiological parameters of the COVID-19 outbreak - Update 03.01.2021 Austria, 2020 AGES (2020). in German

A Sample Study for Determining Energy Consumption Values in Public Buildings: Central Anatolia Region

Selmin Ener Rusen[1](✉) and Aydın Rusen[2]

[1] Karamanoglu Mehmetbey University, Energy Systems Engineering, 70200 Karaman, Turkey
selmin.enerrusen@gmail.com

[2] Springer Karamanoglu Mehmetbey University, Metallurgical and Materials Engineering, 70200 Karaman, Turkey
aydinrusen@kmu.edu.tr

Abstract. Today, it is known that the role of energy efficiency in the development of countries is very crucial. When our country is viewed from this point of view, it is seen that energy efficiency studies aimed to reduce the energy density value have become more and more important and policies are implemented in this direction. Therefore, in this study, an exemplary study was conducted to determine the saving potential of the energy used in Turkey. In order to find energy usage in the public buildings in KOP Region, which includes 7 cities in Central Anatolia Region (Karaman, Aksaray, Kırıkkale, Kırşehir, Yozgat, Nevşehir, and Niğde), the amount of energy consumed per person and per square meter was determined. The calculation was made based on the energy consumption values obtained from public buildings (administrative buildings, schools, dormitories, sports facilities, prisons, and universities) in 7 cities in the KOP Region between 2014–2018 years. According to the obtained data, it is observed that energy consumption is high in mass living areas of public buildings. The results are encouraging for the future works to use energy efficiency in public buildings.

Keywords: Energy saving potential · Public buildings · Central anatolia region

1 Introduction

Sustainable energy is one of the most important issues in the world. Rapidly increased the demand for energy, unconsciously the consumption of fossil resources for this demand, and the negative effects of these fossil resources on the environment seriously cripple sustainable energy. In this sense, renewable energy has an important role in the maintaining sustainable energy. It is known that solar, wind, biomass, hydropower, geothermal, and energy efficiency are described as renewable energy sources. Among these renewable energy sources, energy efficiency could be considered to be the cheapest, best, and one of the most used solutions systems [1, 2]. Today, it is known that energy efficiency and

H. Gökçekuş and Y. Kassem (Eds.): NRSEM 2021, EESCI, pp. 100–109, 2022.
https://doi.org/10.1007/978-3-031-04375-8_12

even the use of zero energy are very important for all countries. In particular, the European Council has been adopted energy efficiency as the keystone of European energy policy, and "Energy Efficiency First" is one of the key principles of the Europe 2030 strategy [2, 3]. In our developing country, on the other hand, environmental problems and dependence on foreign energy are increasing due to energy demand. According to the reports of the Ministry of Energy and Natural Resources, our primary energy demand will continue with high increases in the coming years. In this context, the key factor in the field of sustainable development for our country is "Energy Efficiency" studies [4]. Increasing the efficiency and quality of energy at every point of the production to consumption, reducing energy density by reducing unnecessary use and preventing waste will increase the welfare level of our country [5, 6]. In addition, it is extremely important for our country to increase the energy efficiency, develop studies on sustainability and waste management, use renewable energy, train Energy Managers that our country needs, and realize energy efficiency increasing projects in industry and the public sector. Buildings (residential, commercial, and public) use one-third of Turkey's total final energy consumption [7]. As a result, the building sector consumes the same amount of total final energy as the manufacturing industry in Turkey. With increasing rates of urbanization and population expansion, total demand for energy in buildings is likely to rise even more [1]. This paper summarizes the findings of research conducted on the subject of public building energy benchmarking and applied to the building stock of the KOP Region in Turkey, based on a thorough assessment of the each region's energy performance. In accordance with this purpose, the amount of energy spent per person and per square meter in public buildings in KOP Region, which encompasses 7 cities in Central Anatolia Region (Karaman, Aksaray, Krkkale, Krşehir, Yozgat, Nevşehir, and Nide), was determined. Between 2014 and 2018, energy consumption values from public buildings (administrative buildings, schools, dorms, sports facilities, prisons, and universities) in seven cities in the specified region were calculated.

1.1 Energy Consumption in Public Buildings

The buildings vary greatly in terms of energy consumption and intended purpose, which means that some are considerable energy efficient, on the other hand, others have a lot of potential for improvement. For example, according to a research issued by the Buildings Performance Institute Europe (BPIE), in 2011, buildings which used for education have the third greatest building stock and approximately 10% of total energy consumption by the construction industry in Europe. In brief, the construction sector accounts for roughly 40% of total final energy consumption in the developed countries. Among these buildings, the government ones account for a major share of global energy consumption [1, 2, 10].

The space and water heating leads to more than half of all public buildings' energy demand. Table 1 shows the typical electricity and fuel consumption as a percentage of public buildings in Europe, according to usage areas.

Table 1. Typical electricity and fuel consumption of public buildings [2].

Electricity consumption	Consumption (%)	Fuel consumption	Consumption (%)
Air conditioning & Fans	50	Space heating	75
Lights	15	Refrigeration	12
Refrigeration	13	Generator	10
Office equipment	10	Other	3
Water heater/ Elevator	12		

Field tests on individual energy consumption in the public buildings show that energy consumption on office computers has contributed to about 30% of energy demand in the European service sector over the last decade [11, 12]. In Europe, data and studies show that a large quantity of IT office equipment is underused and kept in overnight in public buildings [13, 14]. Because of the building users leave lighting and some equipment on at the end of the day, nearly 56% of total energy used in public buildings is spent during the non-working hours [15]. Researchers have found that behavioral change could reduce energy consumption in public and commercial buildings by up to 30–40% [16]. The most common barrier to energy saving is that there is no financial benefit from saving energy since the bill is paid by the government and not by the office worker [15, 16]. Existing research on energy usage in public buildings frequently emphasizes the adoption of technological energy-saving solutions in public buildings and the possibility for behavioral change.

1.2 Energy Consumption in Public Buildings in Turkey

Turkey has a high rate of urbanization approaching a growth rate of 2% per year and new construction rates nearly 4%, indicating that the building stock is rapidly expanding. In addition, the developing construction industry is one of the most significant drivers of the Turkish economy. Turkey has approximately 23 million dwellings and 9.1 million structures, also 100,000 new structures have been added to these stocks each year on average. Due to the increase in the energy consumption of the sector and the effect of the pandemic, the construction sector has essentially become Turkey's largest energy user in the last few years [1]. As a result, addressing building energy consumption is an important approach to minimize reliance on imported energy for Turkey and meet EU year 2050 commitments. In addition, energy demand in the housing sector accounts for more than half of the final energy consumption of the entire construction sector. The rest is filled with public and commercial buildings. Nevertheless, there is insufficient data in the available statistics to allow a further analysis of this subject [13].

The majority of Turkey's housing stock is made up of relatively new homes. Nonetheless, old structures constructed before the year 2000 have been exist. Since then, multi-family residential structures have accounted for almost 80% of all new developments. Non-residential structures make up the majority of the remaining 20%. Besides, by the year 2030, the Urban Transformation Plan aims to renew 7.5 million houses in Turkey.

The national figures gathered by the General Directorate of Energy Affairs and the international statistics compiled by the OECD/IEA give a thorough analysis of Turkey's construction industry energy demand [1, 13]. Today, it is known that energy efficiency and even the use of zero energy are very important for the development of countries. Also, it is very essential to research the efficient use of energy in Turkey. An exemplary study was conducted to determine the saving potential of the energy used in Turkey by the Ministry of Energy and Natural Resources, General Directorate of Renewable Energy (YEGM) [16].

Table 2. The potential for savings in electricity and fuel consumption as a result of the energy audit study in public buildings built by YEGM in 2017 in Turkey [16].

Public building	Count		Electricity saving (TEP/Year)	Saving (%)		Fuel saving (TEP/Year)	Saving (%)
Schools (72)	359	Lights	114,12	15	Space heating	1213,36	25
Dormitories (13)	51	Lights	131,09	20	Hot water	1105,22	30
University (9)	180	Lights	792,33	20	Space heating	6163,36	20
Hospital (25)	186	Air conditioning & Fans	2532,84	15	Space heating	4110,26	35
Administrative buildings (36)	175	Air conditioning & Fans	429,86	10	Space heating	2843,09	35
Airports (8)	59	Air conditioning & Fans	843,16	20	Space heating	792,97	25
Prison (3)	35	Lights	110,7	25	Space heating	969,74	60

Here mentioned study, a survey was conducted in 166 public buildings in Turkey between 2014–2015, and the results were shared in 2018. Table 2 summarizing the savings potentials in electricity consumption and fuel consumption in public buildings in Turkey as a result of the energy audit study by YEGM in 2017 [16]. Total energy saving potential to be obtained as a result of energy audit work in public buildings built by YEGM in 2017 in Turkey is given at 20% [16]. In the light of the above information, it can be said that a more thorough examination of the effects of various factors on energy savings in public buildings is required.

2 Data Collection and Analysis

The buildings analyzed under the present case study have different locations in Central Anatolia Region (Karaman, Aksaray, Krkkale, Krşehir, Yozgat, Nevşehir, and Nide) of Turkey. However, these selected cities are similar in terms of climate characteristics and are in Temperate Continental (Dc) climate type according to Threwartha [17] climate classification.

Heating-degree-day (HDD) value is one of the most important parameters in energy calculation in buildings. The fact that the cities selected in this study are of the same climate type is important in terms of ensuring that the HDD value is considered almost the same. Also, these selected cities have similar longitude, latitude and altitude values. A list of the selected stations' names and their geographical locations together are presented in Table 3 and locations are given in Fig. 1.

The available data set contains a five-year period (between 2014 and 2018) for the seven stations and it consists of consumption values obtained from energy bills paid by public buildings and some data to be correlated with energy consumption.

The public buildings have specific characteristics that make the task of consuming energy different from that performed in other types of buildings. These include buildings dedicated to lecturing, offices, hospitals, other research facilities, libraries, air-ports, prisons, and the others. These kinds of public buildings are frequently organized from a single point, and their one-line energy supply infrastructure is sometimes shared. Therefore, the issue of determining individual usage is a significant difficulty because most public buildings are not equipped with partial energy meters.

Table 3. Geography and climate type information on selected stations.

Number	Selected stations	Latitude	Longitude	Altitude (m)
1	Karaman	37.18 N	33.22 E	1039
2	Aksaray	38.36 N	34.03 E	980
3	Niğde	37.96 N	34.68 E	1229
4	Kırıkkale	39.84 N	33.51 E	712
5	Kırşehir	39.15 N	34.16 E	985
6	Yozgat	39.82 N	34.81 E	1300
7	Nevşehir	38.62 N	34.71 E	1224

Each public building's necessary information was determined from the obtained from public staff available on site with the required character information form for the public buildings. Information about Building Name, Intended purpose, Net Floor Area (m2)/Year of build, Number of people(Yearly number of total students/patients), Yearly Electric (kWh), Yearly Natural Gas (Sm3), Yearly Coal (ton), Yearly Lignite (ton), and Yearly Wood Consumption (ton) exist in this form.

Fig. 1. Geographical locations for selected 7 stations in Central Anatolia Region [18].

3 Materials and Methods

The work presented in this paper relies on data obtained during energy consumption to the available set of public buildings of the KOP region. KOP region includes 7 cities in Central Anatolia Region (Karaman, Aksaray, Kırıkkale, Kırşehir, Yozgat, Nevşehir, and Niğde). It was an opportunity to use data from a logical group of buildings as an input to a benchmarking procedure. These energy consumption values were specifically designed to obtain data for scientific research for the energy efficiency potential in public buildings and took place in a limited timeframe. In addition, the study mentions and discusses certain constraints on the availability of public building data.

As a result of energy analysis of existing public buildings, we have properly classified the buildings, according to their main final uses and divided them into schools, dormitories, universities, hospitals, social service buildings (sports facilities, teachers' homes, retirement homes, education centers, etc.), airports and prisons.

As a beginning with approach, the specific energy consumption per net floor area was utilized as the reference energy usage indicator (EUI). The EUI was calculated using data received during each building's energy invoicing for the government buildings of the KOP Region between 2014–2018 years. The EUI is calculated as the amount of end-use energy tons of oil equivalent (TOE) delivered to buildings per square meter of floor area over the course of the selected period of each year [TOE/m^2].

Similarly, the reference energy usage indicator of the person (EUIP) is calculated as tons of oil equivalent (TOE) of end-use energy consumed in public buildings per person using the buildings during the selected period of each year [TOE/person]. To evaluate the potential energy savings for each city in KOP region, the total energy consumed by public buildings in the study was considered. The public buildings are often assigned to performance categories based on their performance in comparison to other buildings of the same kind using energy benchmarking methodologies.

Buildings are typically assigned to performance categories based on their performance in comparison to other buildings of the same type using energy benchmarking methodologies. It is known that the categorizing process can begin once the measured data have a sufficient statistical distribution. In this study, the categorization of public buildings was made by taking into account the results of extensive research conducted by YEGM in 2017 in Turkey.

4 Results and Discussion

In order to analyze the energy performance of the public building, we examined the five-year data for a total of 753 public buildings in 7 provinces in Central Anatolia (KOP Region). Figure 2 presents the number of buildings, by selected cities, ranked by available energy consumption data, which used in the calculation. In addition, The selected 7 stations here have been chosen with especially the same climatic types, so that the heating-degree-day (HDD) value, which is one of the most important parameters of energy work in buildings, can be considered the same.

To find energy analysis, the existing public buildings were classified according to their main final uses. These are schools, dormitories, universities, hospitals, social service buildings (sports facilities, teachers' homes, retirement homes, education centers, etc.), airports, and prisons. As a beginner with the approach, the reference energy consumption per net floor area was utilized as the reference energy usage indicator (EUI) and similar, the reference energy usage indicator of the person (EUIP) was computed for all buildings. The reason for this investigation was to reveal if the reference energy usage indicator of the public building can be used for the energy efficiency estimations where there is no measured energy consumption. The calculation was using data received during each building's energy invoicing for the government buildings of the KOP Region.

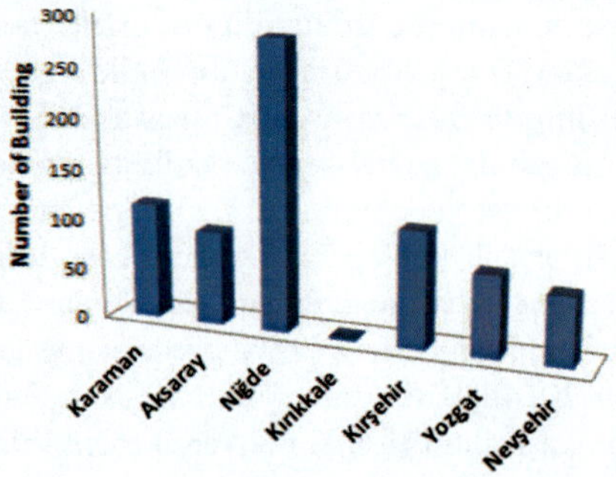

Fig. 2. Total number of available data by province for selected 7 stations.

As mentioned before, total 753 public buildings thermal and electrical energy consumptions (2014–2018) were collected and analyzed before the pandemic years. During the pandemic period, energy consumption values have changed considerably in all kinds of buildings due to the closures.

When public buildings are classified among themselves, it is obvious that energy consumption will differ according to the purpose of use. In this study, the variation of the total energy consumption of public buildings in different provinces has been examined similarly. The heated area m^2 serves as a starting point for calculating the precise energy consumption of buildings. The dispersion of reference (thermal and electrical) energy consumption for analyzed public buildings of each city is presented in Fig. 3.

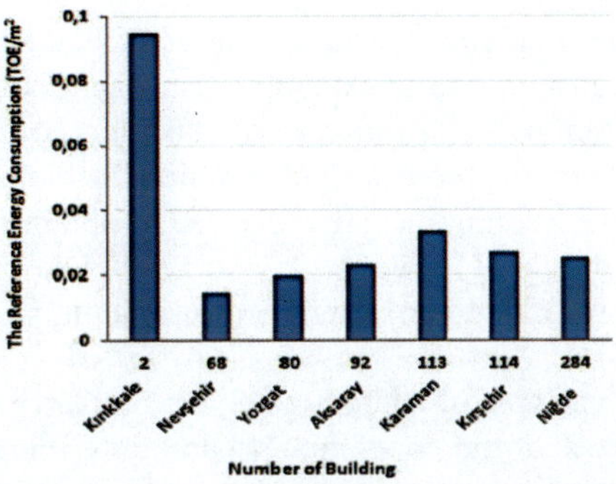

Fig. 3. Scattering of the reference energy consumption of public buildings for 7 stations.

In order to visualise the range of these differences more accurately, Fig. 4 shows the reference energy consumption per net floor area (TOE/m^2) and the reference energy consumption indicator of the person (TOE/Person) in public buildings per person, indicators grouped for each city.

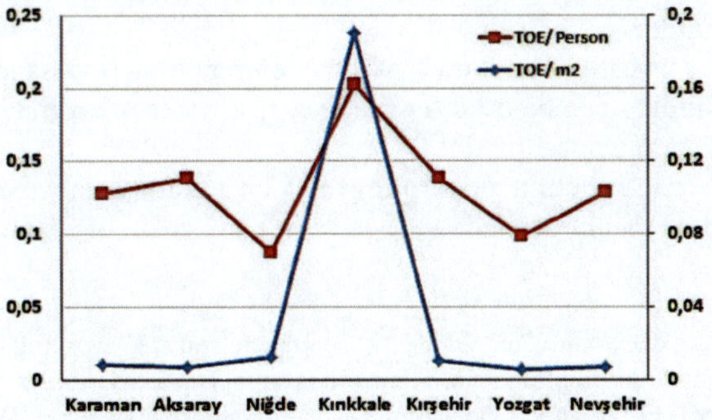

Fig. 4. Total number of available data by province for selected 7 stations

As can be seen in Fig. 3 and 4, the energy consumption for Kırıkkale province is high. Similarly, the highest energy consumption per person is again in this province. Some inconsistencies have been identified in the data received from this station, but we have still included it in terms of subject integrity. In other provinces, the reference energy consumption values per square meter and per person are in harmony. When the graphs are examined, Yozgat and Niğde are the provinces that consume relatively less energy per person.

5 Conclusions

In Turkey, government buildings account for a major share of global energy consumption. While newly constructed buildings may be relatively energy-efficient, the existing old structures, which make up more than 85% of the entire building stock, were erected before energy rating systems were implemented and are, as a result, inefficient to a great extent. Decreasing the energy consumption in existing government buildings is fundamental, because it will not, as it diminished the energy costs, but moreover appear governments' solid commitment towards the decrease of CO2 emission. In this report, calculations were made based on the energy consumption values that can be obtained from public buildings in 7 provinces in the KOP Region between 2014–2018. The following conclusions may be grouped based on an examination of energy consumption in public buildings:

- The KOP Region, with an area of approximately 95,580 km^2, has 12.2% of Turkey's total surface area and 6% of the population. In addition, it is an important region in terms of energy consumption.
- The data obtained are classified according to the purpose of use of the public building. Calculations include information on the number of people reported by public institutions, net floor area (m^2), and consumption in various energy sources (Electricity, Natural Gas, Wood, Coal, Fuel oil, etc.).
- While making the evaluations, all provinces were selected in the same climate type and no climate correction was made. This means that this set of data is more likely to be homogeneous than other constructing databases.
- The development of energy-saving solutions in public buildings is more difficult than in residential buildings.
- As a result of this investigation show that if the reference energy usage indicator (EUI) of the public building can be used for the energy efficiency estimations where there is no measured energy consumption.
- Regular energy consumption monitoring and an energy-saving strategy for each building should be recommended.

Acknowledgements. We would like to thank the managers and staff working in all public buildings for sharing energy consumption data. The work is supported under the project number KOP-2018D000190 by the KOP Regional Development Administration, Republic Of Turkey Ministry of Industry and Technology.

References

1. Enhancing Turkey's policy framework for energy efficiency of buildings, and recommendations for the way forward based on international experiences, SHURA Energy Transition Center Buildings Performance Institute Europe (2019)
2. The European PPP Expertise Centre (EPEC), Guidance on Energy Efficiency in Public Buildings (2018)
3. Zou, P.X.W., et al.: Achieving energy efficiency in government buildings through mandatory policy and program enforcement. Front. Eng Manag. **4**(1), 92–117 (2017). https://doi.org/10.15302/J-FEM-2017101
4. Energy Efficiency in Government Buildings, Low Carbon Communities in the Caribbean Project (LCCC) and the Caribbean Sustainable Energy Program (CSEP) with the financial assistance of the European Union and the United Nations Industrial Development Organization (UNIDO) (2013)

5. Energy, E.U.R.O.S.T.A.T., Transport and Environment Indicators,: Edition, p. 2014. Publications Office of the European Union, Luxembourg (2014)
6. European Commission. European Commission Europe 2020: Integrated Guidelines for the Economic and Employment Policies of the Member States; European Commission: Brussels, Belgium (2010)
7. EİGM, Enerji İşleri Genel Müdürlüğü - 2016 Yılı Ulusal Enerji Denge Tablosu (2017). eigm.gov.tr
8. Bernardo, H., Oliveira, F.: Estimation of energy savings potential in higher education buildings supported by energy performance benchmarking: a case study. Environments **5**, 85 (2018)
9. Cibinskiene, A., Dumciuviene, D., Andrijauskiene, M.: Energy consumption in public buildings: the determinants of occupants' behavior. Energies **13**(14), 3586 (2020).https://doi.org/10.3390/en13143586
10. Murtagh, N., et al.: Individual energy use and feedback in an once setting: a field trial. Energy Policy **62**, 717–728 (2013)
11. Mulville, M., Jones, K., Huebner, G.: The potential for energy reduction in UK commercial uces through ective management and behaviour change. Archit. Eng. Des. Manag. **10**, 79–90 (2014)
12. Masoso, O.T., Grobler, L.J.: The dark side of occupants' behaviour on building energy use. Energy Build. **42**, 173–177 (2010)
13. Yun, G.Y., Kong, H.J., Kim, H., Kim, J.T.: A field survey of visual comfort and lighting energy consumption in open plan oces. Energy Build **46**, 146–151 (2012)
14. Salleh, M.N.M., Kandar, M.Z., Sakip, S.R.M.: Benchmarking for energy eciency on school buildings design: a review. Procedia Soc. Behav. Sci. **222**, 211–218 (2016)
15. Schleich, J.: Barriers to energy eciency: a comparison across the German commercial and services sector. Ecol. Econ. **68**, 2150–2159 (2009)
16. Energy Efficiency Study in Public Buildings, Implementation Monitoring Report – I, EV-2018–02-V1 (2018). www.yegm.gov.tr
17. Threwartha, G.T.: An Introduction to Climate. McGraw Hill, New York (1968)
18. Extracted from Google maps. http://maps.google.com/ (2021)

Adjustment of the Evaporation Pan Coefficient: Case Study of Konya Closed Basin

Alara Cicibiyik[1](✉), Nermin Şarlak[1], and Deniz Üstün[2]

[1] Engineering Faculty, Civil Engineering Department, Karamanoğlu Mehmetbey University, Turkey, Japan
alarabolgenn@gmail.com, nsarlak@kmu.edu.tr
[2] Engineering Faculty, Computer Engineering Department, Tarsus University, Turkey, Japan
denizustun@tarsus.edu.tr

Abstract. Global warming is an accepted fact by everyone, even if climate change still seems to be a controversial issue. The increase in temperature caused by global warming increases the rate of evaporation from open water surfaces. Although it is necessary that the presence of water and removing the saturated vapour layer from the air with the wind to keep on evaporation, temperature is an essential factor. Evaporation is the most difficult component of the hydrological cycle to determine. Evaporation can be estimated by various methods. One of them is the evaporation pan even if the value measured cannot be expected to represent evaporation in large bodies of water. The pan is more easily affected by changes in temperature in the air. Another is the empirical evaporation estimation methods based on measured meteorological parameters. One of these methods is Penman. The Penman method is used as a reference method in studies related to evaporation since it gives the closest results to the measured data in different parts of the world. In this study, evaporation rate was estimated by the Penman method using data measured at five meteorological stations in the Konya Closed Basin. The aim of the study is to recalculate the pan coefficient, which is used as 0.7 in Turkey. The Artificial bee colony algorithm, one of the statistical optimization techniques, was used for this purpose. It was concluded that the measurements would close to Penman estimates if the pan coefficient was taken as 0.79 for the Konya Closed Basin.

Keywords: Evaporation · Penman · Pan coefficient · Konya Closed Basin

1 Introduction

The water varies in quality and quantity on Earth, though it is the most important source of life for living things. The world's population, which is increasing day by day, will probably face with water problems due to overuse of existing sources and pollution. The world is experiencing problems depending on the global warming with clean water due to excessive temperature rise, lack of precipitation, an increase in evaporation, industrial activities and pollution. The greenhouse gases in the atmosphere have increased the temperature of the Earth. It is believed that the most important parameter that has a

H. Gökçekuş and Y. Kassem (Eds.): NRSEM 2021, EESCI, pp. 110–118, 2022.
https://doi.org/10.1007/978-3-031-04375-8_13

direct effect on evaporation is the surface temperature. Therefore, as surface temperatures because of global warming increase, evaporation also increases. For this reason, accurate determination of losses caused by evaporation is of great importance in terms of water resources [1].

Experimental and empirical methods for evaporation estimation have been developed. The evaporation pan is the one of the experimental methods. Although its cost is low, its capability to represent evaporation on land or in larger bodies of water is restricted. The evaporation pan is more easily affected by temperature changes in the air. Heat reflection of the pan, heat exchange around the pan and low humidity of the environment also affects the evaporation. Several different standard types of pans (metal containers) are used by different countries. Class A pan is preferred to use in Turkey [2]. The pan is 1.22 m in diameter by 25.4 cm high and set on a low platform. The pan is usually filled with 20 cm of water and evaporation is measured with a ruler called a limnimeter. The pan measurement values are multiplied by a coefficient to approach the open surface evaporation value. This coefficient, known as the pan coefficient, is taken as 0.7 in Turkey. The limits of change of this value are 0.6–0.8. It is assumed that 70% of the annual evaporation amount from the pan will be equal to the annual evaporation that may be occurred from a lake near the pan [3].

Empirical methods for evaporation estimation are based on mass-transfer, energy balance and approaches in which both are considered together. These methods' equations were derived from meteorological observation parameters [4]. One of these methods is Penman. The Penman method is used as a reference method in studies related to evaporation since it gives the closest results to the measured data in different parts of the world. Many studies related to evaporation have been studied both in our country and around the world from the past to the present [5–15].

In this study, evaporation values estimated by the Penman method for selected stations in the Konya Closed Basin were compared with pan measurements. This basin is faced with water shortage problem in these days. Water is transported from adjacent basin to this basin by Blue Tunnel. Thus, this basin has no chance of losing any drop of water. The aim of the study is to recalculate the pan coefficient used in Turkey to estimate evaporation amounts correctly. The Artificial bee colony algorithm, one of the statistical optimization techniques, was used for this purpose. It was concluded that the measurements would close to Penman estimates if the pan coefficient was taken as 0.79 for the Konya Closed Basin.

2 Material and Method

2.1 Penman Method

In 1948, Penman combined energy balance approaches with mass-transfer (E_a) to develop an evaporation equation that does not require surface temperature data [16]:

$$\lambda E = \frac{\Delta \cdot (R_n - G) + \gamma \cdot E_a}{(\Delta + \gamma) \cdot \rho_w} \tag{1}$$

$$E_a = 6.43\,(a_w + b_w * u_2)\,(e_s - e_a) \tag{2}$$

where E is evaporation rate from open water surface [mm/day]; Δ is slope of the relation between saturated vapor pressure and temperature [kPa /°C]; λ is latent heat of vaporization [MJ/kg]; G is soil heat flux which is not taken into account for open surface evaporation [MJ/m^{-2}day^{-1}]; γ is psychometric constant [kPa /°C]; u_2 is wind speed measurement at 2 m height [m/s]; R_n is net radiation [MJ/m^{-2}day^{-1}]; e_s is saturated vapor pressure [kPa]; e_a is actual vapor pressure [kPa]; E_a is the mass transfer [MJ/m^{-2}day^{-1}]; a_w and b_w are the empirical constants; ρ_w is the mass density of water [1000 kg/m^3].

2.2 Artificial Bee Colony Algorithm (ABCA)

The ABC algorithm which inspires from the honeybee behaviors was developed by Karaboğa [17]. There are some assumptions in the proposed algorithm. The first assumption is that there are food sources in the algorithm and the position of each food source is represented a possible solution in the search space of the problem. Another is that the algorithm has three virtual bees consisting of employed, onlooker and scout. In addition, only one bee is assigned to each food source and the bees try to search a better food source by exhausting their food sources. The search process is iteratively performed by bees at each of these phases. In the search process, the main aim is that food source with most nectar is to find. The model related to the food search behaviors of bees is given in Fig. 1 [17].

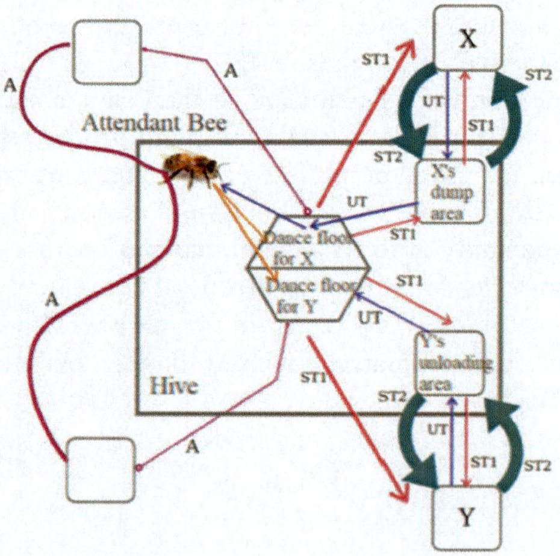

Fig. 1. ABC algorithm flowchart

The initial nectar positions in the search space of the problem are performed by using a mathematical operation that can generate random values among to lower and upper limits of each of its parameters.

$$x_{ij} = x_{minj} + rand * (0, 1) * \left(x_{maxj} - x_{minj}\right) \quad (3)$$

where $i = 1....SN, j = 1...$ D and SN are the number of food sources and D is the number of parameters to be optimized. x_{min_j} is the lower limit of the parameters. The number of food sources is equal to the number of bees on duty. The worker bee determines a new food source in the neighborhood of the food source where it works and evaluates the quality of it. If the new resource is better, it memorizes the new resource:

$$v_{i,j} = x_{i,j} + f_i * \left(x_{i,j} - x_{k,j}\right) \tag{4}$$

For each resource denoted by x_j, a single parameter (randomly selected parameter, j) of this resource, i.e. its solution, is changed, and the v_i resource is found in the neighborhood of x_i. If the $v_{i,j}$ generated as a result of this operation exceeds the previously specified parameter limits, j. moves to the lower or upper bound values that belong to the parameter:

$$x_{ij} = \begin{bmatrix} x_j^{min}, \ v_{ij} < x_j^{min} \\ v_{i,j}, \ x_j^{min} \leq v_{ij} \leq x_j^{maks} \\ x_j^{min}, \ v_{ij} > x_j^{min} \end{bmatrix} \tag{5}$$

The v_i parameter vector produced within the limits represents a new resource, and its quality is calculated and assigned a compliance value:

$$fitness_i = \begin{bmatrix} 1/(1 + f_i) \ f_i \geq 0 \\ 1/abs(f_i) \ \ f_i < 0 \end{bmatrix} \tag{6}$$

3 Application Area

The Konya Closed Basin is located in the Central Anatolia Region of Turkey. The total area of the drainage basin, which covers about 6.4% of Turkey's area, is 49,786 km^2 [18]. Historical data of daily maximum, minimum and mean humidity, maximum, minimum and mean temperature, mean wind speed and pan evaporation data for the period 2000–2019 were provided by Turkish State Meteorological Service (DMI). Figure 2 shows the geographic location of the meteorological stations. Site information including altitude, latitude, longitude, station name and number and observation period for each selected stations are given in Table 1.

Fig. 2. Geographic location of meteorological stations

Table 1. Site information including altitude, latitude, longitude, station name and number and observation period for each station.

Station name	Station No.	Altitude (m)	Latitude	Longitude	Observation period
Aksaray	17192	970	33.59	38.22	2000–2019
Ereğli	17248	1046	34.02	37.31	2000–2019
Karapınar	17902	996	33.31	37.42	2000–2019
Karaman	17246	1026	33.13	37.11	2000–2019
Konya	17244	1018	32.34	37.59	2000–2018

4 Results and Discussion

4.1 Missing Data Completion

Missing daily solar radiation data of five stations in the study were completed by a radial-based function (RBF) surrogate interpolation method [19]. The missing solar radiation data was tried to be estimated by the measured daily humidity, temperature and wind speed data. An independent sample Student-t test was performed to check whether the completed solar radiation data was significant [20]. Since the p values calculated from the Student-t-test at the 5% significance level were greater than 0.05, the null hypothesis assuming that solar radiation data does not have a significant difference between each other was accepted.

4.2 Homogeneity Tests

There are many methods in the literature to test the homogeneity of the data. Pettitt, Buishand, Standard normal homogeneity Test (SNHT) and Von-Neumann tests were used to check the homogeneity of the data in this study [21].

Minimum-mean humidity and mean wind speed data were found to be inhomogeneous in all stations. In the solar radiation data, homogeneity was deteriorated only in the Konya station. It was observed that SNHT gave rejection in all parameters. Before deciding to utilize the original data for inhomogeneous data, double-mass curve analysis was performed since this method is used to detect the deterioration of homogeneity in stations' data (change of station location, errors in the measuring instrument, etc.) [22]. We could not detect systematic error between stations. Therefore, it was decided to use the original measured data to prevent the loss of information.

4.3 Comparison of Pan Evaporation Measurements with Estimated Evaporation Data

Pan evaporation measurements were compared with evaporation data estimated by Penman method. The correlation coefficient (r) can be used to test whether the relationship between two variables exists. If the value of r is greater than 0.8, it can be interpreted that there is a good relationship between two variables, while if it is less than 0.5, there is a weak relationship [23].

Daily comparisons of Penman evaporation estimates and pan evaporation measurements during 2000–2019 in Aksaray, Ereğli, Karapınar, Konya and Karaman stations were given in Fig. 3 as a scattering plot.

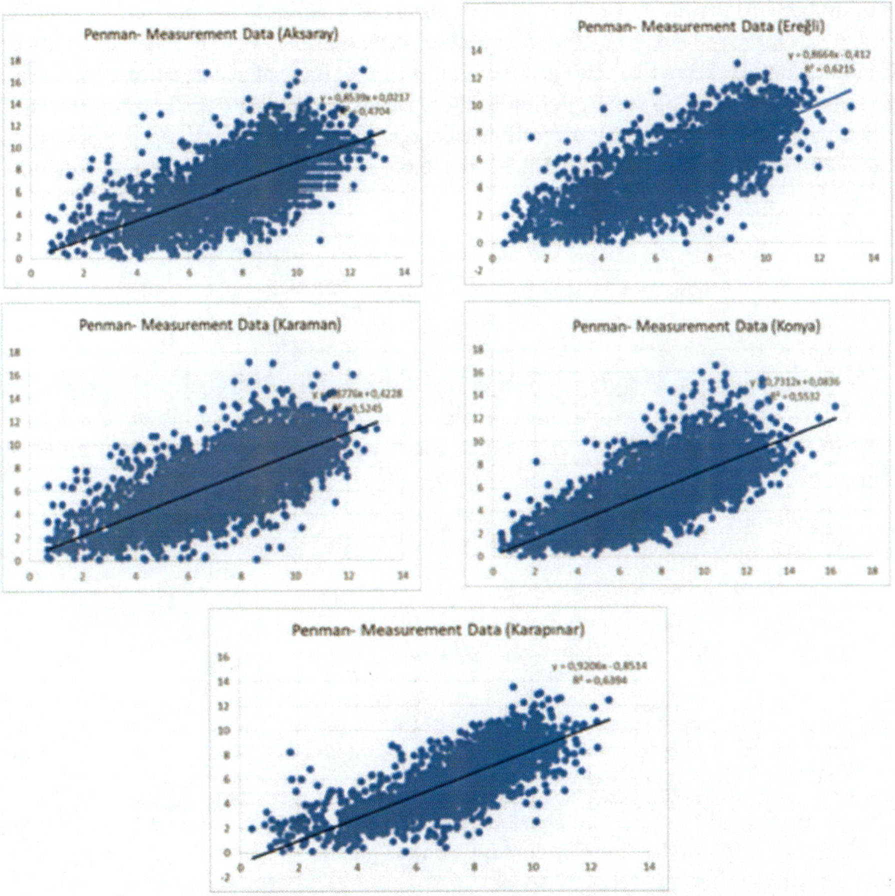

Fig. 3. Penman evaporation estimation versus pan measurement values for stations

There was a linear medium relationship (r = 0.68) between the evaporation pan values and Penman method estimates at Aksaray station. The highest evaporation value of 17 mm observed in the pan measurement was estimated as 12.51 mm by the Penman method. The lowest value of 0.1 mm observed in the pan measurement was estimated as 3.32 mm. It was found nearly strong linear relationships (r = 0.78, 0.79, 0.74 and 0,72) at Ereğli, Karapınar, Konya and Karaman stations. While the highest observed evaporation value of 17 mm was estimated as 10.39 mm, the lowest observed value of

0.1 mm was estimated as 2.74 mm for Ereğli station. Penman method mostly estimated higher value than pan measurements for low measured values.

4.4 Adjustment of the Pan Coefficient with the Artificial Bee Colony (ABC) Algorithm

The number of swarms to be used in the optimization solution was taken as 20. The number of employed bee was 10 and onlooker bee's number was 10. Also, the number of iterations was taken as 20 and the coefficient a (upper limit of acceleration coefficient) was taken as 1 [24]. The stop criterion in the study is the number of iterations. Cost(x) is the difference between the current Penman evaporation estimates and the measured values obtained without multiplying by the pan coefficient (0.7). The cost function, Cost(x) is determined as:

$$\mathrm{Cost}(\times) = 100 \times \left(\frac{\sum_{i=1}^{m} \frac{\left| E_{penreal_i} - E_{penpredict_i} \right|}{E_{penreal_i}}}{m} \right) \tag{7}$$

The development curve graph given in Fig. 4 shows the curves corresponding to the error values obtained by the optimization algorithm in each iteration. When the data of all stations were organized as a single data set, the pan coefficient was determined as 0.79.

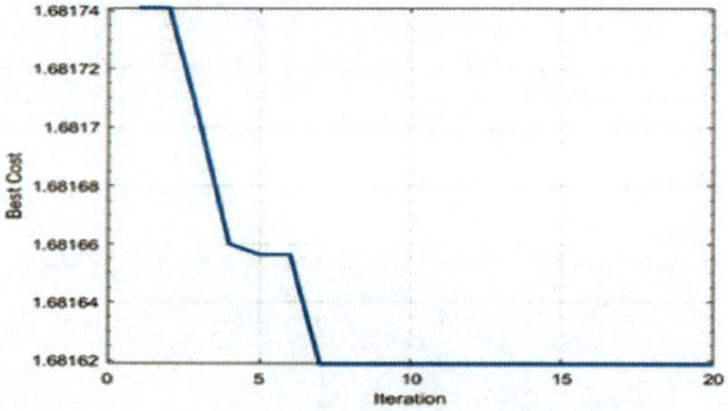

Fig. 4. ABC solution graph

4.5 Evaluation of the Adjustment Pan Coefficient

The original and adjusted pan evaporation measurements obtained by multiplying 0.7 and 0.79 respectively, were statistically compared with Penman estimates in order to see the effect of the adjustment. Regression analysis (R^2), Nash-Sutcliffe (NSE), square root of the mean errors (RMSE) and mean absolute error (MAE) were used for this purpose. The comparisons are presented in Table 2.

Table 2. Statistical comparisons.

	Pan coefficient = 0.7				Pan coefficient = 0.79			
	NSE	RMSE	R^2	MAE	NSE	RMSE	R^2	MAE
All stations' data	0,25	2,63	0,51	1,81	0,48	2,1	0,52	1,62

The pan coefficient improvement was not found to be significant when the results were evaluated. However, it can be said that the results would be more reliable if the pan coefficient was taken into account as 0.79 for the Konya Closed Basin.

5 Conclusion

Evaporation loss plays an important role in the operation of water resources, determining the amount of irrigation water and accurately detecting water losses etc. The Konya Closed Basin, chosen as the application area, constitutes approximately 6% of Turkey. Daily meteorological data were provided from five stations in the Konya Closed Basin. The Penman, an internationally accepted as a reference method, was used for this study. Daily Penman evaporation estimates were calculated for each station from 2000–2019. The Artificial Bee Colony Algorithm was utilized to recalculate the pan coefficient. This value was found as 0.79 for this basin. Although little improvement was detected in this study, it is hard to say that the improvement is significant. However, it is still recommended to multiply adjusted pan coefficient with unprocessed measured data taken from the station to calculate daily evaporation.

References

1. Yüksel, İ., Sandalcı, M., Çeribaşı, G., Yüksek Ö.: Küresel Isınma ve İklim Değişikliğinin Su Kaynaklarına Etkileri, Ulusal 7. Kıyı Mühendisliği Sempozyumu, 51–58 (2011)
2. Meteoroloji Genel Müdürlüğü (MGM), Meteoroloji Genel Müdürlüğü (2020)
3. Bayazıt, M.: Hidroloji, Birsen Yayınevi. 219 s, İstanbul (2003). ISBN: 975–511–364–9
4. Şarlak, N., ve Bağçaci, S. Ç.: The assessment of empirical potential evapotranspiration methods: a case study of Konya closed basin. Teknik Dergi/Tech. J. Turk. Chamb. Civil Eng. **31**(1), 9755–9772 (2020)
5. Şorman, A.Ü.: Van Gölü su bütçesinin uzaktan algılama tekniklerinin kullanımı ile hesaplanması. TÜBİTAK-İNTAG-830 Araştırma Projesi. Proje No: 199I016, Ankara (2002)
6. Terzi, Ö.: Eğirdir Gölü'ne Ait Buharlaşma Modellerinin Geliştirilmesi Ve Uygu-lanması, Süleyman Demirel Üniversitesi, Fen Bilimleri Enstitüsü, Doktora Tezi, 124s, Isparta (2004)
7. Aydın, H., Düzen, H.: Comparison of measured and calculated evaporation values of open water surface: The case of lake Van (Turkey)/Serbest su yüzeyinde ölçülen ve hesaplanan buharlaşma değerlerinin karşılaştırılması: Van gölü örneği (Türkiye). Atmosfer Bilimleri Sempozyumu, 275–284 (2011)
8. Öztekin, T., Sezer, Ç.Ö.: A Sınıfı Buharlaşma Kabından olan Günlük Buharla-şmanın Penman ve linacre modelle. J. Agri. Fac. Gaziosmanpasa Univ. **33**(2016–3), 137 (2016). https://doi.org/10.13002/jafag1055

9. Öztekin, T., Sezer, Ç.Ö., Kutlu Sezer, E.: A-sınıfı Buharlaşma Kabından Olan Buharlaşma Miktarının Penman ve Priestley-Taylor (PT) Modelleri ile Tahmini. Ege Ün-iversitesi Ziraat Fakültesi Dergisi, 55, 11–20 (2018)
10. Özel, A., ve Büyükyıldız, M.: Aylık Buharlaşma Tahmininde Yapay Zeka Yöntem-lerin Kullanılabilirliği. Ömer Halisdemir Üniversitesi Mühendislik Bilimleri Dergisi **8**(1), 244–254 (2019). https://doi.org/10.28948/ngumuh.516891
11. Keskin, M.E., Şorman, Ü., Terzi, Ö.: Aylık Tava Katsayılarının Penman Metodu Kullanılarak Belirlenmesi, ĐMO Teknik Dergi, 3395–3403, Ankara (2005)
12. Wossenu, A.: Evaporation estimation for lake Okeechobee in South Florida. J. Irrig. Drain. Eng. **127**(March), 140–147 (2001)
13. Irmak, S., ve Haman, D. Z.: Evaluation of five methods for estimating class a pan evaporation in a humid climate. HortTechnology **13**(3), 500–508 (2018). https://doi.org/10.21273/horttech.13.3.0500
14. Althoff, D., Rodrigues, L.N., ve Silva, D.D. da.: Evaluating evaporation methods for estimating small reservoir water surface evaporation in the Brazilian savannah. Water (Switzerland), **11**(9) (2019). https://doi.org/10.3390/w11091942
15. Jasmıne, M.: A Comparatıve Study On Predıctıon Of Evaporatıon İn Arıd Area Based on Artıfıcıal Intellıgence Technıques, University of Ottawa, Master Thesis, 92s (2020)
16. Markovnikov, N., Kipyatkova, I., Karpov, A., Filchenkov, A.: Deep neural networks in russian speech recognition. In: Filchenkov, A., Pivovarova, L., Žižka, J. (eds.) AINL 2017. CCIS, vol. 789, pp. 54–67. Springer, Cham (2018). https://doi.org/10.1007/978-3-319-71746-3_5
17. Karaboğa, Derviş, B.B.: An artificial bee colony algorithm for solving con-strained optimization problems. In: Proceedings of 2018 2nd IEEE Advanced Information Management, Communicates, Electronic and Automation Control Conference, IMCEC 2018, 10 (2007). https://doi.org/10.1109/IMCEC.2018.8469371
18. Tarım ve Orman Bakanlığı. Konya Kapalı Havzası Yönetim Planı (2018)
19. Halıcı, U.: Artıfıcıal Neural Networks, Chapter IX- Radyal Basic Function. Orta Doğu Teknik Üniversitesi Ders Notu, s139–147, Ankara
20. Çelebi, H.: Mühendisler için İstatistik, Yöntemler ve Uygulamalar, Mersin Üniversitesi Ders Notu, 34 (2007)
21. Agha, O.M.A.M., Bağçacı, S.Ç., ve Şarlak, N.: Homogeneity analysis of precipitation series in North Iraq. IOSR J. Appl. Geol. Geophys. **05**(03), 57–63 (2017). https://doi.org/10.9790/0990-0503025763
22. Usul, N.: Mühendislik Hidrolojisi (ODTÜ Geliş.) Ankara: ODTÜ Yayıncılık, 418 (2009)
23. Şenyay, L.: İstatistik II- Regresyon Ve Korelasyon Analizi, Dokuz Eylül Üniversitesi Ders Notu, s1–23, İzmir
24. Fındıklı, M.A., ve Kocamaz, A.F.: Yapay Arı Koloni Algoritmalarının Sürü Robot Hedef Bulma Problemine Uygulanması. SETSCI Conf. Index. Syst. **3**, 1034–1039 (2018)

AI-Based (ANN) Model for Predicting Electrical Conductivity Using Lysimeter Experiments

Aida H. Baghanam[1](✉), Amirreza Tabataba Vakili[1], Vahid Nourani[1], and Dominika Dąbrowska[2]

[1] Department of Water Resources Engineering, Faculty of Civil Engineering, University of Tabriz, Tabriz, Iran
hosseinibaghanam@gmail.com, hosseinibaghanam@tabrizu.ac.ir
[2] Faculty of Earth Sciences, University of Silesia, Bedzinska 60, 41-200 Sosnowiec, Poland

Abstract. Prediction of the leachate pollutants is of prime importance to detect nearby water resources pollution. In this way, electrical conductivity (EC) as a physicochemical water pollution parameter with the possibility of portable measurement can be used as an indicator of the leachate quality. For this purpose, two lysimeter experiments were carried out simultaneously to simulate the Tychy-Urbanowice landfills. During the tests, the EC, waste temperature, and the moisture data were measured by the installed sensors. This study aims to develop an artificial neural network (ANN) model to determine the parameters affecting the EC value and subsequently, predict the EC parameter of the lysimeter employing the developed ANN model. The performance of the model was evaluated by determination coefficient (DC) as well as root mean square error (RMSE). The study results declared that the moisture content of the lysimeters had a significant contribution to the EC value prediction.

Keywords: Landfill leachate · Lysimeter · Electrical conductivity · Artificial Neural Network (ANN) Introduction

1 Introduction

Disposal of the solid waste in a sanitary landfill, as the last option of the MSW management hierarchy, is one of the safest and economic approaches [1]. Waste composition, liquid infiltration, chemical, physical, and biological reactions within the landfill are the parameters affecting the leachate generation. Adjacent soil and water resources pollution would be resulted in case of uncontrolled leachate migration [2]. Hence, the leachate must be monitored and treated within the landfill.

Lysimetric experiment can be carried out as a dynamic test to simulate the landfill performance; alongside the statistical approaches. Since, the study of the surface and groundwater supplies pollution caused by the leachate are costly and time consuming [3]. In this way, physical parameters such as EC, moisture content, and temperature of the waste were recorded during the lysimetric tests.

H. Gökçekuş and Y. Kassem (Eds.): NRSEM 2021, EESCI, pp. 119–128, 2022.
https://doi.org/10.1007/978-3-031-04375-8_14

Several factors can influence the quality and quantity of the leachate such as waste quantity landfill age, initial moisture, landfill type, and climatological conditions [4]. EC as an indication for the leachate contaminant, is resulted from the presence of the soluble substances (ions), which can be affected by water content in the landfill.

Various studies have been conducted to estimate the EC of the landfill leachate. Li and Zeiss [5] noted that one of the key factors to optimize the landfill performance is moisture content; since it could affect the EC and biodegradation rate. Electrical resistivity (or EC in reverse) in diverse studies has been utilized to assess the relationship between moisture content, temperature, and EC. Guérin [6] mentioned that the moisture content plays an essential role in the bioreactors design and control. Subsequently, the EC could be employed to monitor the leachate diffusion and affected area of the landfill by the leachate recirculation. Considering the important role of the moisture content in bioreactors, Grellier [7] measured the ER of the two leachate samples under the conditions of saturated medium and only leachate. Both tests proved that there was a relationship between temperature, moisture content, and EC.

Artificial intelligence methods can be utilized to determine the statistical relation between diverse physiochemical factors in leachate as well as leachate generation rate [8]. Data-based methods like Artificial intelligence model (ANN) was proved to be practical in environmental engineering problems with non-linear and unclear physics [9]. An ANN model was developed by Karaca and Özkaya [10] to estimate and predict the leachate generation in landfill by implementation of the recorded leachate rate and daily meteorological data. SVM and Multi-Linear Perceptron (MLP) ANN were used by Abunama [8] to predict the leachate amount of a landfill. The results declared that the ANN model outperformed the SVM. In this study, (ANN) will be employed alongside the lysimetric tests; since, there was a few studies in modeling the leachate properties via AI based models. The aim of this study is to develop an ANN model to predict the EC as contamination indicator of the leachate using lysimetric and meteorological data.

2 Material and Methods

2.1 Study Area

The Tychy-Urbanowice landfill is located in southern Poland. The landfill system entails two parts: an operating one (I site) and an abandoned one (II, III sites). Prior to 1988, the old uncovered site had been used to dump building materials, and later, it was turned into a sanitary landfill. Since 1994, the mentioned site has been closed. More than 70000 m^2 of the new active area is comprised of two sealed landfills. The lysimeter tests was conducted adjacent to the landfill system in Tychy-Urbanowice (Fig. 1). Municipal waste with European Waste Code of 190599 was filled into two lysimeters.

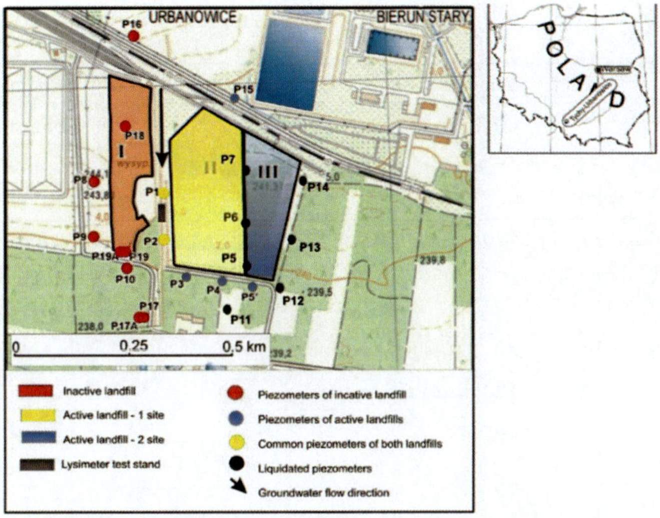

Fig. 1. Study area (based on Sołtysiak [11] with modifications)

From November 2018 until December 2019, the lysimetric tests conducted for about 400 days. The initial moisture content of waste was 69.5%. A stand – pipe lysimeter test was run in Tychy-Urbanowice (Southern poland), on the property of MASTER Waste and Energy company, the owner of the landfill complex. Based on rainfall volume measurements for this landfill for the period 2004–2009, the first lysimeter was recharged by rainwater or by using distilled water with an equivalent volume to the average monthly rainfall [3]. The open lysimeter was equipped with two GS3 soil moisture, temperature, and EC sensors. After adding distilled water to the second lysimeter, leachate was added to fill it up to the volume of monthly precipitation. Each of the lysimeters had two sensors, one at 0.6 m and one at 1.2 m depth. Using the Em50 basic logger, sensor's data was recorded hourly. Table 1 and Fig. 2 show the datasets used in the current study, which entail the Pszczyna synoptic station and the lysimeter data. There were four different ports installed on the mentioned lysimeters (Port-1 and Port-2 on the first lysimeter and Port-3 and Port-4 on the second lysimeter) from which the data were collected. Port-1 is comprised of lysimeter's inner Temperature (T_1), Moisture content (M_1), and Electrical conductivity (EC_1); Port-2 consists of, lysimeter's inner Temperature (T_2), Moisture content (M_2), and Electrical conductivity (EC_2); Port-3 and Port-4 presented Moisture content (M_3) and (M_4), respectively. Pszczyna synoptic station provided the local Minimum temperature (T_{Min}), Maximum temperature (T_{Max}), Minimum temperature at the ground (T_{Ground}), Mean temperature (T_{Mean}), and Precipitation (P) data. Soil temperature and moisture content were derived from four various bands of the GLDAS satellite for the region 19.0448E, 50.0974N, 19.0532E, 50.1014N, namely Soil temperature 0–10 cm, Soil temperature 10–40 cm, Soil temperature 40–100 cm, Soil temperature 100–200 cm, and Soil moisture content for the same depths. Soil moisture and temperature for the mention depths used in this study were produced with the Giovanni online data system.

Table 1. Data information

Data source	Port	Depth [cm]	Time period [hour]	Parameter
First lysimeter	1	60 (inside lysimeter)	1	Moisture (M_1)
				Temperature (T_1)
				EC (EC_1)
	2	120 (inside lysimeter)	1	Moisture (M_2)
				Temperature (T_2)
				EC (EC_2)
Second lysimeter	3	60 (inside lysimeter)	1	Moisture (M_3)
	4	120 (inside lysimeter)	1	Moisture (M_4)
Synoptic data	–	Surface	24	Tmax
				Tmean
				Tmin
				Tmin at the ground
				Precipitation
GLDAS satellite	–	0–10	3	Soil moisture (SM_{10})
				Soil temperature (ST_{10})
		10–40	3	Soil moisture (SM_{40})
				Soil temperature (ST_{40})
		40–100	3	Soil moisture (SM_{100})
				Soil temperature (ST_{100})
		100–200	3	Soil moisture (SM_{200})
				Soil temperature (ST_{200})

2.2 Proposed Methodology

To estimate the EC and Temperature of the second lysimeter using the first lysimeter data, the relation between the available data must be indicated. An AI-based model was implemented considering the nonlinear relationship between the mentioned parameters.

First Step (Data Preparation): Since the field data were recorded in a daily time series, in the first step, all lysimetric hourly data were converted to daily data. Considering the tangent sigmoid as the activation function, to increase the performance of AI-based models, the parameters of the models were normalized to be mapped between 0 and 1. To prevent the overtraining phenomenon, all the parameters were randomized due to the limited lysimetric test data (parameters gathered for less than one year).

Second Step (Data Assessment): Using the Correlation Coefficient (CC) method, the effect of synoptic, and satellite data on EC, Temperature, and Moisture of lysimeters

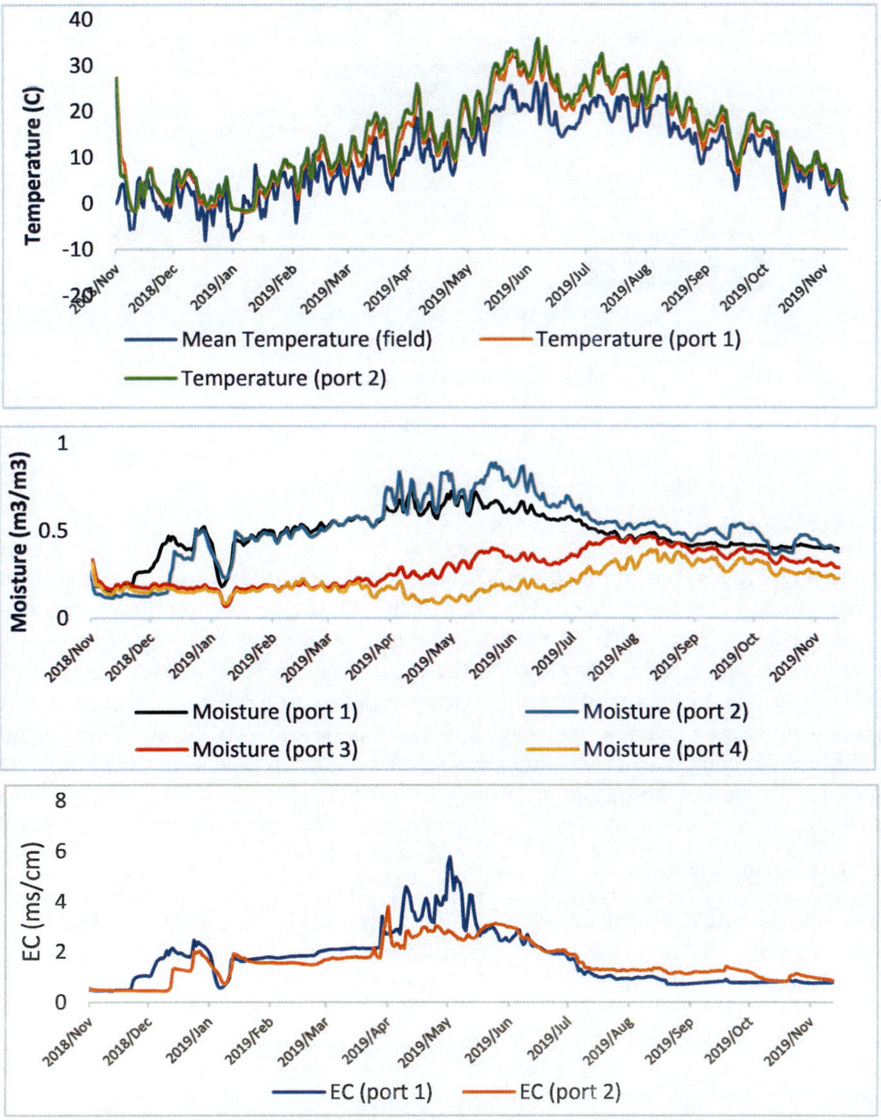

Fig. 2. Temperature time series, Moisture time series and EC time series

was then explored. Alongside this method AI method was applied to analyze the relation between the inner parameters of the lysimeter.

Third Step (EC Prediction): Finally, EC of the both ports in the first lysimeter were predicted utilizing the parameters in the last step of this research as the inputs of the ANN model. According to schematic figure (Fig. 3) the following steps present the proposed methodology.

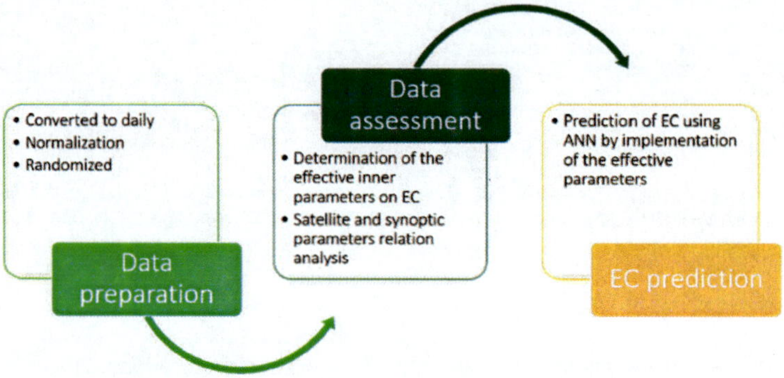

Fig. 3. Proposed methodology

2.3 Artificial Neural Network (ANN)

Data processing algorithms like the ANN were developed by drawing inspiration from biological neural networks. It can be said that the ANN is made up of interconnected processing units (nodes). The ANN structure can be described by three layers. The initial layer of the ANN is correlated to the input data; the second layer is called the hidden-layer which consists of hidden nodes parallel to each other, and the number of hidden-layer can be different considering various issues; the last layer, known as the output layer, is comprised of the summation of the weighted inputs passed through the activation function in the second layer [9].

2.4 Evaluation Criteria

In both the training and verification phases, the developed models were evaluated based on the following statistical measure, namely Determination Coefficient (DC), and Root Mean Square Error (RMSE):

$$RMSE = \sqrt{\frac{\sum_{i=1}^{N} (R_i - Z_i)^2}{N}} \quad (1)$$

$$DC = 1 - \frac{\sum_{i=1}^{N} (R_i - Z_i)^2}{\sum_{i=1}^{N} \left(R_i - \overline{R}\right)^2} \quad (2)$$

where R_i is the observed data, Z_i is the model output, $\overline{R}$ is the mean of observed data, and N is the total number of outputs.

3 Results and Discussion

Our research implemented ANN models to accurately estimate the second lysimeter's EC. As the proposed methodology consisted of three steps, we present and discuss the results in three steps as follows:

3.1 Results of Data Assessment

The correlation coefficient method was employed in the first step of this paper to analyze the linear relationship between the current external and the lysimeter parameters (Table 2). The results demonstrated that the internal moisture content of the lysimeters was well associated with the soil moisture acquired from the satellite at depths ranging from 100 to 200 cm. Furthermore, as depth lowers, the linear correlation gradually decreases, since numerous factors, such as evaporation, alter the amount of moisture at this depth, reducing this linear correlation. Thus, soil moisture in the depth of 0 to 10 and 10 to 40 cm were ruled out. The lysimeter's Moisture content can alter not only by infiltrated water, but also by elements such as initial waste moisture and internal chemical and biological reactions (Tables 2 and 3).

Table 2. The results of CC method

Parameters	SM100	SM200	Precipitation	ST100	ST200	T_{mean}
M_1	0.36	0.68	0.09	0.39	0.24	0.59
M_2	0.15	0.54	0.06	0.58	0.78	0.72
T_1	−0.35	0.11	0.09	0.87	0.78	0.95
T_2	−0.32	0.15	0.09	0.87	0.78	0.95
EC_1	0.58	0.54	0.12	0.02	−0.12	0.22
EC_2	0.58	0.75	0.12	0.02	−0.12	0.22

Table 3. The results of sensitivity analysis by ANN

Inputs	Output	DC		RMSE	
		C[a]	V[b]	C[a]	V[b]
M_1	EC_1	0.75	0.75	0.10	0.10
M_1, T_1	EC_1	0.79	0.76	0.09	0.10

(continued)

Table 3. (*continued*)

Inputs	Output	DC		RMSE	
		C[a]	V[b]	C[a]	V[b]
M_2	EC_2	0.85	0.79	0.08	0.09
M_1	EC_2	0.79	0.62	0.10	0.12
M_2, T_2	EC_2	0.88	0.86	0.07	0.09

[a]Calibration , [b]Verification

By knowing the significant correlation between Moisture, EC, and external Temperatures, and Temperature of the lysimeter, the links between the external parameters and the moisture content of the lysimeters were analyzed to find the optimal combination to estimate the moisture content. The results showed that (Table 4), similar to CC method, the link between soil moisture at depths of 100 to 200 cm is better than the ones at lower depths, and then by including the precipitation parameter, this relationship improves slightly. When all Moisture related parameters (i.e., synoptic and satellite moisture) is present in the input set, it is optimal to estimate the Moisture content of the lysimeter, exclusively using external moistures. The results of temperature effect showed that when all the effective parameters are present within the input combination of the model, the ANN reaches its highest accuracy.

Table 4. The results of sensitivity analysis by ANN

Inputs	Outputs							
Combination	M_1				M_2			
	DC		RMSE		DC		RMSE	
	C[a]	V[b]	C[a]	V[b]	C[a]	V[b]	C[a]	V[b]
P	0.02	0.03	0.21	0.21	0.05	0.04	0.24	0.20
P, SM_{100}	0.35	0.12	0.18	0.19	0.22	0.17	0.21	0.22
P, SM_{200}	0.78	0.73	0.10	0.12	0.64	0.53	0.15	0.15
P, SM_{100}, SM_{200}	0.85	0.83	0.06	0.08	0.90	0.91	0.07	0.07
P,T_{min}, T_{max},T_{mean}	0.46	0.44	0.15	0.17	0.62	0.53	0.15	0.14
SM_{100}	0.27	0.13	0.19	0.19	0.11	0.12	0.23	0.22
SM_{200}	0.72	0.70	0.12	0.10	0.62	0.55	0.14	0.16
T_{min}	0.33	0.25	0.18	0.19	0.42	0.47	0.18	0.17
T_{max}	0.40	0.39	0.17	0.17	0.60	0.50	0.15	0.16
T_{mean}	0.40	0.40	0.17	0.16	0.53	0.47	0.17	0.15
T_{min}, T_{max},T_{mean}	0.43	0.48	0.16	0.15	0.64	0.61	0.14	0.16

(*continued*)

Table 4. (*continued*)

Inputs	Outputs							
T_{mean}, ST_{100}, ST_{200}	0.81	0.79	0.09	0.09	0.83	0.85	0.07	0.09
ST_{100}	0.30	0.29	0.18	0.18	0.52	0.41	0.17	0.16
ST_{200}	0.16	0.17	0.19	0.20	0.32	0.18	0.20	0.19
ST_{100}, ST_{200}	0.79	0.82	0.08	0.08	0.84	0.81	0.07	0.09
ST100, ST200, SM100, SM200	0.93	0.94	0.04	0.04	0.94	0.96	0.04	0.05
All parameters	0.95	0.93	0.03	0.06	0.96	0.97	0.03	0.04

[a]Calibration , [b]Verification

Table 5. The results of EC prediction by ANN

Inputs	Outputs							
Combination	EC_1				EC_2			
	DC		RMSE		DC		RMSE	
	C[a]	V[b]	C[a]	V[b]	C[a]	V[b]	C[a]	V[b]
P, SM_{100}, SM_{200}	0.87	0.85	0.06	0.07	0.87	0.90	0.08	0.07
T_{mean}, ST_{100}, ST_{200}	0.85	0.83	0.07	0.08	0.86	0.85	0.07	0.07
ST_{100}, ST_{200}, SM_{100}, SM_{200}	0.86	0.87	0.04	0.03	0.87	0.88	0.04	0.03
All parameters	0.90	0.89	0.03	0.03	0.91	0.90	0.02	0.03

[a]Calibration , [b]Verification

Considering the results, it can be seen that external parameters can be used to obtain the EC parameter on which the lysimetric Moisture and temperature have the greatest effect (with DC above 0.80). For this reason, the best combinations obtained in this step were used to predict EC, which was a satisfactory result and could confirm the hypothesis because the accuracy of all obtained models is above 0.80.

3.2 Results of EC Prediction

Finally, it can be considered that there is a strong correlation between the parameters of Moisture and internal temperature and EC, which needed to be weighed to convert this proportion to an equation, which is what the ANN does. On the other hand, it was observed that satellite and synoptic data have a significant effect on the most important parameter that affects EC, namely moisture. Therefore, in case the existing landfill data is destroyed or the installed sensors had a problem, it is possible to use external parameters, which include meteorological stations and satellites, to predict the EC parameter, which is known as a leachate pollution indicator (Table 5).

4 Conclusions

In the current study, the EC parameter was estimated employing ANN model. In this way, the Tychy-Urbanowice landfills were simulated via a lysimetric experiment including two lysimeters. Despite the first lysimeter, which measured the Temperature, Moisture Content, and EC values, the second lysimeter's sensors recorded the Moisture content of the waste only. The moisture content was indicated to be in a potent relation with the EC parameter. On the other hand, the precipitation data declared a weak relation to the mentioned parameter. Moreover, the soil temperature and the field temperature indicated a strong relation with the temperature of the lysimeter. Considering the sensitivity analysis and CC, several combinations were proposed to simulate both ports of the first lysimeter's EC. Comparing the results of the simulation, it was concluded that the ANN model resulted in an acceptable performance.

References

1. White, P., Franke, M., Hindle, P.: Landfilling, in integrated solid waste management: a lifecycle inventory. In: White, P., Franke, M., Hindle, P., (eds.), pp. 271–302. Springer US Boston, MA (1995). https://doi.org/10.1007/978-1-4684-6705-5_11
2. Xie, H.-J., et al.: Investigation of migration of pollutant at the base of Suzhou Qizishan landfill without a liner system. J. Zhejiang Univ.-Sci. **10**(3), 439–449 (2009). https://doi.org/10.1631/jzus.a0820299
3. Dąbrowska, D., et al.: Application of hydrogeological and biological research for the lysimeter experiment performance under simulated municipal landfill condition. J. Mater. Cycles Waste Manage. **21**(6), 1477–1487 (2019). https://doi.org/10.1007/s10163-019-00900-x
4. Zhang, Q.-Q., Tian, B.-H., Zhang, X., Ghulam, A., Fang, C.-R., He, R.: Investigation on characteristics of leachate and concentrated leachate in three landfill leachate treatment plants. Waste Manage. **33**(11), 2277–2286 (2013). https://doi.org/10.1016/j.wasman.2013.07.021
5. Li, R., Zeiss, C.: In situ moisture content measurement in MSW landfills with TDR. Environ. Eng. Sci. – Environ. Eng. Sci. **18**, 53–66 (2001). https://doi.org/10.1089/109287500750070252
6. Guérin, R., et al.: Leachate recirculation: moisture content assessment by means of a geophysical technique. Waste Manage. **24**(8), 785–794 (2004). https://doi.org/10.1016/j.wasman.2004.03.010
7. Grellier, S., Robain, H., Bellier, G., Skhiri, N.: Influence of temperature on the electrical conductivity of leachate from municipal solid waste. J. Hazard. Mater. **137**(1), 612–617 (2006). https://doi.org/10.1016/j.jhazmat.2006.02.049
8. Abunama, T., Othman, F., Ansari, M., El-Shafie, A.: Leachate generation rate modeling using artificial intelligence algorithms aided by input optimization method for an MSW landfill. Environ. Sci. Pollut. Res. **26**(4), 3368–3381 (2018). https://doi.org/10.1007/s11356-018-3749-5
9. Nourani, V., Kalantari, O., Baghanam, A.H.: Two semidistributed ANN-based models for estimation of suspended sediment load. J. Hydrol. Eng. **17**(12), 1368–1380 (2012). https://doi.org/10.1061/(asce)he.1943-5584.0000587
10. Karaca, F., Özkaya, B.: NN-LEAP: a neural network-based model for controlling leachate flow-rate in a municipal solid waste landfill site. Environ. Model. Softw. **21**(8), 1190–1197 (2006). https://doi.org/10.1016/j.envsoft.2005.06.006
11. Sołtysiak, M., Dąbrowska, D., Jałowiecki, K., Nourani, V.: A multi-method approach to groundwater risk assessment: a case study of a landfill in southern Poland. Geolog. Q. **62**(2) (2018). https://doi.org/10.7306/gq.1411

AI-Based Statistical Downscaling of Precipitation and Temperature via Convolutional Neural Network Using Nonlinear Predictor Screening Approach

Aida H. Baghanam[1] (✉), Vahid Nourani[1], and Mohammed Bejani[2]

[1] Faculty of Civil Engineering, University of Tabriz, Tabriz, Iran
Hosseinibaghanam@gmail.com, {Hosseinibaghanam, nourani}@tabrizu.ac.ir

[2] Environmental Engineering, Faculty of Civil Engineering, University of Tabriz, Tabriz, Iran
m.bejani99@ms.tabrizu.ac.ir

Abstract. Statistical downscaling with affective variables which stem from screening techniques is required as part of studies project climate conditions in the future, based on climate change situation. The application of Artificial Intelligence (AI) methods for downscaling has become very common in the recent decade. However, deep learning methods are the new generation of AI methods that are gaining more attention due to better training compared to older generations. Convolutional Neural Network (CNN) is one of the novel deep learning methods which was used to downscale daily precipitation and temperature (predictand) data in this study. To this end, large-scale climate variables (predictor) were extracted from Coupled Model Inter Comparison Phase 6 (CMIP6) General Circulation Models (GCMs) for Tabriz synoptic stations in the northwest of Iran. To diminish the dimension of variables, eliminate redundant information and obtain the most accurate downscaling and projection results, predictor screening methods must be employed. Here in this study Random Forest (RF) and Mutual Information (MI) as the two non-linear screening methods were used for screening purpose predictors. Then, the dominant predictors fed into the CNN downscaling model and Results of the study in terms of evaluation criteria denotes that not only does the performance of CNN-based downscaling surpass ANN, but also the conjunction with screening methods also improves the efficiency. Therefore, models that have not benefited from predictor screening methods, performed weakly.

Keywords: Convolutional neural networks · Artificial neural networks · Climate change · Predictor screening method · Random forest (RF) · Mutual information · General circulation models · Statistical downscaling method

H. Gökçekuş and Y. Kassem (Eds.): NRSEM 2021, EESCI, pp. 129–138, 2022.
https://doi.org/10.1007/978-3-031-04375-8_15

1 Introduction

One of the most important environmental challenges is to assess future changes in precipitation and mean temperature due to climate change. General Circulation Models (GCMs) are mathematical representations of atmospheric, oceanic, and continental processes with their interactions. Downscaling is the specific name for the procedure to take large-scale information to make predictions at local scales. This procedure makes the relationship between predictor and predict and based on their nature. Hence, various downscaling techniques such as statistical [1] and dynamical models were developed.

The main deficiencies of the dynamical downscaling present with high computational resources, expertise and needs a huge volume of input data [2]. Statistical downscaling was developed to make the statistical relationship between coarse resolution and local resolution and bridge the spatial scale gap between GCM outputs and catchment scale climatic parameters [3]. In this investigation, not only deep learning built-in feature extraction was used, but also another feature selection method such as Mutual Information (MI) and Random Forest (RF) (as a novel predictor screening) were used to enhance the accuracy of the CNN model in terms of evaluation metrics. The last concept of the downscaling procedure is Weather Generators such as Long Ashton Research Station-Weather Generator (LARS-WG).

In this study, the CNN method from Deep Learning and ANN from machine learning-based were used to downscale large-scale climatic information (predictor) to local scale (predictand). The architecture of CNN has the most similarity with classical ANN but this method follows an end-to-end design, in which a fully connected dense layer is used at the output layer to recover the dimensions of the input data [4]. According to studies, models which have a non-linear-based form such as AI-based models contains involve noise and redundant information from input data and also need huge data to train [5]. In this way, to cover all deficiencies of non-linear models and enhance model accuracy, predictor screening techniques must be employed before downscaling. The application of predictor screening methods is in various sciences but there is a common purpose that led to diminish dimensionality of inputs and select prominent variables. According to studies, precipitation patterns and mean temperature changes are major concerns about the future. Therefore, in this investigation, these two parameters are assessed and discussed in the last section.

2 Study Area and Data

The domain of interest of this research is the Northwest of Iran-Tabriz with a latitude of 38.0962 °N, a longitude of 46.2738 °E, and 1361 m above sea level.

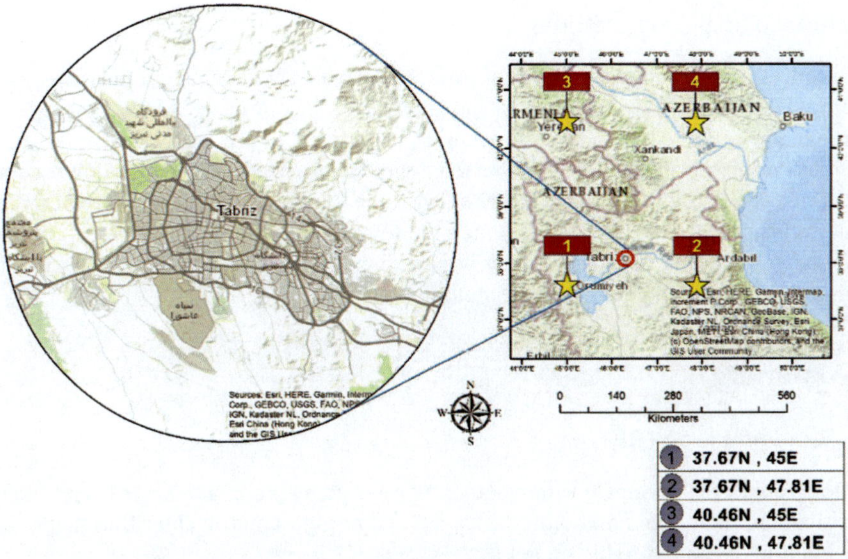

Fig. 1. Study area, Tabriz, Iran, with four different grid points around synoptic station

To calibrate and validate downscaling models,188 large-scale atmospheric variables were extracted from a single GSM (Can_ESM5) used as input predictor from the 6th Coupled Model Inter Comparison Phase (CMIP6) for the period 1951–2015. Figure 1 illustrates the location and situation of each grid point. For this purpose, totally 188 different variables used as the input of each model that Table 2 shows extracted variables.

3 Proposed Methodology

This research contains two steps to reach downscaled precipitation and temperature from 1951 through 2014. The first step is the screening process to select dominant predictors and the second step is the downscaling procedure to calibrate input data.

3.1 The First Step (Input Screening)

To obtain the best downscaling results, the following techniques must be employed to screen all input variables which reduces dimensionality and diminish the noise of outputs. in this study, we presented two non-linear predictor screening methods which could be analyzed and select variables as dominant features. At first, RF constructed decision trees that each decision tree contains nodes and leaves. In the internal nodes, features are selected with variance reduction criteria. Then, each feature is sorted by average in a decrease in impurity. In the end, RF polls for extracted most important features to select dominant variables.

The framework of feature extraction based on MI could be described as a given set of initial data X with T variables and Y defines as output label. Then, find subset $S \subset X$ with N variables that minimizes MI $H(C, X)$ and maximizes MI $I(C, X)$.

3.2 Second Step (Downscaling)

In this investigation, two downscaling process such as 1D CNN and a classical ANN based on DL and ML were employed respectively. Hence, a single GCM was employed as a screening procedure to enhance the accuracy of outputs. In this way, extracted features and local observation data were collected together, then applied to the downscaling process. In this process, each method calculates the relation of large-scale predictor and local predictand, extract special net which model will produce.

The results revealed that techniques which inherited from deep nature, produce high-resolution outputs compared to other methods.

4 Materials

4.1 Random Forest (RF)

Random forests (RF) is an algorithm that first proposed by breiman 2001 [6] that used for classification, regression, and other tasks. RF is a non-parametric algorithm that builds decision trees as an ensemble model [7].

The mathematical equations of RF follow as:

$$RFfi_j = \frac{\sum_{j \in alltrees} fi_{ij}}{T} \tag{1}$$

$RFfi_i$ = the importance of feature I calculated from all trees in the RF model
Norm fi_{ij} = the normalized feature importance for I in tree j
T = total number of trees

4.2 Mutual Information

The theory of information, MI describes as: measures information between two random variables based on statistical dependencies and entropy. The entropy of a variable is the amount of information contained in the variable.

To quantity the information that is storage in a variable, Shannon entropy (1948) is being used.

Shannon entropy defines as:

$$H(X) = -\sum_{i=1}^{n} P(X_i)\log[P(X_i)] \tag{2}$$

X: random variable with length of n
P: corresponding probabilities

4.3 Convolutional Neural Network (CNN)

CNN is a subset of Deep Learning that uses multi-layered classical ANNs to result state-of-the-art accuracy in various sciences Generally, CNN construction is similar to. Other NNs consisting of input layer, hidden layers, and output layer. the architecture of

CNNs contains three types of layers: 1st convolutional layer 2nd pooling layer 3rd fully connected layer.

It should be mentioned that one layer is situated between the convolutional layer and the fully connected layer which is named pooling layer.

The pooling layer is another building block of a CNN that employs next to activation function (non-linearity). This layer reduces dimensions of feature maps and computations and helps to overcome overfitting.

The main purpose of the activation function is to transform linearity into complex form. In other words, the activation function introduces non-linearity to the model. This node situates at the end (except the last layer) or in between layers. RELU is the typical activation function which used in regression issues mostly but recently, it uses on deep learning issues.

The last few layers are fully connected layers or Multi-Layer Feed Forward Network (MLFF) which could be flattened and connected to the output layer. Neurons which situated at this layer, connected with neighboring neurons to flatten the matrix by adding this layer. Features, which are extracted from previous layers, helps to make the final output.

4.4 Evaluation Criteria

To get feedback from the model, to do an operation to improve accuracy, evaluation criteria must be called. In this research, two evaluation metrics used to retrieve each model performance consist of Root Mean Squared Error (RMSE) and Determination Coefficient (DC or R^2).

In this way, to assess non-linear models, DC is also useful. The range of this metric varies from 0.00 to 1.00 or 0 to 100%.

$$RMSE : \sqrt{\frac{\sum_{i=1}^{n}(O_i - C_i)^2}{N}} \tag{3}$$

$$DC = 1 - \frac{\sum(O - C)^2}{\sum(O - \overline{O})^2} \tag{4}$$

5 Results and Discussion

5.1 Results of the First Step (Input Screening)

Selected two non-linear predictor screening methods (i.e. MI and RF) were used as the extraction of dominant meteorological variables to predict the pattern of precipitation and changes of mean temperature using single GCM with daily data from 1950 to 2014. To do that, 188 different variables were extracted from Can_ESM5 from four different grid points around Tabriz synoptic station and applied on screening methods to find non-linear relationships between predictor and predictand.

The most ranked variables, which were extracted from each method, were selected as dominant features. According to the screening procedure, it was found that RF resulted in a good performance in selecting the most effective atmospheric variables in terms of evaluation metrics.

Precipitation: *MI.* Selected four different grid points located near water bodies (e.g. Sevan, Urmia and Caspian Sea), so the humidity type variables transfer humidity by the wind (e.g. ua) at low-pressure level to the study area. Thus, the humidity type predictors got the dominancy of input to the next step. Geopotential height (e.g. zg_S) is one of the other dominant predictors that relate to local precipitation observation in non-linear way. Also, it could mention that geopotential height represents the weight of the atmosphere. The highlighted variable that correlates highly with precipitation, announce with zg at 1000 pressure level. This means increasing from the surface causes decreasing in pressure level. So that, with increasing from the surface, air particles become decreasing and getting cooler than the surface. In this way, particles that contain humidity (e.g. vapor) become condensing and affect local precipitation.

RF. The most reliable variables, which gained from RF method are humidity and temperature type variables (e.g. tas-huss-prc). Surface temperature at first grid point (e.g. tas) cause to increase sea surface temperature (e.g. Urmia lake) and led to evaporate waters from the surface of Urmia lake and transfer to the low-pressure level of atmosphere that finally global circulation generates. With this interaction and circulation, eastward winds (e.g. ua) at low-pressure levels, blow these humidity and temperatures to the study site and affects precipitation pattern. Eastward winds from the first and second grid point (e.g. ua) not only correlate in a non-linear way but also connects linearly. At this point, humidities which raised from the Caspian Sea and Urmia lake, transfers by winds at high-level of the sea. This is because of situated mountains which prevent transferring at high-pressure levels (e.g. shallow layer).

Temperature: *MI.* During the screening process, three different types of variables contain humidity type (e.g. hus), temperature type (e.g. tas-tasmax), and height (e.g. zg_S) selected as dominant predictors. At first, hus at middle-pressure levels, which are believed to be related to temperature in a non-linear way. In view of this, humidity that transfers to the study area, are those evaporations that stand from the Black sea and Urmia lake. Moreover, humidity-type particles which resulted from the evaporation procedure, has greenhouse gas specifics. Thus, these particles absorb emitted infrared waves from the sun and earth then affects local temperature. The selection for the third predictor was most common in various studies. Hence, the surface temperature is another important item in local temperature formation. In this way, during the day, radiation that emits from the sun impacts surface temperature and becomes warmer. In essence, the air is such a poor heat conductor that in calm weather, the hot ground only warms a shallow layer of air a few centimeters thick by conduction [8]. Hence, resulted temperatures are converted to thermals and raises to higher levels by the lightness of created thermals and causes circulations named convective circulation. Thus, tas and tasmax are getting two other variables that have a relationship with local temperature.

RF. At the top of dominant variables with RF algorithm, humidity type variables such as relative humidity and specific humidity got the most ranked variables related to local temperature. It is obvious that Sevan lake and Black sea humidity transfers at low pressure level to study site and impacts local temperature which discussed above. The eastward wind which is coming from the east and transfers Caspian sea humidity and warm weather toward the west is another most rated variable through RF algorithm. To this end, as discussed above, The air temperature (e.g. Tasmax) at the first grid point, which affects on the creation of local temperature not only in a non-linear way, but also it affects in a linear way.

5.2 Results of the Second Step (Downscaling)

After passing the screening step and extracting important features based on two non-linear algorithms to eliminate outliers, redundant information, and overcome over fitting the model, the downscaling procedure was employed to simulate the sequence of precipitation and temperature pattern. Before downscaling step, the whole data normalized and distributed in the range of 0 to 1. To downscale predict ands, two non-linear based (i.e. AI-based) models such as classical ANN and Deep learning-based CNN with activation function named RELU was employed to identify the strongness of the deep learning method over the classical machine learning method. To obtain the robust result of modeling, whole data were split to Train, Validation and test datasets by the amount of 0.7, 0.2, and 0.1 respectively from 1950 to 2014 (Fig. 2) (Table 1) and (Table 3).

Table 1. Results of evaluation criteria for precipitation based on CNN model

			Precipitation			
	RF				MI	
RMSE		DC		RMSE		DC
1.07		0.1038		1.18		0.0563

Table 2. Results of evaluation criteria for precipitation based on ANN model

			Precipitation			
	RF				MI	
RMSE		RF		RMSE		DC
2.5845		0.0198		2.6114		0.0125

Furthermore, there must be a benchmark model to assess the new model (i.e. CNN). In this way, a classical ANN model with three FFNN, 500 epochs, and optimum hidden neurons (based on each feature extraction) was selected as a benchmark model (Fig. 3) (Table 4).

Table 3. Results of evaluation criteria for temperature based on CNN model

			Temperature			
	RF				MI	
RMSE		DC		RMSE		DC
0.1941		0.9672		0.2268		0.8967

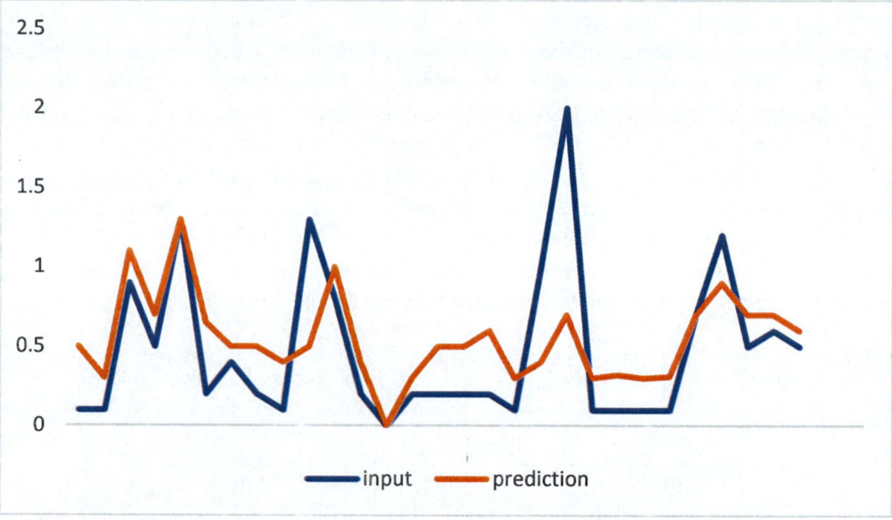

Fig. 2. Result of predicted precipitation (a month) based on CNN

Table 4. Results of evaluation criteria for temperature based on ANN model

			Temperature			
	RF				MI	
RMSE		DC		RMSE		DC
4.9621		0.7729		5.6874		0.6534

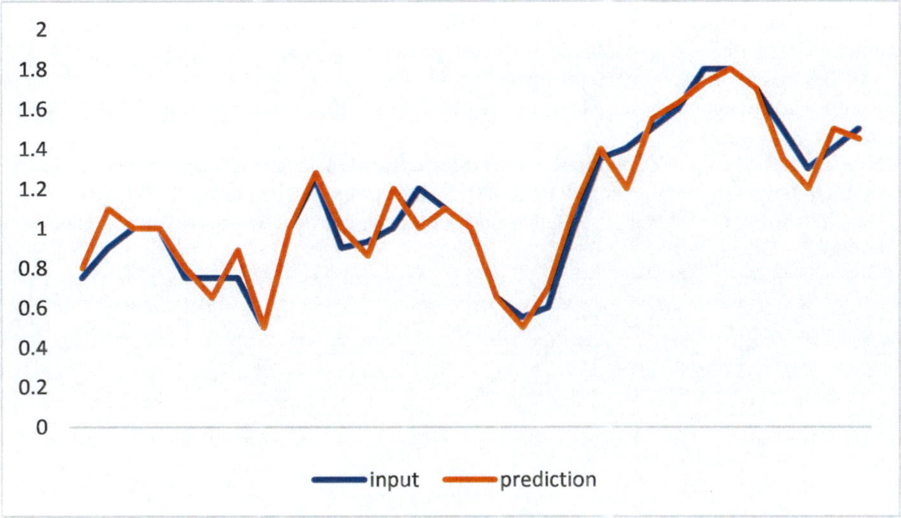

Fig. 3. Result of predicted temperature (a month) based on CNN

6 Conclusions

In this study, in order to assess the combining of novel feature extraction with deep learning and machine learning model accuracies on precipitation and temperature over the study area, a single GCM (Can_ESM5) from CMIP6 was employed at four different grid points near Tabriz synoptic station. At first, two predictor screening methods were used to extract prominent features that most correlated with predictand, eliminate extra features and prevent overfitting models which were named Random Forest (RF) and Mutual Information (MI). After that, variables which extracted from previous techniques were employed to models which named CNN and ANN that known as Deep Learning and machine learning methods (i.e. non-linear nature) respectively. Then, it was observed that RF-CNN based-downscaling models resulted low RMSE and high R^2 for the precipitation and temperature variables in both calibration and validation periods over other methods such as MI-CNN, MI-ANN, and RF-ANN.

References

1. Wilby, R.L., Dawson, C.W.: The statistical downscaling model: insights from one decade of application. Int. J. Climatol. **33**(7), 1707–1719 (2013). https://doi.org/10.1002/JOC.3544
2. Trzaska, S., Schnarr, E.: A Review of Downscaling Methods for Climate Change Projections. https://www.researchgate.net/publication/267097515
3. Benestad, R.E., Hanssen-Bauer, I., Chen, D.: Empirical-Statistical Downscaling, pp. 1–215 (2008). https://doi.org/10.1142/6908
4. Sun, A.Y., Tang, G.: Downscaling satellite and reanalysis precipitation products using attention-based deep convolutional neural nets. Front. Water **2** (2020). https://doi.org/10.3389/frwa.2020.536743

5. Nourani, V., Baghanam, A.H., Gokcekus, H.: Data-driven ensemble model to statistically downscale rainfall using nonlinear predictor screening approach. J. Hydrol. **565**, 538–551 (2018). https://doi.org/10.1016/j.jhydrol.2018.08.049
6. Breiman, L.: Random forests. Mach. Learn. **45**(1), 5–32 (2001). https://doi.org/10.1023/A:1010933404324
7. Nguyen, T.T., Huang, J.Z., Nguyen, T.T.: Unbiased feature selection in learning random forests for high-dimensional data. Sci. World J. (2015). https://doi.org/10.1155/2015/471371
8. Donald Ahrens, C., Henson, R.: Meteorology today : an introduction to weather, climate, and the environment

A Comparative Study of a Small-Scale Solar PV Power Plant in Nahr al-Bared, Lebanon

Youssef Kassem[1,2](✉), Hüseyin Gökçekuş[2], Hüseyin Çamur[1], and Engin Esenel[1]

[1] Faculty of Engineering, Mechanical Engineering Department, Near East University, 99138 Nicosia, North Cyprus
{yousseuf.kassem,huseyin.camur,engin.esenel}@neu.edu.tr

[2] Faculty of Civil and Environmental Engineering, Near East University, 99138 Nicosia, North Cyprus
huseyin.gokcekus@neu.edu.tr

Abstract. The ability and accuracy of machine learning techniques have been investigated for predicting the power of PV output. The objective of this paper is to predict the PV power output of a 0.5 kW off grid-connected PV system located in Nahr al-Bared, Lebanon. In this paper, the PV output of the proposed system was evaluated through Multilayer Feed-Forward Neural Network (MFFNN), and Cascade Feed-forward Neural Network (CFFNN) based on experimental data. Additionally, the proposed models were compared with multiple linear regressions (MLR) to show the ability and accuracy of the proposed models. The results indicated that the MFFNN model has higher predictive accuracy compared to other models.

Keywords: Machine learning models · PV output · Offgrid-connected · Nahr al-Bared · Lebanon

1 Introduction

In recent years, energy consumption and demand have been increased due to the growing populations and number of construction [1]. The increase in fossil fuel consumption leads to an increase the greenhouse gas emissions and air pollutions [2]. Several scientific studies showed that renewable energy is considered a powerful energy source for reducing the energy demand for urban regions [3]. Solar energy is one of the most promising renewable energy. It is clean energy and environmental-friendly energy source. Solar energy can be converted into electricity using solar photovoltaic (PV).

Lebanon is located on the eastern coast of the Mediterranean Sea and the climate of Lebanon is the Mediterranean climate. Besides, the main energy source is fossil fuel (diesel and heavy fuel oil). Based on the atlas solar map, Lebanon has huge solar energy compared to wind energy. Many studies have focused on solar energy in different locations in Lebanon [4–8]. For instance, Kassem et al. [5] investigated the feasibility of 100 MW grid-connected wind/solar systems in the Rayak region in Lebanon. The results indicated wind power is considered more efficient than solar power in the selected region.

H. Gökçekuş and Y. Kassem (Eds.): NRSEM 2021, EESCI, pp. 139–146, 2022.
https://doi.org/10.1007/978-3-031-04375-8_16

Moreover, several techniques such as regression techniques, soft computing sys techniques have been used to predict the energy production from renewable energy technologies. The machine learning models have been widely used for estimating the required properties [9].

Therefore, this study's goal is to predict the power production of a 0.5 kW off grid-connected PV system in Nahr al-Bared, Lebanon using two machine learning tools, namely, Multilayer Feed-Forward Neural Network (MFFNN) and Cascade Feed-forward Neural Network (CFFNN). In addition, the accuracy of models is compared with multiple linear regressions (MLR).

2 Material and Methods

2.1 PV System Description

In the present study, a 0.5 kW grid-connected rooftop PV system is proposed for generating enough energy to power the residential building in a coastal Palestinian camp (Nahr al-Bared) in Lebanon. For the proposed 0.5 kW grid-connected solar PV system, the mono-Si - SF160-24-1M170 module manufactured by Hanwha Solar One is selected as it is an efficient PV module that is currently available in the market. The specification of the used module is shown in [10]. Besides, three modules are required to build the proposed system. Moreover, the BSM-600W-OFF inverter is used for the PV system with a total capacity of 600 kW. The specification of the used inverter is shown in [11]. The descriptive statistics of the variables are presented in Table 1. Figure 1 shows the mean hourly AC output from the PV system during January 2017–December 2017.

Table 1. Statistical parameters of mean hourly data during 2017.

Variable	Mean	Standard deviation	Coefficient of variation	Minimum	Maximum
Ambient temperature [°C]	16.52	9.518	57.62	−6	43.4
Wind speed [m/s]	3.1482	2.5375	80.6	0	16.4
Solar radiation [W/m^2]	162.81	237.49	145.87	0	852.29
AC system output [W]	59.568	86.683	145.52	0	317.998

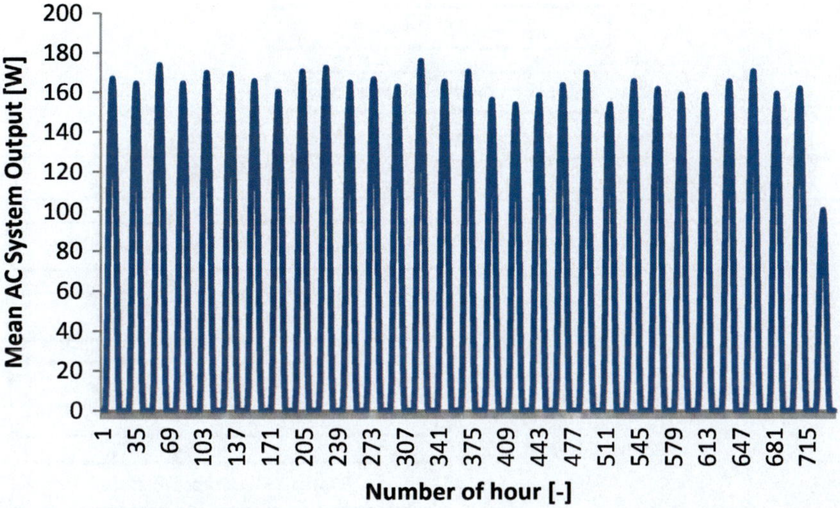

Fig. 1. Mean hourly AC system output during the period of January2017-December 2017.

2.2 Empirical Models

Many models and techniques are such as machine learning models and mathematical models are used as alternative tools to descript a complex system [12–15]. They are utilized in a wide variety of applications [12–15]. In this study, three empirical models (Multilayer Feed-Forward Neural Network (MMFFNN), Cascade Feed-forward Neural Network (CFNN), and multiple linear regression (MLR)) are developed to estimate the hourly power generation from the PV system. In this work, the empirical models have utilized wind speed (WS), solar radiation (SR), ambient temperature (AT), and the number of hours (NH) as input. For machine learning models, TRAINLM is utilized as a training function. Also, Mean squared error (MSE) is estimated to find the best performance of the training algorithm. The descriptions of developed models in detail were given in Ref. [7–9]. MATLAB software was used to develop the proposed models. Figure 2 shows the structure of the ANN models (MMFFNN and CFNN) used in this study.

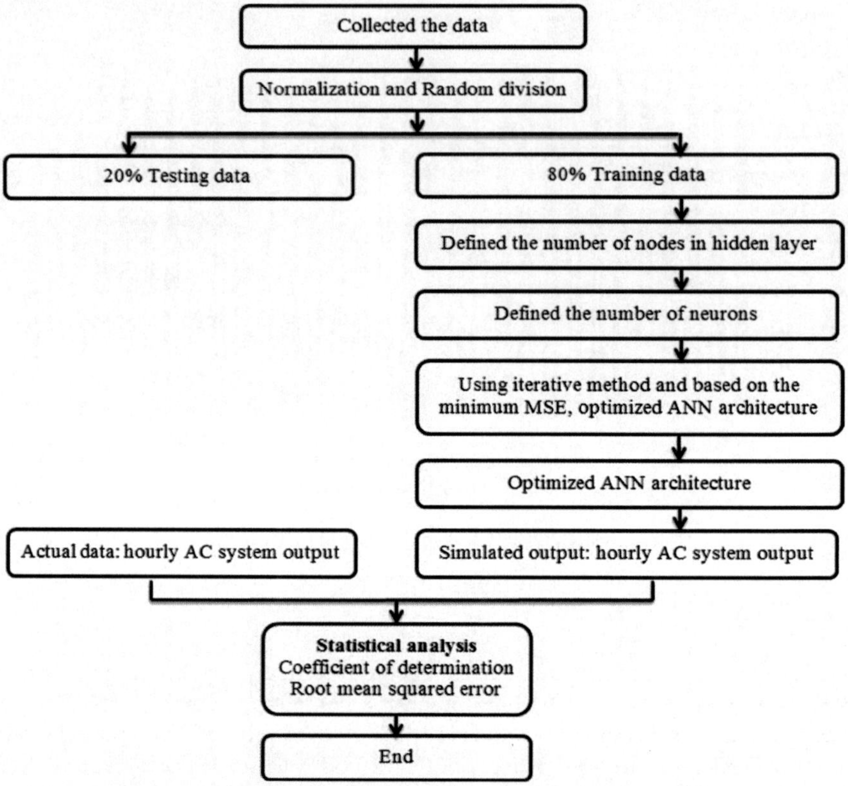

Fig. 2. Flowchart of the ANN-based method prediction procedure

3 Results and Discussions

3.1 ANN Models

In this research, The data were divided into training and test groups and the results of the models were compared with each other. The optimum network architecture for all models was determined through the trial and error method. It should be noted that the optimum number of HLs and NNs in the MFFNN and CFFNN models were estimated based on the minimum value of MSE.

It is found that the best transfer function for the hidden neurons is the tangent-sigmoid function. Based on the value of MSE, it is found that one hidden layer and 5 neurons are selected as the best for the MMFFNN model (4:1:1) with an MSE value of 8.742×10^{-5}. While it found that 1 hidden layer and 8 neurons are chosen as an optimum number for the CFNN model (4:1:1) with an MSE value of 1.35×10^{-4}.

3.2 MLR Model

The developed mathematical model including MLR was implemented to predict the hourly AC system output (AC-output). The data of ambient temperature (AT), wind speed (WS), solar radiation (SR), and the number of hours (NH) were used to generate a mathematical equation as given in Eqs. (1).

$$AC - output = A + B \cdot Hour + C \cdot AT + D \cdot WS + E \cdot SR \tag{1}$$

where A = 2.026, B = −0.000133, C = −0.2628, D = 0.7317 and E = 0.369518.

3.3 Performance Evaluation of Empirical Models for Testing Data

To select the best model for estimating the MP of the rotors, R-squared and root mean squared error (RMSE) is determined as shown in Table 2. The comparison of the predicted and actual values of the PV output for all models is shown in Fig. 3. It is found that the highest R-squared value of 0.9988 and lowest RMSE value of 3.0152 are obtained from the MFFNN model.

Table 2. Performance evaluation of the proposed models.

Performance criteria	MFFNN	CFNN	MLR
R-squared	0.9988	0.9986	0.9951
RMSE	3.0152	3.2796	7.7523

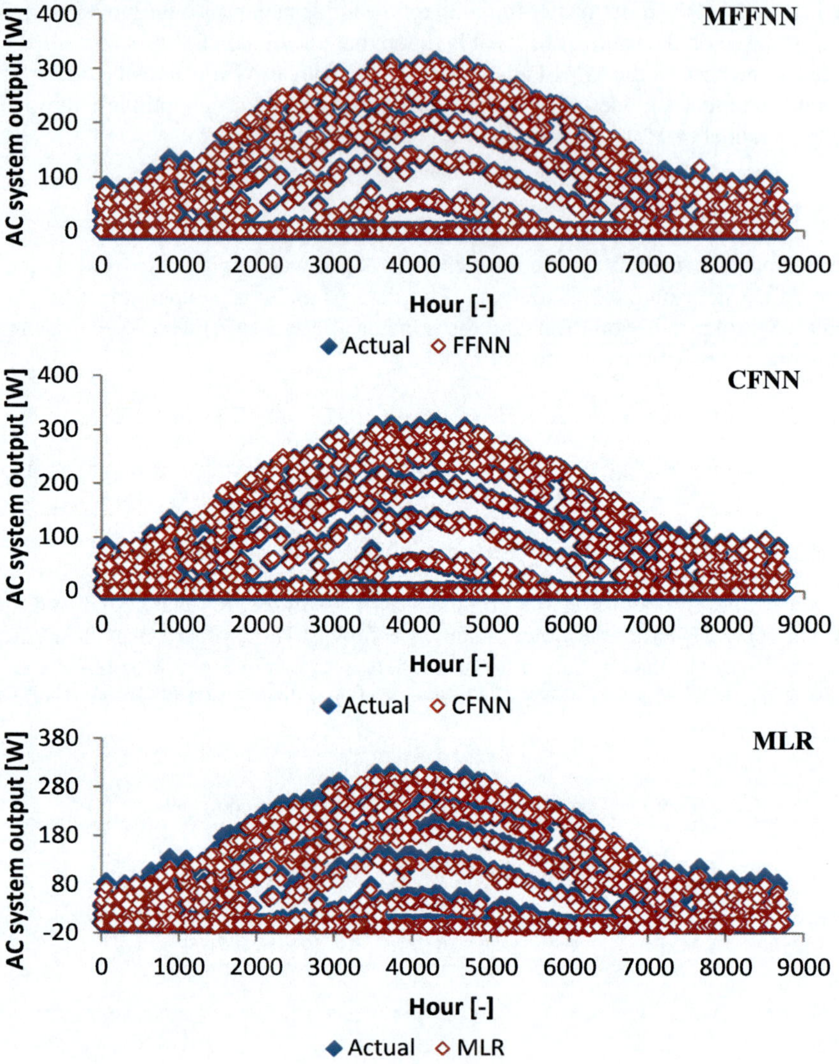

Fig. 3. Comparison of the predicted and observed values of all models

4 Conclusions

The main objective was to examine the application of artificial neural network models (Multilayer Feed-Forward Neural Network and Cascade Feed-forward Neural Network) for predicting the PV power output of the proposed system. The ANN models were also compared with multiple linear regression (MLR) to show the predictive accuracy of the proposed model. In this work, the impact of ambient temperature (AT), wind speed (WS), solar radiation (SR), and the number of hours (NH) on the PV power output was investigated and the experimental data were used to develop the proposed models.

Moreover, the coefficient of determination (R^2) and root mean squared error (RMSE) were used to assess the best empirical model. It is found that the MFFNN model was found to be the best model for estimating the PV power output of the PV system and more precise compared to CFFNN and MLR models.

References

1. Alayat, M., Kassem, Y., Çamur, H.: Assessment of wind energy potential as a power generation source: a case study of eight selected locations in Northern Cyprus. Energies **11**(10), 2697 (2018)
2. Kassem, Y., Çamur, H., Alhuoti, S.M.A.: Solar energy technology for Northern Cyprus: assessment, statistical analysis, and feasibility study. Energies **13**(4), 940 (2020)
3. Kassem, Y., Çamur, H., Aateg, R.A.F.: Exploring solar and wind energy as a power generation source for solving the electricity crisis in Libya. Energies **13**(14), 3708 (2020)
4. Berjawi, A.H., Najem, S., Faour, G., Abdallah, C., Ahmad, A.: Assessing solar PV's potential in Lebanon. Issam Fares Institute for Public and International Affairs (2017)
5. Kassem, Y., Gökçekuş, H., Janbein, W.: Predictive model and assessment of the potential for wind and solar power in Rayak region. Lebanon. Model. Earth Syst. Environ. **7**, 1475–1502 (2020)
6. Houri, A.: Solar water heating in Lebanon: current status and future prospects. Renew. Energy **31**(5), 663–675 (2006)
7. El-Jamal, G., Ghandour, M., Ibrahim, H., Assi, A.: Technical feasibility study of solar-pumped hydro storage in Lebanon. In International Conference on Renewable Energies for Developing Countries 2014, pp. 23–28. IEEE, November 2014
8. Tannous, S., Manneh, R., Harajli, H., El Zakhem, H.: Comparative cradle-to-grave life cycle assessment of traditional grid-connected and solar stand-alone street light systems: a case study for rural areas in Lebanon. J. Clean. Prod. **186**, 963–977 (2018)
9. Jeong, H., Obaidat, M.S., Yen, N.Y., Park, J.J.: Advances in Computer Science and Its Applications: CSA 2013. Springer, Heidelberg (2014). https://doi.org/10.1007/978-3-642-41674-3
10. Renugen: Hanwha Solarone SF160-170 Watt Solar Panel Module. https://www.renugen.co.uk/hanwha-solarone-sf160-170-watt-solar-panel-module/. Accessed 8 Mar 2018
11. Bluesun off grid 600w DC to AC Power Inverter Pure Sine Wave Inverter 0.6KW. https://www.bluesunpv.com/bluesun-off-grid-600w-dc-to-ac-power-inverter-pure-sine-wave-inverter-0-6kw_p242.html. Accessed 6 Feb 2018
12. Kassem, Y., Çamur, H.: Prediction of biodiesel density for extended ranges of temperature and pressure using adaptive neuro-fuzzy inference system (ANFIS) and radial basis function (RBF). Procedia Comput. Sci. **120**, 311–316 (2017)
13. Kassem, Y., Gökçekuş, H., Çamur, H.: Analysis of prediction models for wind power density, case study: Ercan area, Northern Cyprus. In: Aliev, R.A., Kacprzyk, J., Pedrycz, W., Jamshidi, M., Sadikoglu, F.M. (eds.) ICAFS 2018. AISC, vol. 896, pp. 99–106. Springer, Cham (2019). https://doi.org/10.1007/978-3-030-04164-9_16

14. Kassem, Y., Gökçekuş, H., Çamur, H.: Prediction of kinematic viscosity and density of biodiesel produced from waste sunflower and canola oils using ANN and RSM: comparative study. In: Aliev, R.A., Kacprzyk, J., Pedrycz, W., Jamshidi, M., Babanli, M.B., Sadikoglu, F.M. (eds.) ICSCCW 2019. AISC, vol. 1095, pp. 880–887. Springer, Cham (2020). https://doi.org/10.1007/978-3-030-35249-3_117
15. Kassem, Y., Gökçekuş, H., Çamur, H.: Artificial neural networks for predicting the electrical power of a new configuration of Savonius rotor. In: Aliev, R.A., Kacprzyk, J., Pedrycz, W., Jamshidi, M., Babanli, M.B., Sadikoglu, F.M. (eds.) ICSCCW 2019. AISC, vol. 1095, pp. 872–879. Springer, Cham (2020). https://doi.org/10.1007/978-3-030-35249-3_116

Application of WASP8 Deterministic Water Quality Model to Acısu Creek in Antalya, Turkey

Pelin Orhan[1], Secil Tuzun Dugan[2], Murat Yesiltas[2], Mehmet Ali Turan Kocer[2], Hicran Coban[3], Ayse Muhammetoglu[1](✉), and Habib Muhammetoglu[1]

[1] Department of Environmental Engineering, Akdeniz University, Antalya, Turkey
aysemuh@akdeniz.edu.tr, muhammetoglu@usa.net

[2] Mediterranean Fisheries Research Production and Training Institute, Antalya, Turkey
{seciltuzun.dugan,murat.yesiltas}@tarimorman.gov.tr,
mehmetalituran.kocer@tarim.gov.tr

[3] 13th Branch General Directorate, State Hydraulic Works, Antalya, Turkey
hbaykam@dsi.gov.tr

Abstract. Increasing urbanization, tourism and agricultural activities threaten the quality of water resources. In order to decrease this environmental threat, research studies are carried out to improve water quality status around the world. Consequently, various water quality analysis and simulation programs are used to evaluate and predict water quality for different management options. A research study has been initiated for Acısu Creek located in Antalya province of Turkey, which discharges directly to the Mediterranean Sea and impairs the sea water quality. There are intense agricultural and tourism areas within the watershed of Acısu Creek. In this study, WASP8, an open access water quality analysis and simulation program developed by US-EPA, was applied to simulate water quality at Acısu Creek. An intense monitoring program was initiated to collect the required data sets for modelling where the water quality measurement and analysis studies were conducted monthly at twelve monitoring stations for one year. This study presents the initial findings of the model calibration study which was realized using six months of monitoring data collected from September 2020 till March 2021. The predicted results of flow rate and water quality parameters were in good agreement with observations. The results of the ongoing monitoring study will be used for model verification and then several management scenarios will be tested for improving water quality.

Keywords: Water quality model · WASP · Deterministic model · Acısu Creek

1 Introduction

Due to availability of diverse job and education opportunities, health services and improved social life, there is a remarkable shift of population from rural to urban areas. The increase in the level of economic development and welfare in urban areas causes

H. Gökçekuş and Y. Kassem (Eds.): NRSEM 2021, EESCI, pp. 147–154, 2022.
https://doi.org/10.1007/978-3-031-04375-8_17

environmental pollution and creates a pressure on natural resources. Uncontrolled point and non-point sources of pollution adversely affect the water quality of surface waters. When the amount of waste input exceeds the waste assimilation capacity of the receiving environment, contamination of water resources becomes obvious. Therefore, protection of natural resources and sustainable management issues gain importance to prevent pollution of surface waters and maintain a good status [1].

Intense anthropogenic activities, which are concentrated close to rivers and urban settlements, create a water pollution problem and threaten the existence of the river ecosystem. The Water Framework Directive implemented by the European Union to assess, monitor and protect the quantity and quality of water in river basins also directs the poor environment to be improved [2]. Determining the main reasons that cause adverse spatial and temporal changes in the quality of rivers is very important for sustainable water management. In this context, use of water quality simulation models gains an importance. With the advances in computation and processing technology in the last decades, several advanced surface water quality simulation models were developed, such as AQUATOX, QUASAR, WASP, QUAL2E/K, CE-QUAL/W2 and others. These models could be used for different aims such as strategy development and decision-making for different types of water bodies and they are continuously improving based on the latest innovation and studies [3–5].

In this study, the water quality of Acısu Creek, located in Serik district of Antalya province in the south of Turkey, is analyzed using the latest version of WASP model. Acısu Creek directly discharges to the Mediterranean Sea and adversely effects the coastal sea water quality and hence the intense tourism activities in the region. The advanced eutrophication module of WASP8 was applied to simulate the spatial and temporal variations in the quality and quantity of Acısu Creek. The aim of this study is to develop an integrated approach for assessment and management of the river water quality. The integrated approach includes a detailed monitoring study and application of a deterministic water quality simulation model for conventional pollutants. In this paper, the details of monitoring study and the initial findings of model calibration are presented [6].

2 Materials and Methods

2.1 Study Area

Acısu Creek is located in the south of Turkey and it discharges directly to the Mediterranean Sea, as shown in Fig. 1. There are intense agricultural and tourism activities in the watershed of Acısu Creek which are the main economic resources of the region. Belek town, which is connected to Serik district of Antalya province, has many hotels where tourism is very active. There are three wastewater treatment plants in the watershed which are important sources of point pollution. As a result, point sources of pollution from wastewater treatment plants and non-point sources of pollution caused by agricultural activities adversely affect the river water quality.

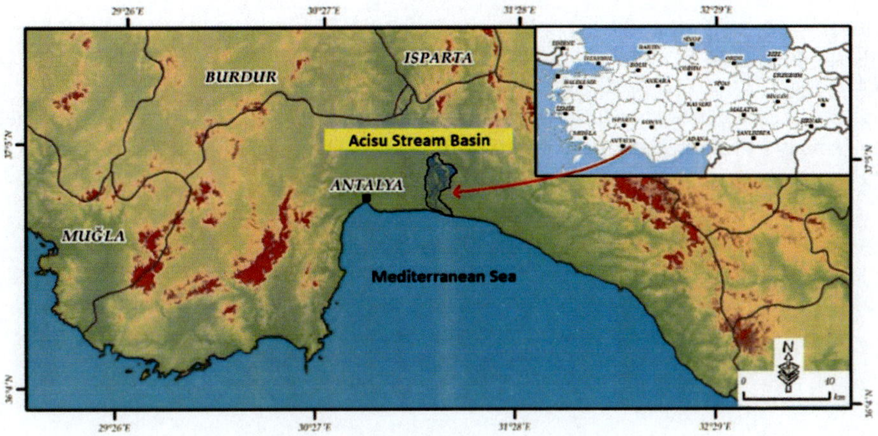

Fig. 1. Locations of Antalya province & watershed of Acisu Creek in Turkey [adapted from 7]

2.2 Monitoring Studies and Data Collection

Twelve measurement and sampling points, located along the stream network of about 40 km length, were selected for the monitoring studies at Acısu Creek (Fig. 2). Nine of the monitoring points are on the mainstem of Acısu Creek and three monitoring points are on the tributaries. Measurement and analysis of many physico-chemical and bacteriological water quality parameters including pesticides were realized at all the monitoring stations with monthly intervals. The monitoring study started in September 2020 and it continued for a period of one year.

Chemical oxygen demand (COD), 5-day biochemical oxygen demand (BOD_5), bacteriological parameters, alkalinity were performed on the same day of sampling. Temperature (T), pH, electrical conductivity, salinity, dissolved oxygen (DO) saturation and concentration, chlorophyl-a (Chl-a) values were measured in-situ using a multi parameter probe set. The color value was determined at wavelengths of 436, 525 and 620 nm with a UV/VIS spectrophotometer [8]. In order to determine the suspended solids (SS) concentration, water samples were filtered through tare weighted glass fiber filter papers and then dried in an oven at 103–105 °C for 4 h. Titrimetric methods were used for total alkalinity and bicarbonate analysis. The COD test was done with the open reflux method whereas BOD_5 values were determined by measuring DO consumption during five-day incubation period [9]. Water samples were collected for fecal coliform (FC), total coliform (TC), fecal streptococcus (FS) and Escherichia-coli (E-coli) analyses in sterile 100 mL amber glass sample bottles. Pesticide analyses were conducted by an accredited private laboratory.

The measurements of flow rate at three monitoring stations on the mainstem of Acısu Creek and two main tributaries were conducted by the 13th Regional Directorate of State Hydraulic Works in Antalya. The discharge flow rates and the characteristics of wastewater treatment plant effluents were obtained from Antalya Metropolitan Municipality, Water and Wastewater Administration (ASAT) and the responsible operating company namely, Turas Tourism. The required meteorological data sets (wind speed, solar radiation, air temperature, humidity and cloud cover) were obtained from Antalya Regional Meteorology branch. Information on agricultural activities in the basin area (such as agricultural production, type and amount of used fertilizers and pesticides) was obtained from Antalya Provincial Directorate of Agriculture. The collected data sets were used for deterministic modeling of water quality in Acısu Creek.

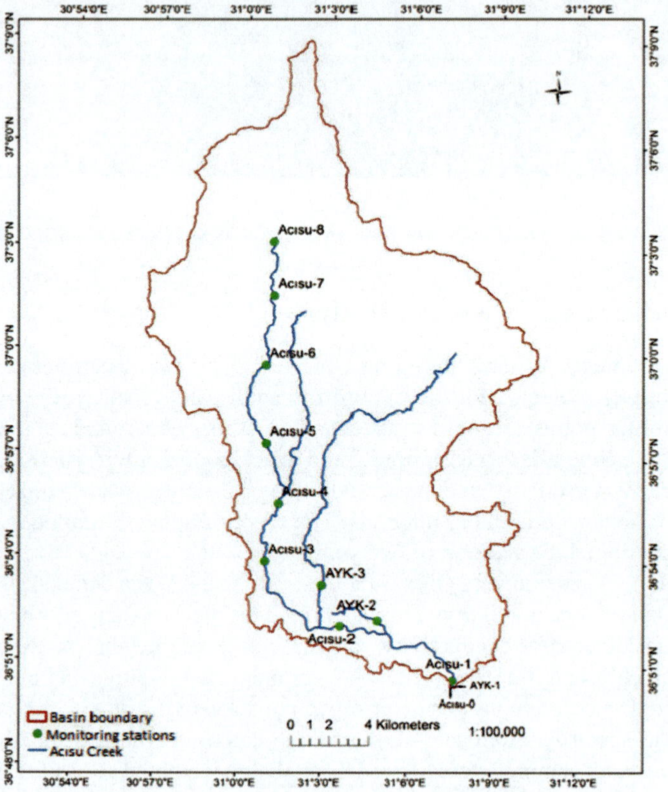

Fig. 2. Locations of monitoring stations [6]

2.3 Water Quality Analyses and Simulation Program, WASP

In this study, the Advanced Eutrophication module of the WASP8 program is used to simulate water quality at Acısu Creek. Water Quality Simulation Program (WASP), developed by The United States Environmental Protection Agency (USEPA), is a differential, spatially-resolved, mass balance, fate and transport modeling framework structured to allow users to simulate concentrations of environmental contaminants in surface waters and sediments [10–13]. During the last few decades, several versions of WASP model were applied at many different water systems to provide feasible solutions for improving the water status in terms of quality and quantity [14]. Model calibration study is performed using the monitoring results of the first six months by adjusting the predicted and observed values of model state variables (T, BOD_5, DO, NH_4-N, NO_3-N, inorganic phosphorus (Inorg-P) and SS) using trial and error method.

3 Results and Discussion

The statistical summary of the monitoring results of selected parameters for the initial six months of monitoring are presented in Table 2 for all stations. The water quality is generally good at the upstream stations (Acısu-8 and Acısu-7) and it is deteriorated due to point sources (discharges of three wastewater treatment effluents) and diffuse pollution sources from agricultural activities. The high numbers of investigated bacteriological parameters indicate the presence of fecal contamination from septic tanks and animal wastes and impose a threat on the coastal water quality at the direct river discharge to the Mediterranean Sea. The model predictions of WASP8 model are presented in Fig. 3 for the downstream station, Acısu-0. The model prediction errors were computed using Root Mean Square Error (RMSE) analysis and coefficient of determination (R^2). The model prediction errors are presented in Table 1 for all monitoring stations and predicted parameters. The model predictions were in good agreement with the measurements and the model prediction errors were acceptable. The model validation will be performed using the monitoring results of the additional six months of monitoring following the calibration period.

Table 1. WASP8 model prediction errors for calibration

Station	Error analysis	T (°C)	BOD_5 (mg/L)	DO (mg/L)	NH_4-N (mg/L)	NO_3-N (mg/L)	Inorg-P (mg/L)	Chl-a (μg/L)	SS (mg/L)
Acısu-6	RMSE	1.18	2.14	0.44	0.0361	0.04	0.003	0.15	2.53
	R^2	0.99	0.90	0.99	1.00	1.00	0.97	0.94	0.94
Acısu-5	RMSE	1.00	0.41	0.20	0.02	0.20	0.001	0.54	9.18
	R^2	0.99	0.99	0.99	0.99	0.99	0.99	0.96	0.99
Acısu-4	RMSE	0.85	1.52	0.70	0.03	0.90	0.01	0.38	13.71
	R^2	0.99	0.80	0.99	0.99	0.93	0.98	0.62	0.86
Acısu-3	RMSE	0.40	3.73	0.81	0.13	1.22	0.01	0.21	8.23
	R^2	0.99	0.82	0.99	0.01	0.90	0.97	0.73	0.88
Acısu-2	RMSE	1.22	1.07	1.03	0.82	1.81	0.11	0.11	3.26
	R^2	0.99	0.87	0.99	0.49	0.66	0.89	0.84	0.96
Acısu-1	RMSE	1.24	0.36	1.67	0.17	0.66	0.04	2.04	3.04
	R^2	0.99	0.99	0.99	0.98	0.94	0.98	0.99	0.99
Acısu-0	RMSE	1.31	2.79	1.70	0.17	0.96	0.03	3.46	8.44
	R^2	0.99	0.88	0.99	0.98	0.91	0.98	0.69	0.95

Table 2. Monitoring results of selected parameters for six months of monitoring (TB: Tributary)

Parameters		Monitoring stations											
		Acısu-0	TB-1	Acısu-1	TB-2	Acısu-2	TB-3	Acısu-3	Acısu-4	Acısu-5	Acısu-6	Acısu-7	Acısu-8
T (°C)	min	14.52	15.95	14.45	15.73	14.18	13.53	13.23	12.98	12.71	13.70	12.74	11.81
	max	26.04	20.79	25.21	23.10	21.86	19.14	19.66	19.18	19.23	21.37	14.72	13.05
	mean	17.98	17.60	18.48	17.78	16.21	15.23	15.17	14.88	15.13	15.94	13.92	12.30
	SD	4.45	2.17	5.06	2.82	2.96	2.16	2.40	2.31	2.24	3.13	1.05	0.66
DO (mg/L)	min	6.59	6.98	6.38	1.33	7.60	9.02	8.44	8.56	9.25	7.98	10.18	10.21
	max	9.28	10.59	9.25	4.61	10.39	10.75	9.68	9.86	10.55	9.87	10.46	10.50
	mean	7.79	9.38	7.46	3.17	8.93	10.17	9.27	9.31	10.00	9.31	10.36	10.38
	SD	0.87	1.64	1.03	1.45	1.05	0.67	0.47	0.43	0.49	0.77	0.16	0.15
TN (mg/L)	min	1.119	1.380	1.060	1.983	0.890	1.204	1.021	0.986	0.840	0.923	2.750	0.131
	max	6.520	12.877	10.471	24.752	4.808	5.685	4.547	5.884	5.771	6.905	6.069	0.529
	mean	3.817	4.689	3.787	9.050	2.314	3.116	2.144	3.398	2.663	3.742	4.465	0.316
	SD	2.138	5.476	3.556	8.571	1.768	1.947	1.386	2.008	1.904	2.682	1.662	0.200
TP (mg/L)	min	0.001	0.084	0.001	0.530	0.038	0.052	0.040	0.038	0.035	0.025	0.023	0.023
	max	0.465	0.186	0.335	5.335	0.108	0.161	0.088	0.092	0.076	0.054	1.059	0.581
	mean	0.195	0.116	0.145	1.615	0.074	0.090	0.061	0.064	0.054	0.037	0.369	0.213
	SD	0.187	0.047	0.135	1.860	0.026	0.043	0.019	0.018	0.015	0.013	0.597	0.319
BOD_5 (mg/L)	min	0.10	6.58	4.15	3.80	0.10	0.10	3.36	0.10	2.48	0.10	1.67	1.82
	max	9.80	13.15	4.67	16.12	3.67	8.61	6.15	5.61	3.82	11.50	2.21	3.55
	mean	5.38	9.86	4.41	8.86	2.01	3.56	4.76	2.94	3.15	4.90	1.94	2.68
	SD	3.23	4.65	0.36	4.55	1.65	2.81	1.97	1.81	0.94	4.79	0.39	1.22
SS (mg/L)	min	6.8	6.8	7.6	4.8	4.4	1.2	7.7	6.4	2.0	1.2	0.4	1.5
	max	27.5	29.5	19.9	41.0	21.9	105.2	49.8	40.4	45.7	20.4	7.3	2.3
	mean	16.4	21.9	13.0	19.6	12.8	29.2	18.2	26.1	11.5	7.8	2.9	1.9
	SD	8.5	10.3	4.6	14.8	6.7	38.6	15.9	13.4	17.3	7.4	3.8	0.4
Chl-a (µg/L)	min	0.09	4.46	0.07	0.13	0.02	0.02	0.03	0.04	0.09	0.04	0.02	0.05
	max	12.43	30.92	14.14	1.02	0.24	0.23	0.17	0.16	3.21	0.73	0.44	0.06
	mean	2.25	13.44	2.54	0.48	0.15	0.12	0.09	0.09	1.08	0.23	0.17	0.06
	SD	4.99	11.89	5.69	0.36	0.09	0.09	0.05	0.04	1.14	0.29	0.23	0.01
TC (cfu/100 mL)	min	190	400	450	733	130	60	10	80	10	15	67	0
	max	24267	48000	20480	18240	27867	9013	21813	29280	12000	6240	6347	187
	mean	9256	15783	8231	12649	8813	3349	6908	7131	5331	1832	2191	62
	SD	9061	22403	7938	6440	10260	3477	8299	11030	5113	2601	3599	108
FC (cfu/100 mL)	min	0	0	0	0	0	0	0	0	0	0	0	0
	max	6000	10613	6000	6000	3627	1227	1513	2300	6000	340	33	13
	mean	1121	2685	1072	2156	689	247	571	497	1035	160	11	4
	SD	2400	5286	2416	2490	1442	484	604	900	2433	156	19	8
FS (cfu/100 mL)	min	0	0	0	0	0	0	0	0	0	0	0	0
	max	383	70	238	880	102	33	609	218	38	113	0	0
	mean	87	23	116	351	28	8	144	44	16	36	0	0
	SD	167	33	110	446	44	14	264	97	20	54	0	0
E-coli (cfu/100 mL)	min	0	0	0	0	0	0	0	0	0	0	0	0
	max	5767	4480	5700	5640	740	40	67	33	5320	15	0	0
	mean	967	1122	954	1120	131	14	25	11	890	3	0	0
	SD	2352	2239	2325	2233	299	18	29	14	2170	7	0	0

After model calibration and validation, WASP model will be applied to test several water quality management scenarios. Acısu Creek faces both water quantity and quality problems and the management scenarios will address mitigation to these pressures.

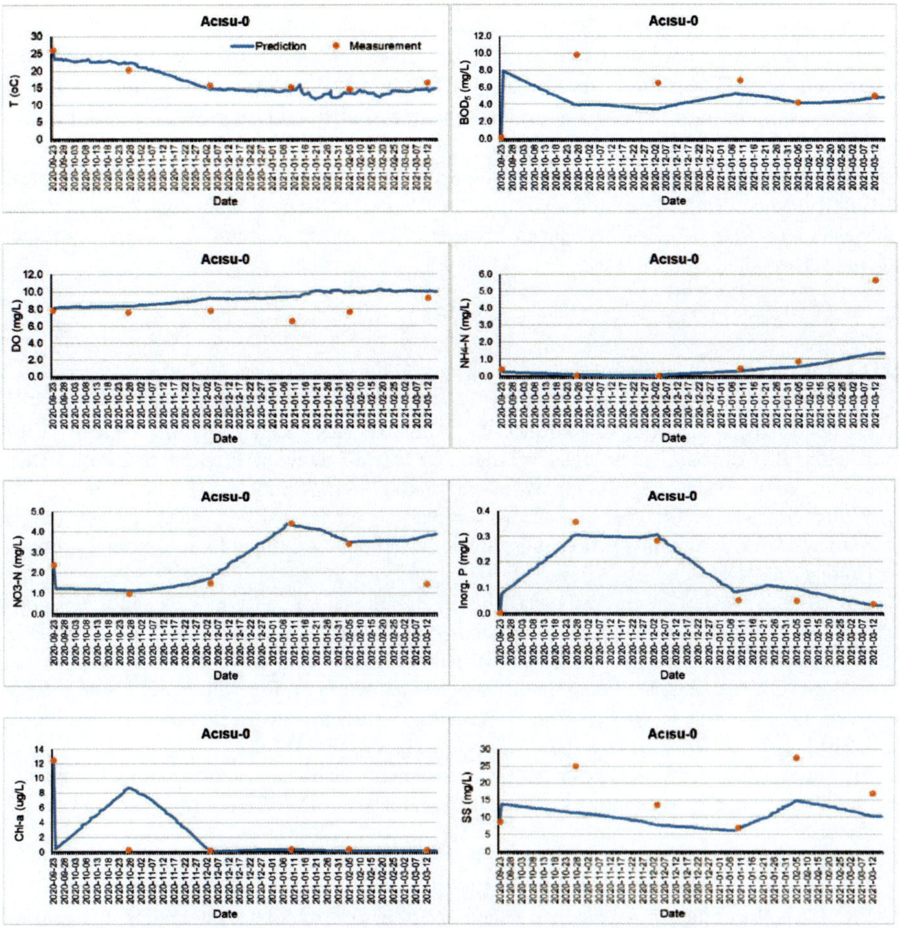

Fig. 3. The model simulation results of downstream station Acısu-0

4 Conclusion

The effluents discharged from three wastewater treatment plants and the diffuse pollution from agricultural activities, animal husbandry and septic tanks in the watershed of Acısu Creek deteriorate the water quality. The excessive air temperatures, increased evaporation rates and irrigation activities in summer season have an adverse effect on Acısu Creek by reducing the waste assimilation capacity of the stream due to reduced flow rates. The stream directly discharges to the Mediterranean Sea and the coastal bacteriological water quality is impaired. Deterministic modelling of Acısu Creek water quality is implemented to find best management options to improve water status. The applied water quality analysis and simulation program, WASP8, is a powerful tool to understand and analyze environmental systems and make predictions for future. The advanced eutrophication module of WASP8 model has proven itself for modeling direct or diffuse pollution inputs to aquatic environments such as rivers, lakes and reservoirs. The model calibration study of Acısu Creek will continue with model validation and scenario analysis. The model predictions will be used to develop an integrated approach for the assessment and management of river water quality and quantity in Acısu Creek.

Acknowledgments. Thanks to the Scientific and Technological Research Council of Turkey, TÜBİTAK (Project No: 119Y267), State Hydraulic Works 13th Branch in Antalya, Antalya Water and Wastewater Authority, Mediterranean Fisheries Research Production & Training Institute, Akdeniz University.

References

1. Burigato Costa, C., da Silva Marques, L., Almeida, A.K., Leite, I.R., de Almeida, I.K.: Applicability of water quality models around the world—a review. Environ. Sci. Pollut. Res. **26**(36), 36141–36162 (2019). https://doi.org/10.1007/s11356-019-06637-2
2. WFD: Directive 2000/60/EC of the European Parliament and of The Council. Establishing a framework for Community action in the field of water policy. Official Journal of the European Union (2000)
3. Cox, B.A.: A review of currently available in-stream water-quality models and their applicability for simulating dissolved oxygen in lowland rivers. Sci. Total Environ. **314**, 335–377 (2003). https://doi.org/10.1016/S0048-9697(03)00063-9
4. Tsakiris, G., Alexakis, D.: Water quality models: an overview. Eur. Water. **37**, 33–46 (2012)
5. Wang, Q., Li, S., Jia, P., Qi, C., Ding, F.: A review of surface water quality models. Sci. World J. **2013**, 1–7, 231768 (2013). https://doi.org/10.1155/2013/231768
6. Muhammetoglu, A.: River water quality assessment and management using deterministic modelling, water quality indices and statistical analysis. TUBITAK Project No: 119Y267 (2020)
7. Fural, S.: Morphometric analysis of drainage characteristic of Acısu Stream (Serik-Antalya). J. Acad. Soc. Sci. Stud. **72**, 541–556 (2018)
8. ISO 7887: Water quality – examination and determination of colour. International Organization for Standardization (2011)
9. APHA: Standard Methods for the Examination of Water and Wastewater, 20th edn. American Public Health Association. Washington, DC (1998)
10. Di Toro, D.M., Fitzpatrick, J.J., Thomann, R.V.: Documentation for water quality analysis simulation program (WASP) and model verification program (MVP) (1983)
11. Ambrose, R.B., Wool, T.A., Connolly, J.P., Schanz, R.W.: WASP4, a hydrodynamic and water-quality model-model theory. User's Manual and Programmer's Guide; Environmental Protection Agency, Environmental Research Lab, Athens, GA, USA (1988)
12. Wool, T.A., Ambrose, R.B., Martin, J.L.: The Water Analysis Simulation Program, User Documentation for Version 6.0, Distributed by USEPA Watershed and Water Quality Modeling Technical Support Center, Athens, GA (2001)
13. Wool, T.A., Ambrose, R.B., Martin, J.L., Comer, E.A.: Water Quality Analysis Simulation Program (WASP), Version 6.0 DRAFT: User's Manual, US Environmental Protection Agency, Environmental Research Laboratory, Athens, GA (2006)
14. USEPA: Literature Review on Nutrient-Related Rates. Constants, and Kinetics Formulations in Surface Water Quality Modeling. U.S. Environmental Protection Agency. Office of Research and Development. Regions 6 and 10 (2019). 101 pages

Blue Growth, a Key for Sustainable Development of Islands; the Potentials of Turkish Republic of Northern Cyprus

Selin Deliceirmak[1,2](✉) and Ilkay Salihoglu[2]

[1] Molecular Biology and Genetics Department, Faculty of Arts and Sciences, Near East University, Near East Boulevard, 99138 Nicosia, Turkish Republic of Northern Cyprus
selin.kucukavsar@neu.edu.tr

[2] Biosphere Research Center, University of Kyrenia, Yahya Bakir Street, Karakum, Kyrenia, Turkish Republic of Northern Cyprus
ilkay.saglihoglu@kyrenia.edu.tr

Abstract. Oceans possess 97% of the water found on earth. Life evolved in oceans, and they shaped the environmental conditions we thrive and depend on. In addition, oceans support the global population with food and livelihoods. Over-exploitation of goods and destruction of natural systems are now threatening the sustainability of economic growth and the health of the marine ecosystems, which requires holistic management strategies, especially under the global problem of climate change. Among various geographies, islands are more prone to climate change, natural disasters and human perturbations. Therefore, Small Island Developing States suggested a holistic approach to managing economic growth and its substantial harm to natural resources and emphasised the economic and social importance of oceans and inland waters in the Rio + 20 conference. The new concept, namely "blue growth", provides a framework for sustainable developmental goals for those economies based on marine and coastal activities. Turkish Republic of Northern Cyprus (TRNC) is a unique example where blue growth practices might benefit economic sectors while conserving its pristine nature. This study examined the strengths, weaknesses, opportunities, and threats of politically unrecognised countries to discover their blue growth potentials. We suggest that the integrated management of current economic sectors and finding potential ones can benefit the TRNC to self-sustain without compromising the health of the island ecosystem.

Keywords: Blue growth · Blue economy · Island ecosystems · Sustainable development · Conservation

1 Introduction

In August 1967, the first full-disk colour picture of the earth was taken from space by the DODGE satellite [1] and might be the first time we realise that we inhabit a "blue planet". The colour blue is given to our planet by the oceans representing the 72% of

H. Gökçekuş and Y. Kassem (Eds.): NRSEM 2021, EESCI, pp. 155–163, 2022.
https://doi.org/10.1007/978-3-031-04375-8_18

the earth's surface. Throughout the earth's history, life evolved in oceans and shaped the environmental conditions we thrive and depend on. Today, oceans support all life on earth by oxygen generation, carbon dioxide absorption, nutrient cycling and climate regulation. Although oceans play a significant role in sustaining life on earth, after the industrial revolution, the concept of "growth" often results in substantial harm to natural resources, especially underrated marine ecosystems.

The concept of "Blue Economy" was introduced by a group of pacific Small Island Developing States (SIDS), emphasising the economic and social importance of oceans and inland waters in Rio + 20 conference held in Rio in 2012. SIDS questioned the green growth concept being the main focus during the Rio + 20 preparation process. Although the green economy concept promise "economic growth and development while ensuring that natural assets continue to provide the resources and environmental services on which our well-being relies" [2], the growth was mainly based on the sustainable use of terrestrial resources. Therefore, SIDS positioned the importance of a new concept, "blue economy" and "blue growth", as it has broader relevance as we inhabit the blue planet [5].

Turkish Republic of Northern Cyprus is a politically unrecognised island country, having approximately 420 km long coastline. In this study, we discussed the established and potential blue economy sectors of the TRNC and defined the strengths, weaknesses, opportunities, and threats of these sectors.

2 Blue Economy and Blue Growth

The Blue Growth concept is the newest among other calls aiming to integrate maritime activities in an environmentally friendly manner [3]. Even though the "blue economy" term was used more frequently during the Rio + 20, the term was not mentioned in the final report of the congress [4]. The principles of the term left undefined and different usages of both terms by various institutions were inevitable. Lately, blue economy acknowledged as "improved human well-being and social equity, while significantly reducing environmental risks and ecological scarcities" and defining the main aim to "de-couple socio-economic development from environmental degradation" in the Blue Economy Concept Paper [5]. In 2015, The Economist Intelligence Unit conspired the blue economy as a sustainable ocean economy. Regarding the definitions of the Blue Economy concept given above, sustainability is intrinsically declared. However, Mulazzani and Malorgio [4] stressed that the blue economy and blue growth concepts defined in these reports are far beyond desiring the conservation and sustainable use of resources but public policy aspirations [4]. For the first time, *blue growth* was defined as "a smart, sustainable and inclusive growth from the oceans, seas and coasts" by Ecorys et al. [7] in their research for EU Commission. A year later, the Committee on Transport and Tourism of the European Parliament underlined the hidden degradation potential of maritime activities and suggested that blue growth must be consistent with the objectives of the ecosystem approach used for the Marine Strategy Framework Directive aiming to reach Good Environmental Status of the EU marine environment and protect the resources [4]. Before quantifying the value of the blue economy, describing the sectors of the blue economy and blue growth is a priority [6]. Blue economy sectors include maritime trade and transport; food, nutrition, health; energy and raw materials; living, working, and leisure in coastal regions and at sea; coastal protection and nature development; and maritime security (Fig. 1). Gross value added (GVA) by working in coastal areas is estimated to be £ 4.108 bln in the EU. Among these leading maritime sectors, leisure, working and living provide the higher GVA and employment [7]. In addition to the established sectors, blue bioeconomy, ocean energy, desalination, maritime defence, marine observations, research and education were introduced as the emerging sectors [8].

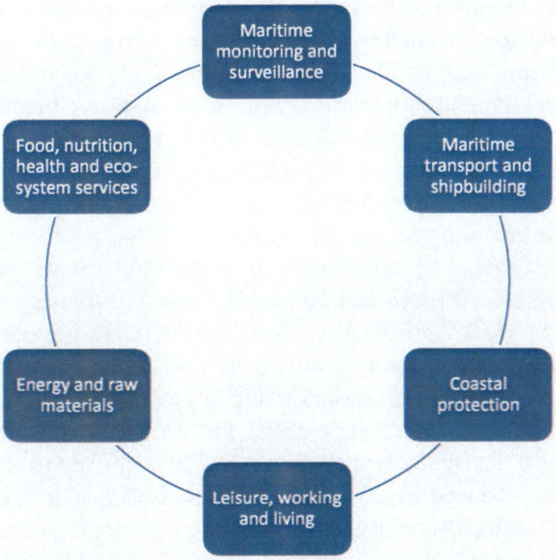

Fig. 1. Blue economy functions defined by Ecorys et al. [7].

3 Blue Growth and Blue Economy Capacity of TRNC

TRNC is located in the northern part of Cyprus and is politically recognised solely by Turkey. The country's economic development is mainly restricted due to the international embargo; therefore, it depends on Turkish financial support and monetary transfer. The country's real economic growth rate declined sharply in 2020 compared with previous years [9]. A decrease in the growth rate might be due to global pandemic conditions since the economy mainly comprises the service sector. Tourism and education is listed on top of the vital economic sectors in TRNC. Following, we discuss the potentials of TRNC blue economy sectors to suggest the blue growth concept as a tool for independent growth while conserving its pristine nature.

The blue economy's maritime trade and transport function include deep-sea shipping, shortsea shipping, passenger ferry services, and inland waterway transport [7]. TRNC has four main ports where maritime trade and transportation are pursued. Cargo transportation is mainly controlled by the Famagusta Port, whereas the Port Kyrenia carries out tourist and passenger transportation. The other two ports are in Kalecik and Teknecik, constructed for baggage cargoes and petroleum unloading [10]. In addition to the built ports in TRNC, Gemikonağı Port in Morphou Bay is proposed to be constructed for container transhipment of Turkey [11]. Yetkili et al. [11] stressed in their study that various markets in the Eastern Mediterranean, Black Sea, Adriatic Sea and Balkan countries can be accessed from Cyprus. Additionally, OECD projected an increase in container traffic by 2030 [12]. An increase in maritime activities provides potential economic growth for TRNC. However, the blue growth concept promotes the development of maritime economic functions sustainably [7]. Therefore, environmental drawbacks of maritime and transport activities should be determined, and appropriate measures should be taken. Protecting natural beauties is critically essential, specifically for TRNC, since its economy and livelihood depend mainly on coastal tourism.

Seas support the human population with food and livelihoods. These natural resources for human use are strictly related to the provisioning ecosystem services [4, 13]. Extraction of natural resources for human and animal consumption, farming marine aquatic products, blue biotechnology, and agriculture on saline soils are the sub-functions of this blue economic sector [7]. By nature, TRNC is expected to be supported economically by the fisheries sector. According to the Ministry of Agriculture and Natural Resources of TRNC, fish production in 2019 was 510 tonnes, and the number of active fishing boats was 680. Additionally, according to the data provided by the Ministry, only 976 fishermen reported. Capturing wild fisheries is not a growing sector due to the oligotrophic nature of the Mediterranean Sea and limited fish production [14]. However, one of the most prominent current and potential blue economic sub-function is the aquaculture applications on the island. Only two cage facilities are reported in the region, and the number of fishes produced in the aquaculture facilities is increasing and exceeding the number of wild catch fishes since 2015 [15]. During the preparation process of this paper, a literature search using the keywords "fisheries" and "TRNC" brought only two academic studies, one of which is a master thesis. It is critically important to stress that academic studies conducted in the country related to fishing activities surrounded by sea are almost nill. These are the most significant weaknesses defined here, constraining the economic growth potential of TRNC. On the other hand, positive efforts to determine issues and possible solutions for the fisheries are present. One particular example can be the seminar organised by the Ministry of Agriculture and Natural Resources of TRNC 2016. Stakeholders participated from a wide range of sectors (fishermen, recreational fishers, universities, governmental agencies) to determine issues and solutions in the fisheries sector, demanding holistic management approaches and practices. Apart from wild capture fishing and aquaculture sub-sectors of the blue economy, blue biotechnology is among the emerging sectors in the near future. Blue biotechnology means "using wild and farmed aquatic living resources as precursors of bio-molecules used for high-value products" [7]. The global market value of blue biotechnology is projected to be the US $6.4 billion by 2026 [16]. Briefly, blue biotechnology requires the usage of living marine resources, which can also be considered extracting biodiversity sustainably, having potential high market values. These organisms include micro/macro-algae, bacteria, fungi and invertebrates. Blue biotechnological applications require integrated study of science and biotechnology to convert marine living resources into food, feed, nutraceuticals, pharmaceuticals, cosmetics, energy and much more [17–20]. Here we strongly suggest developing research infrastructure to investigate marine biodiversity by strengthening universities and encouraging researchers. It is critical to understand the physical, chemical and biological properties of the TRNC marine environment. One of the most significant weaknesses we observe as obstacles to blue biotechnological developments is the limited number of research institutions conducting marine studies and biotechnological applications. As far as we can tell, a limited number of studies are trying to investigate the physical, chemical or biological properties of the TRNC marine environment. On the other hand, one comprehensive project provides a detailed investigation of natural and anthropogenic impacts on the Cilician Basin marine ecosystem, led by the Near East University of TRNC. The project describes the physical and chemical dynamics and biological components of the marine environment of TRNC [21]. The project contributes to the education of young researchers on the topic and attracts the government's attention to support marine environment-related studies. However, scientific studies should be continuously supported and given priority by universities, governments, decision makers and private sectors.

Marine mineral mining includes extracting the minerals from the sea bed, such as iron ore, tin, copper, manganese, cobalt, beryllium, germanium, graphite, gold, sulphides, phosphorites, diamonds and lime [5, 7]. Deep-sea mining raises debates on its negative environmental impacts. The literature survey did not bring any research on the potential of TRNC deep sea bed regarding the potential minerals to be mined. In addition to the lack of scientific studies, we do not foresee that the mineral mining sector will be considered a TRNC blue growth potential due to its political conflicts and lack of the required technological infrastructure. On the other hand, marine energy should be investigated as a potential blue growth sector and compensate for the energy requirements in TRNC. Ocean energy exploits energy sources such as wave energy, tidal energy, ocean thermal energy conversion, blue energy (osmosis) and marine biomass [7]. TRNC is mainly based on fuel oil to generate electricity [22], and raw materials are imported [23]. In 2021, electricity shortages experienced in households around the TRNC show urgent requirements for alternate energy sources. Literature search regarding the ocean energy sources and TRNC did not bring any result. However, renewable energy sources were investigated as a solution to increasing energy demands [22–24]. In addition to the increasing energy demand, supporting freshwater requirements is vital, especially under changing climatic conditions. Potable water supplied to the households contain salt between 1000 and 2500 ppm in TRNC [25]. Desalination is one of the sub-functions of the blue economy and is an alternative source for freshwater production. Agboola et al. [25] investigated four different desalination systems and found that all the systems are economically and technically feasible for TRNC [25]. Here we summarised the potential of blue energy and mineral mining. However, these activities present environmental concerns to be considered [5]. Their environmental impacts should be investigated carefully before consideration.

Last but not least, leisure, working, and living are among the main blue economic functions that present the higher value-added to the EU economy [7]. This sector comprises the following sub-functions: coastline and cruise tourism, yachting and marinas, working, and living. Except for cruise tourism, other functions are an active and essential contribution to the economic development of TRNC. However, tourism brings challenges in terms of increased greenhouse gas emissions, water diversity and damage to ecosystem services [5]. Orhon [26] stressed the environmental pressure of tourism activities in TRNC and proposed wastewater discharge limitation, especially for sensitive areas. Additionally, physical oceanographic features offshore Kyrenia were investigated to determine the effects of wastewater dumping [27]. Adverse impacts of the tourism sector are known, and solutions were presented by the scientific community of TRNC. Public authorities should integrate monitoring activities into their agenda since the pristine environment is the TRNC's asset that cannot be valued. Blue growth ensures balancing economic growth, social development, and sustainable use of marine living resources. The unrecognised country has no other option to holistically manage its blue resources to ensure aquatic habitat protection.

Based on the literature surveys, we highlighted the blue growth potential (Table 1) to take the attention of scientific, public/private sectors and government in the TRNC for sustainable future and economic development. Intrinsically, the country holds marine and coastal biodiversity assets, ready to be extracted for any function of the blue economy discussed earlier. It is critical to stress that nature extraction should not conflict with sustainable usage of the resources and the blue growth concept ensures this. The existence of higher educational institutions is one of the critical strengths of the country. With an appropriate and careful determination of future developmental goals, universities can channel their scientific concentrate on areas lacking knowledge. Universities can also support the education of professionals in the required areas, such as engineers who are expected to develop and apply new technological developments. To the best of our knowledge, the potentials of seas and coast for job, value and sustainability are not recognised by the public, private sectors and universities. One of the main weaknesses is the lack of funding opportunities for scientific studies and even for the government to restructure and frame the legislations and laws required. Blue growth presents the opportunities to the unrecognised country by promoting socio-economic development, sustainability, integrated management of marine and coastal areas and protection of biodiversity and natural resources. Global threats such as global warming and biodiversity losses should be addressed urgently by the government, and appropriate actions and adaptation strategies should be developed. Intense tourism and port activities already put pressure on the ecosystem and ecosystem services; additional negative impacts imposed by climate change and biodiversity losses will adversely affect the socio-economic status of the people inhabiting TRNC. Different stakeholders use the Marine and coastal areas of TRNC, and conflicts between them might obstruct the blue growth initiatives. It is critical to include all stakeholders in decision-making processes as it's a must for all holistic management approaches.

Table 1. Strengths, weaknesses, opportunities and threats of TRNC for promoting blue growth in the country.

Strengths	Weaknesses	Opportunities	Threats
Island country (intrinsically full of marine and coastal biodiversity)	Lack of knowledge and information of public sector, private sector and universities on "blue growth" activities	Holistic management of coastal areas	Marine and environmental pollution
Adequate number of coastal activities (tourism, transportation)	Lack of funding to support scientific studies	The potential of socio-economic development	Conflicts between different uses of coastal areas
Aquaculture potential	Environmental degradation due to touristic and port activities	Promoting sustainability goals	Lack of knowledge and will to encourage ecosystem-based managements
Existence of higher education institutions	Inadequacies in the legal framework	Protecting environment and biodiversity	Biodiversity crisis
Existence of SEPAs	Embargo status of the unrecognised country		Climate change and lack of adaptation strategies
	Inadequate number of institutions working on the marine environment		

4 Conclusions

TRNC is a politically unrecognised country and requires monetary support for its socio-economic wealth from Turkey. However, this pristine island country possesses huge blue growth potential that will promote its social and economic development and ensure its ecosystems' health. Universities should take the first step to investigate the blue economic prospects of the country, and the local government should consider these pieces of knowledge to apply the blue growth approach. Developing countries are under various economic and environmental pressures that cause sustainable development not to be the priority. However, these countries are more prone to natural and human-made catastrophes. Therefore, blue growth is a must to ensure the balance between socio-economic growth and sustainable use of marine resources.

References

1. Forbes: Viewing The Earth From Space Celebrates 70 Years (2021). https://www.forbes.com/sites/startswithabang/2016/08/22/viewing-the-earth-from-space-celebrates-70-years/?sh=6301149552ab. Accessed 07 Oct 2021

2. OECD: Green growth and sustainable development (2021). https://www.oecd.org/greengrowth/. Accessed 07 Oct 2021
3. Eikeset, A.M., et al.: What is blue growth? The semantics of "Sustainable Development" of marine environments. Mar. Policy **87**, 177–179 (2018)
4. Mulazzani, L., Malorgio, G.: Blue growth and ecosystem services. Mar. Policy **85**, 17–24 (2017)
5. U. N. S. D. K. Platform: Blue Economy Concept Paper (2014). https://sustainabledevelopment.un.org/index.php?page=view&type=111&nr=2978&menu=35
6. Katila, J., Ala-Rämi, K., Repka, S., Rendon, E., Törrönen, J.: Defining and quantifying the sea-based economy to support regional blue growth strategies–case Gulf of Bothnia. Mar. Policy **100**, 215–225 (2019)
7. Ecorys, Deltares, and Oceanic: Blue Growth, Scenarios and drivers for Sustainable Growth from the Oceans, Seas and Coasts (2012). https://webgate.ec.europa.eu/maritimeforum/system/files/Blue_Growth_Third_Interim_Report_130312%20Clean_0.pdf
8. European Commission: The EU Blue Economy Report. Publications Office of the European Union, Luxembourg (2021). file:///Users/selinkucukavsar/Downloads/KLAR21001ENN.en.pdf
9. T. C. C. of Commerce: Economic and Social Data. (in Turkish). https://www.ktto.net/ekonomik-gostergeler-ve-veriler/
10. North Cyprus Economy - Transport and Communications. http://www.cypnet.co.uk/ncyprus/economy/econ06.htm. Accessed 20 Oct 2021
11. Yetkili, E., Doğan, E., Baltaoğlu, S., Salihoğlu, İ: Economic analysis of container transhipment in the Eastern Mediterranean region. Int. J. Environ. Geoinform. **3**(1), 12–21 (2016)
12. OECD: The Ocean Economy in 2030 (2016). https://read.oecd-ilibrary.org/economics/the-ocean-economy-in-2030_9789264251724-en#page4. Accessed 20 Oct 2021
13. Koundouri, P., Giannouli, A.: Blue growth and economics. Front. Mar. Sci. **2**, 94 (2015)
14. Turley, C.M., et al.: Relationship between primary producers and bacteria in an oligotrophic sea–the Mediterranean and biogeochemical implications. Mar. Ecol. Prog. Ser. **193**, 11–18 (2000)
15. Akbora, H.D.: General status and growth potential of fisheries sector in Northern Cyprus. Nat. Eng. Sci. **5**(2), 73–81 (2020)
16. SMITHERS: The Future of Marine Biotechnology for Industrial Applications to 2025 (2015). https://www.smithers.com/services/market-reports/energy/marine-biotechnology-for-industrial-applications
17. Barre, S.L., Bates, S.S.: Blue Biotechnology (2018). https://doi.org/10.1002/9783527801718
18. Müller, W.E.G., Wang, X., Schröder, H.C.: New target sites for treatment of osteoporosis. In: Müller, W.E.G., Schröder, H.C., Wang, X. (eds.) Blue Biotechnology. PMSB, vol. 55, pp. 187–219. Springer, Cham (2017). https://doi.org/10.1007/978-3-319-51284-6_6
19. Erakovic Haber, V., Spaventi, R.: Discovery and development of novel drugs. In: Müller, W.E.G., Schröder, H.C., Wang, X. (eds.) Blue Biotechnology. PMSB, vol. 55, pp. 91–104. Springer, Cham (2017). https://doi.org/10.1007/978-3-319-51284-6_3
20. He, F., Mai, L.H., Gardères, J., Hussain, A., Erakovic Haber, V., Bourguet-Kondracki, M.-L.: Major antimicrobial representatives from marine sponges and/or their associated bacteria. In: Müller, W.E.G., Schröder, H.C., Wang, X. (eds.) Blue Biotechnology. PMSB, vol. 55, pp. 35–89. Springer, Cham (2017). https://doi.org/10.1007/978-3-319-51284-6_2
21. Salihoğlu, İ.: İnsan Kaynaklı ve Doğal Değişimlerin Kilikya Baseni (KKTC-TC Arası) Deniz Ekosistemi Üzerindeki Etkilerin Belirlenmesi (2018). (in Turkish)
22. Onder, H.H.: An expert system to estimate the capacity of harvesting energy from biogas and its capacity in TRNC. In: 2009 International Conference on Machine Learning and Applications (2009). https://doi.org/10.1109/icmla.2009.141

23. Al-Turjman, F., Qadir, Z., Abujubbeh, M., Batunlu, C.: Feasibility analysis of solar photovoltaic-wind hybrid energy system for household applications. Comput. Electr. Eng. **86**, 106743 (2020). https://doi.org/10.1016/j.compeleceng.2020.106743
24. Vafaei, L.E.: Lighting the TRNC flag on the Besparmak mountain with solar energy. In: 2020 4th International Symposium on Multidisciplinary Studies and Innovative Technologies (ISMSIT) (2020). https://doi.org/10.1109/ismsit50672.2020.9254618
25. Agboola, O.P., Al-Mutaz, I.S., Orfi, J., Egelioglu, F.: Economic investigation of different configurations of inclined solar water desalination systems. Adv. Mech. Eng. **6**, 925976 (2014). https://doi.org/10.1155/2014/925976
26. Orhon, D., Gökçekuş, H., Sözen, S.: Environmental Basis of sustainable tourism along sensitive coastal areas – principles and applications. In: Gökçekus, H., Türker, U., LaMoreaux, J. (eds.) Survival and Sustainability. Springer, Heidelberg (2010). https://doi.org/10.1007/978-3-540-95991-5_23
27. Deliceırmak, Ç., Salihoğlu, I.: Physical oceanographic features of the Cilician Basin and wastewater dumping at the coastal zone of Kyrenia. Desalin. Water Treat. **177**, 290–296 (2020). https://doi.org/10.5004/dwt.2020.25182

Challenges in Managing Water Crisis and Regulatory Instruments: A Case Study of South Asian and Middle Eastern Countries

Mirza Mohammed Abdul Basith Baig(✉) and Bertuğ Akıntuğ

Middle East Technical University, Northern Cyprus Campus, Guzelyurt, 99738 Mersin 10, Turkey
{mirza.baig,bertug}@metu.edu.tr

Abstract. The water crisis is now a global issue and is growing parallelly with the increasing population, urbanization, and industrial progress. Those increases in the water-stressed regions are hampering the economic prospects of the nations. Along with the water scarcity, lack of hygiene, poor sanitation, and water pollution is already hindering the developments of the countries. The most afflicted ones are the developing countries which are at greater risk due to their sizeable population, inadequate infrastructure, and impoverished policies. Accompanying the developing countries many developed nations which are naturally water deficit are also struggling to manage the water crisis. The escalating effects of climate change may disturb the water demand and supply. By extensively reviewing various literature, this paper will assess the current situation of water scarcity and policies of both developed and developing countries from South Asian and Middle Eastern parts of the world, alongside reviewing their policies towards managing the water crisis. The outcome of the research paper is inclusion of lacking provisions which can further strengthen the existing regulatory framework in a way that doesn't affect economic growth meanwhile improving water sustainability.

Keywords: Sustainability · Policy development · Water crisis · Water treatment technologies · Water economics

1 Introduction

Water is the only chemical compound which distinguishes earth from other planets, and is responsible for all the life on earth, there is approximately 71% of Earth's surface covered with water [1], and most of it is in oceans which is around 96.5%, some of it is consolidated in the ice caps, and a very little proportion of it is available for us to use which is in the form of ground and surface water. Water resources are the source of life and one of the essential components of the human development, but due to the climate change and population explosion, the water resources are getting under intense pressure. The urbanization and raising demand of water are straining the availability and the quality of water. It is estimated that, a number of millions of people are struggling to get access to safe drinking water, according to one of the reports published by World

H. Gökçekuş and Y. Kassem (Eds.): NRSEM 2021, EESCI, pp. 164–172, 2022.
https://doi.org/10.1007/978-3-031-04375-8_19

Health Organization (WHO), about 785 million people are lacking the basic drinking water services, and at least 2 billion of world population is using the drinking water source which is contaminated by feces, this promotes the transmission of diseases and may also lead to epidemic kind of situations, by the year 2025 the half of the world's population will live under water stressed regions under acute water shortages [2]. The Fig. 1 shows the estimated regions which are going to be water stressed by the year 2040.

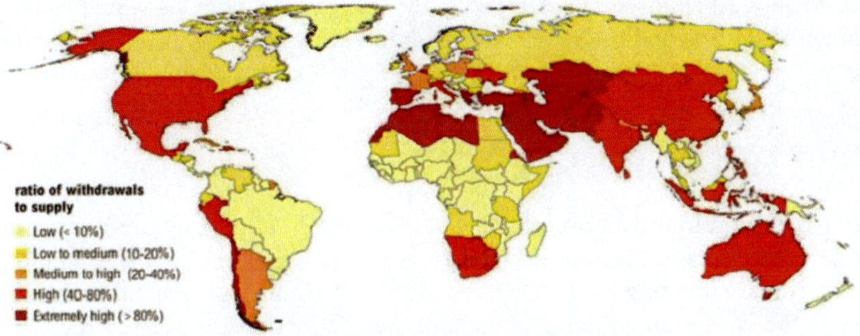

Fig. 1. Water stressed countries by 2040 [3].

The difference between water demand and supply is increasing day by day and creating imbalances. These consequences are further worsening due to climate change. As seen from Fig. 1, most of the middle eastern countries such as Saudi Arabia, Yemen, and UAE will be water deficient countries by the year 2040. For example, Yemen which is affected by water crisis, is also economically unstable and is now a zone of conflict for both water and land rights. Even countries like United States which are more stable may face water shortage [4].

Growing water demand is one of the important problems of developing countries. For example, countries like India and Pakistan are the developing nations with gradually increasing economies. The water demand and supply were not a problem in these countries because of their monsoon patterns, as both the countries receive 4 to 5 months of precipitation which was enough to feed the rivers and other surface water bodies. However, recently due to increasing population and climate change, there is a growing gap between supply and demand. Supply is inadequate or contaminated because of poor water infrastructures and the lack of strong regulatory policies and substantial technological advancement in water treatment processes. All together, these are contributing towards disturbing the economic growth and prospects. Water is the key element in the development of the country's infrastructure as well as it contributes towards all the sectors such as energy, industry, and agriculture. Now a days, water is called as the modern-day fuel, because it is the major component in the production of hydrogen gas [5].

When it comes to the development level of a country or the gross domestic product (GDP) of a country, it is seen in many such cases that there is a correlation between the average rainfall in a country and its GDP growth. Because agricultural and energy sectors both can immensely affected by drought or flood conditions which in turn affects the employment in the country [6]. Additionally, the degradation in water quality would reduce the touristic activities in the country which again impedes the GDP of the country. This paper focused on the water challenges in both the Middle Eastern and South Asian countries where these regions are more effected by the water crisis, mainly because of their population or natural weather patterns. The countries which have been considered in this paper are India as the South Asian country and Saudi Arabia and Yemen from Middle Eastern region. In the paper, their existing policies are assessed and an attempt is made to suggest some of the changes in the present-day policies which can help in managing the present and future water crises with achieving sustainability in the water sector.

2 Water Policy in India

As India is an Agrarian Nation, the demand for water and energy are raising for the irrigation purposes and because of an increase in population, the water supplies are limited and thus hampering the development efforts. As India is the second most populous country in the world, the growing population has its merits as parallelly the economy expands which creates financial stability but at the same time competition for the water limitation will most likely to intensify resulting in conflicts. The major consumers of water are the farmers, industries, and large metropolitan cities. The regulation in these sectors or communities are not strong enough which creates mismanagement of water resources. The deteriorating surface water quality is evident because of lack of proper regulations which gives rise to large industries disposing their waste into the water bodies without maintaining proper environmental regulations. With the polluted surface water bodies, the quality of ground water of India is worse than ever. As given in Fig. 2, it is

Fig. 2. Water stressed regions in India [7].

clearly estimated that 54% of India is facing high to extremely high-water stress. The most of the northern states of India are dependent on agriculture for their revenues, thus the ground water withdrawal rates are higher relative to the northern states (see Fig. 2).

To manage the water crisis in India, The Ministry of Water Resources came up with a new national water policy in 2012. In the following sections, the key elements stated in the water policy and the critical assessment of the policy are explained.

2.1 An Overview of National Water Policy 2012

The PREAMBLE of the policy clearly states that, the water is the fundamental to life and is the natural resource. It is also a key element in food security and sustainable development. The main objective of the policy is to take the awareness of the present water situation of the country and devise a framework for managing the negative impacts of water crisis.

2.1.1 Water Pricing

According to the policy, the water pricing will gradually increase to promote the efficient use and at the same time maximizing the value of water which will lead to the population treating water as an economic good. Recycling was also mentioned in the policy and the treatment of the water will have to strictly follow the specific standards. There will be pricing for over usage or withdrawal of water. The pricing of water and energy will affect both the electricity system of the nation parallelly with the water [8].

A water regulatory board or authority will be setup in every states. In order to regulate the tariffs, this board will work towards the pricing and monitoring the ground water table of the states, regulate the allocations of water to different communities, review the tariffs and change the pricing, if necessary. The board will also work towards resolving inter-state water conflicts and disputes.

2.1.2 Water Framework Law

The state governments must be given the rights to frame their water policies accordingly. Giving the states necessary authority will leave the governance of water issues with the states and the over influence of the central government towards the states will be minimized. As the northern states are highly affected by water stress, the water infrastructure must be improved in those states and the states which are water rich must be partner with the deficit states to form a policy in order to evolve the agricultural sector of the northern states for effective utilization of water. An attention to the preservation of surface water must be paid. The water bodies must not be polluted. The periodic checks and inspection of water bodies must be performed. A penalty system for any industry that pollutes the water must be initiated. The strict rules and regulations and more focus towards polluter pays principle must be implemented. Newer and larger water treatment facilities and systems should be proposed and financed by the government accordingly.

2.2 Assessment of Different Provisions of National Water Policy 2012

The various provisions of the policy are having both significant to no variations in the water sector. The policy includes the recharging of ground water using some advance technologies. However, there is nothing about community involvement in recharging the aquifers and nothing related to involving large water consumers such as farmers, industries and corporates in ground water recharging. The major investment which is going to be used in the water related infrastructure improvement are lacking behind and is not mentioned clearly [9]. The policy doesn't include any concrete steps for local level water management, which can include watershed management and rain water harvesting. Before setting up the tribunal, the central government must work on inter-state relations and transparency should be maintained within the policies of different states, at least the states which shares the common boundaries. The policy must have included a strong view of including private players into the sector, instead there is nothing mentioned related to the integration of private players. Private players can bring efficiency in the water systems and they should have included with certain partnership agreements which shouldn't bother the pricing of water to the end consumers [10]. Improving or enhancing the infrastructure in water sector, such as laying of new water distribution networks, widening of water channels, expansion of lakes, building of new reservoirs may require additional land, acquiring of land from the people and relocation and resettlement provisions are missing in the policy. There is also no adequate information on compensation to be paid to the affected people.

3 Water Policy in Kingdom of Saudi Arabia

The middle eastern regions suffer from acute water scarcity. The major reasons behind this are the climate change, growing population, and incapacitated policies. Most of the drinking water is extracted from the deep aquifers [11]. The average annual rainfall in the kingdom is very less and is estimated to be around 59 mm. During the summer seasons, the temperatures are too harsh and can go up to 55 °C. Because of such dry weathers, the country has no rivers or lakes and nearly 90% of the kingdom is covered in deserts. According to the UNESCO Water Scarcity Index (WSI), the kingdom falls under the extreme water shortage condition [12]. The increasing population of the country will further worsen the situation of water. It is expected that by the year 2040 the population of Saudi Arabia will reach to 42 million [13]. The kingdoms agricultural activity is completely dependent on the ground water sources. If necessary effective steps are not taken, Saudi Arabia may run out of ground water completely. This will have direct effect on the food security which will in turn effect the economic growth. Ministry of Environment, Water and Agriculture (MEWA) has developed a strategy to manage the water scarcity issue in the country. The key objective of the strategy is to address the challenges and restructuring of the water sector. Following sections discusses more about the strategy of MEWA and its evaluation.

3.1 National Water Strategy of Saudi Arabia

The policy's vision is bringing sustainability in water sector and safeguarding the natural water resources. The main priorities are given to the continuous supply of water even under emergency situations, cost-effective water supply to the end consumers, and optimized use of water resources.

3.1.1 Water Law and Resource Management

The policies included will be more focused towards implementing a legal and adequate framework to manage the water crises. The MEWA will be responsible for the implementation of different set of policies and regulation. The top priority will be given to integrated resource management, which aims to bring more sustainability in water sector by efficiently using the available water resources. Ground water revival will be implemented and the monitoring of activities related to this will be taken care or managed by MEWA. The reduction in consumption rates in both the urban and agricultural sectors will be prioritized. The building of more water and waste water treatment facilities will be initiated in order to ensure the continual supply of water without any disruptions. Research and development in water management sector will be sponsored and supported by the MEWA [14].

3.1.2 Water Services Regulations

Electricity and Cogeneration Regulatory Authority (ECRA) is the regulatory authority in Saudi Arabia which will ensure the economic regulations, such as to regulate water services, parallelly with the electricity and cogeneration. It also oversees the licensing of service provider and reviewing of tariffs in the sector with implementation of agreements with the private players. Further privatization of distribution sector will be overseen by private players, presently distribution sector is managed by National Water Company (NWC).

3.2 Evaluation of National Water Strategy

Despite of mentioning the sustainability objective in the National Water Strategy, there isn't any concrete plan proposed in the policy, the integration of renewables in the energy sector which is in turn fueling the water sector of the country is still unknown. Ground water revival and recharging requires proper planning and execution. The methods and technologies used in the revival are not mentioned. There are no provisions provided for water recycling. The reused water can be diverted towards the irrigation sector, but there is no adequate information mentioned related to recycling of water. Water efficient agricultural practices can be a key breakthrough in the agricultural sector of the kingdom. However, there aren't any policies stated related to this. Information regarding investments and funding for both infrastructure improvement and research and development on water management is not adequately stated.

4 Water Policy in Yemen

The country is facing with critical development problems and the most serious challenges are the water scarcity and over-utilization of the aquifers. Together with the climate change and its impacts, the whole situation of the country is worsening day by day. Antagonizing by the continual differences between the annual recharge of water and the growing demand, because of the gradually growing population, has led the country towards the alarming depletion of the groundwater resources which are the major source of the fulfilling the nations demand. According to the rough estimation, only 32% of the Yemen's population is having access to public drinking water system and around 21% is having access to the sanitation networks [15].

As of today, Yemen comes under the world's most water scarce countries and the disputes and conflicts did strongly affect the water sector of the country. Nearly 18 million people are in the urgent necessity of water assistance [16]. The country is already affected due to the increasing number of conflicts in the coastal areas and also with the natural disasters, food security, and other epidemic outbreaks have crumbled the economic structure of the country and the revival of it is problematic. During the early 2000's, the water sector of the country was re-organized, the Ministry of Water and Environment (MWE) of Yemen has initiated a combined strategy which included an action plan, multi-stakeholder process of preparing consolidated strategy for the water sector in whole. The program named as National Water Sector Strategy and Investment Program (NWSSIP), whose primary focus was to improve the water sector of the country. Further sections explain the major characteristics of the strategy.

4.1 National Water Sector Strategy and Investment Program

This strategy puts forward certain set of institutional and financial measures, which are focused towards improving the water sector and its sub sectors in order to ensure the interests of stakeholders in those resources. This strategy enables to protect the water resources and to stop the unsustainable use of water. If the improper usage of water continues, this will eventually reduce the levels of water resources, and will impact and harm many sectors in the country which includes the agricultural sector which will be the first victim of water crisis. This will result in economic decline and impact the food security of the country. The NWSSIP had setup some objectives in order to sustain the water sector management, firstly, NWSSIP ensure the coordination amongst the urban and rural water supply sector which comes under the MWE. The finances in the sector are monitored and warranted that it effectively supports the growth in water sector, the strategy focuses on the sustainability aspect to support the continual meeting of water needs of both domestic and industrial sector parallelly maximizing economic benefits. The policies bring attention towards the pollution of the ground water resources and the ways to recover from the problem.

4.2 Evaluation of NWSSIP and Humanitarian Assistance in Yemen

Apart from all the policies and strategies drafted, Yemen is still under crisis. Because of the number of crises in the last decade, the country is known as the worst humanitarian crisis, the nation is far away from the sustainability goals and is now undergoing a five-year long war. Because of the growing population and unmanageable famine and poverty, the extremely harsh living conditions have severely affected the water sector. With the

Yemen governmental organizations, there are some other organizations which are supporting the country with different aids and funds. Some of the known organizations are UNCIEF, UN, and Baitul Maal and Mona [17].

UNICEF is working more towards the water sector of Yemen by supporting and maintaining various supply systems in the country, and providing safe drinking water access to nearly 5 million people in the country. Despite having a prodigious strategy, the implementation is very arduous. Because of the instability in the nation, the reviving through the situation takes plenty of time. As the NWSSIP strategy does not include in concrete steps in recharging the ground water resource, the water recycling is not included in the program. As stated, the coordination with the urban and rural water sector may bring some positive outcome, but still the financial investments are not clearly mentioned in the program, and the community involvement in aquifer recharge is not adequately explained.

5 Conclusion

We are aware of how vital the role of water is too human and animal life, for maintaining both the economic and ecological developments in a nation. Considering the importance of water and its scarcity, the planning and management of the water resources is utmost importance and making sure of equitable use in an economic manner should be the priority. The water policy in India is very much dependent on the development of its infrastructure. However, the distribution losses and its considerable impacts must also be considered with strategies to overcome the water theft which is also a very common problematic case which should be more enlightened. Without going for the newer and much larger water projects which is mentioned in the policy, it is better to increase the capacities of the existing ones with minimizing the distribution losses. Private player involvement is not much entertained in the policy, the Public private partnership must be promoted with proper agreements and more room should be given to the private players in operations and maintenance side of the water supply systems. Regulations must be strengthened on the resettlement and reallocation of lands to the people who are affected by the expansion of water bodies and other infrastructural works. Therefore, the proper compensation strategies must be developed. Special subsidies and incentives must be provided on installations of rain water harvesting systems and proper education programs must be set-up across the country for encouraging the people towards water harvesting, as the runoff water can be minimized which will reduce the load on the existing stormwater networks. Trainings must be given to the farmers as most of the water used is in irrigation sector. In order to bring sustainability in the water sector, the trainings must include adaptation to new water efficient agricultural methods, usage of efficient pumps, and water recycling advantages.

The countries like Saudi Arabia, which is considered as the developed nation, should focus more on renewable energy integration within the water sector. As the country is heavily dependent on the technologies like desalination, the usage of renewables will bring sustainability in the field. There must be provisions with sufficient information on water recycling, which can support the agricultural practices in the country. The monitoring of existing aquifers must be accurate and continuous. In order to maintain water availability, the recharging of the aquifers must be prioritized. As Yemen is also a part of Middle East and facing not only a similar water crisis as the other neighboring countries but also the wars, epidemics, economic instability with some other governance issues, only the external financial assistance and end of disputes and wars can help country to revive from the ongoing crises. The NWSSIP strategy can only be possible with sufficient investments and collaboration between governments, donors, and civil societies under a single common agenda.

References

1. Usgs.gov: How Much Water is There on Earth? (2021). https://www.usgs.gov/special-topic/water-science-school/science/how-much-water-there-earth?qt-science_center_objects=0#qt-science_center_objects
2. Who.int: Drinking-water (2021). https://www.who.int/news-room/fact-sheets/detail/drinking-water
3. ReliefWeb: Water Stress by Country: 2040 - World (2021). https://reliefweb.int/map/world/water-stress-country-2040
4. The Water Project: Water Scarcity - The U.S. Connection (2021). https://thewaterproject.org/water-scarcity/water_scarcity_in_us
5. Benavente, C.: Water: the future's fuel. UC Merced Undergrad. Res. J. **7**(1) (2014). https://doi.org/10.5070/M471024999. https://escholarship.org/uc/item/31q17300
6. Media.rff.org (2021). https://media.rff.org/documents/RFF20WP-18-17-rev.pdf
7. 2021. https://www.researchgate.net/publication/325890909_Water_Policy_in_India_Building_Blocks_for_Synergy_with_Science_Technology_and_Innovation_STI_Policy_for_Inclusive_Growth
8. Jalshakti-dowr.gov.in (2021). https://jalshakti-dowr.gov.in/sites/default/files/NWP2012Eng6495132651_1.pdf
9. Mpra.ub.uni-muenchen.de: A Review of Indian Water Policy - Munich Personal RePEc Archive (2021). https://mpra.ub.uni-muenchen.de/45230/
10. Singh, K., Singh, R., Meena, M., Kumar, A.: Water Policy in India: A Review. Academia.edu (2021). https://www.academia.edu/11542452/Water_Policy_in_India_A_Review
11. Frenken, K.: Irrigation in the Middle East region in figures AQUASTAT Survey-2008. Water Reports 34 (2009)
12. Abderrahman, W.A.: Groundwater resources management in Saudi Arabia. In: Special Presentation and Water Conservation Workshop, Al Khobar, Saudi Arabia (2006)
13. Data.worldbank.org: Indicators | Data (2021). https://data.worldbank.org/indicator
14. Mewa.gov.sa: National Water Strategy (2021). https://www.mewa.gov.sa/en/Ministry/Agencies/TheWaterAgency/Topics/Pages/Strategy.aspx
15. Extwprlegs1.fao.org (2021). http://extwprlegs1.fao.org/docs/pdf/yem147103.pdf
16. Unicef.org: Water, Sanitation and Hygiene (2021). https://www.unicef.org/yemen/water-sanitation-and-hygiene
17. Chrol, J.: 5 Organizations Helping During the Yemen Crisis - The Borgen Project. The Borgen Project (2021). https://borgenproject.org/5-organizations-helping-during-the-yemen-crisis/

Cyprus Beaches in the Context of Parabolic Bay Shaped Beach Model

Ramin Layeghi[1], Amin Riazi[2], and Umut Türker[1(✉)]

[1] Eastern Mediterranean University, Famagusta, North Cyprus, Turkey
{ramin.layeghi,umut.turker}@emu.edu.tr
[2] Cyprus International University, Nicosia, North Cyprus, Turkey
ariazi@ciu.edu.tr

Abstract. The importance of shoreline protection has attracted researchers in terms of finding new and applicable methods. The equilibrium planform concept is being widely used and modified over more than two decades. Cyprus has relatively high economical dependency on its sandy shorelines, which are mostly natural headland bays. The fact that most of the rivers located in the island are dried out, duo to careless dam construction over the years; seems to have compromised the main sediment inflow resource for the coastal regions. Hence the existing compulsion for more forecasting and planning, concerning shoreline protection is very clear. Application of parabolic bay shape approach to coastal environments of east of Cyprus have been carried out, using parabolic bay shape approach and two different modifications of the approach. Concerning the shoreline equilibrium, the tested coastlines are in general following a static conditions. However, due to unreliable diffraction points associated with headland bays, in some cases, the coastline is not in a desirable stable situation. Accordingly various suggestions for headland extension are presented at various regions.

Keywords: Parabolic bay shape · Beach · Headland bay · Shoreline · Beach stability

1 Introduction

Waves approaching to nearshore zone causes longshore and cross-shore sediment transport. Both bring about changes in coasts mainly erosion or accretion. Various approaches are developed to protect coastlines from longshore [1–3] and cross-shore sediment transport [4–6]. In particular, for both cases soft engineering applications become a preferred option to implement ecological protection for coastal changes [7–9].

The static equilibrium planforms and equilibrium beach profiles at nearshore zone defines that the equilibrium in sediment mobility is achieved, in other terms net sediment transport within the coastal region has been zeroed. A vast amount of research has been conducted to predict equilibrium conditions [10–12].

The static headland-bay beach concept provides a good approach to issues such as beach stability assessment, erosion prediction, and coastal protection. Many scholars

H. Gökçekuş and Y. Kassem (Eds.): NRSEM 2021, EESCI, pp. 173–182, 2022.
https://doi.org/10.1007/978-3-031-04375-8_20

have compared the advantages and disadvantages of the various models, their applicability to various coasts, the influence of waves, their practical applications, and other factors that influence the models [13, 14]. Four static equilibrium headland bay beach models are the logarithmic spiral model [15], the parabolic model [16], the hyperbolic tangent model [17] and the Elliptical Model [18]. Among them, the parabolic bay shape equation (PBSE) proposed by Hsu and Evans [16] has been recognized for its success and is widely used for global coastal engineering and management applications. This method have become applicable and beneficial when [19] linked it with stability determination and the method was presented as a tool to assess stability of the headland bays. This method provides the ability to measure the full erosion or accretion potential of headland bays and it is highly recommended to be used in order to predict the coastal changes prior to any artificial disruption of the natural processes.

Cyprus is an island in Eastern Mediterranean Sea and is the third most populated and the third largest island. It is located south of Turkey, west of Syria and Lebanon, northwest of Israel, north of Egypt and east of Greece. Because of its critical and strategic location in Mediterranean Sea, beside its political importance it has a great touristic attraction because of its beautiful landscapes and sandy beaches. Maintaining these sandy beaches are critical to Cyprus's economy which is one of the income sectors of its overall income. To study the present situation of these beaches and find any patterns governing the sediment supply status of these coastal regions there are different methods. Among these methods, the most popular one is to trace the input and output of sediment sources in quantitative manner to decide on the movement patterns of sand and gravel particles to then theoretically predict the future condition of these beaches. An alternative method is to determine the headland bay's status by studying its shape using aerial pictures. The first option can be done by very long-term, expensive and difficult field observation to elucidate the sediment movement and beach state by contrasting the historical aerial photographs of usually two or three decades. Therefore, the study area is an ideal location for testing the method performed by Hsu and Evans [16] due to its strongly embayed natural beaches. Hence, the method could be used to show that the sediment volume around the island is finite and if its beaches are not protected, they would be eroded.

The main sediment source of coastal regions is the rivers which are perennial and most of them dried out due to irresponsible dam constructions at their upstream tributaries [20]. Due to the short fetch length of coasts, the waves that are approaching to the coastal areas are not strong enough to continuously erode coastal cliffs to supply sediment source to the coastal zones. Therefore, it is merely rivers that have the ability to carry sediment particles through their watershed area and act as the main source of sand and sediment for sandy headland bays. However, due to the construction of artificial water resources structures on rivers and decrease in the rainfall pattern of Cyprus, sediment supply is minimized recently. Therefore at the coastal zone areas appears to be a lot of sandy headland bays with no stream based sediment supply for the replacement of available sediment losses. In this state if an embayment is in a dynamic equilibrium it loses its sediment inflow and would be eroded to a static state.

Considering the aforementioned concepts and methods, this study will evaluate the equilibrium status of three different beaches, given in Fig. 1, around the eastern coasts of Cyprus. Accordingly the sediment balance system of the different coastal locations will be evaluated to emphasize the importance of the coastline protection issues in present time and for the future. Furthermore, parabolic models were used to suggest artificial coastal structures like jetties and breakwaters while proposing correct location and size of these structures for minimizing environmental concerns of natural habitat at coastal areas.

Fig. 1. Shoreline locations at the eastern coast of Cyprus.

2 Methodology

Hsu and Evans [16] developed the parabolic bay shape equation (PBSE), as a distinct revolutionary method after fitting 27 different bay shaped beaches, accepted to be in static equilibrium. 14 beaches were prototypes and 13 of them were model bays in static equilibrium. The result was a second-order polynomial empirical equation.

$$\frac{R_n}{R_\beta} = C_0 + C_1\left(\frac{\beta}{\theta}\right) + C_2\left(\frac{\beta}{\theta}\right)^2 \tag{1}$$

where R_β is the length of the control line which is the distance between the upcoast diffraction point and a point approximately in the end of the downdrift straight section of the beach, R_n is a radius from updrift control point, stretched to a point on the static equilibrium, corresponding to the angle θ. Angle θ is the angle between the main wave crest tangent and R_n radii and β is the angle between the assumed wave crest line and the control line, R_β. The parameters are illustrated in Fig. 2.

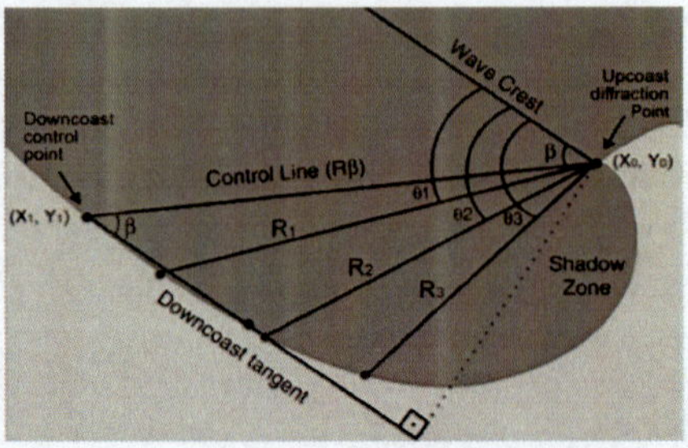

Fig. 2. Parabolic bay shape model sketch [21]

The constants in Eq. (1), as cited in [22] have been determined originally by Hsu and Evans [16] through regression analysis over 27 beaches:

$$C_0 = 0.0707 - 0.0047\beta + 0.000349\beta^2 - 0.00000875\beta^3 + 0.00000004765\beta^4 \quad (2)$$

$$C_1 = 0.9536 + 0.0078\beta - 0.00004879\beta^2 + 0.0000182\beta^3 - 0.000001281\beta^4 \quad (3)$$

$$C_2 = 0.0214 - 0.0078\beta + 0.0003004\beta^2 - 0.00001183\beta^3 + 0.00000009343\beta^4 \quad (4)$$

The range that these constants can obtain value within is from −1 to 2.5 when β is in the usual range of 10 to 80° which is common in normal wave conditions [23]. On the other hand, they have tabulated the R_n/R_β values corresponding to different values of β ranging from 20 to 80° with 2° interval. The modifications of PSBE are generally based on simple boundary condition which can be easily found through geometrical analysis [24]:

$$C_0 + C_1 + C_2 = 1 \quad (5)$$

$$C_1 + 2C_2 = \beta\cot(\beta) \quad (6)$$

The first modification applied to PBSE subsequently was by Tan and Chiew [24], in which the equation was transformed into Eq. (7). In Eq. (7) coefficient α differs for static equilibrium and dynamic equilibrium conditions, as shown in Eq. (8) and Eq. (9), respectively.

$$\frac{R_n}{R_\beta} = \{1 + \alpha - \beta cot(\beta)\} + \{\beta cot(\beta) - 2\alpha\}\left[\frac{\beta}{\theta}\right] + \alpha\left[\frac{\beta}{\theta}\right]^2 \quad (7)$$

For static equilibrium:

$$\alpha = 0.277 - 0.0785 \times 10^{\beta} \quad (8)$$

For dynamic equilibrium:

$$\alpha = -0.004 - 0.0113 \times 10^{\beta} \quad (9)$$

In another study carried by [25], the constants were modified to reduce error and become applicable in two different manners; one for sandy beaches and the other one for gravel beaches. They examined 52 Mediterranean embayed beaches located along Italian, French, Spanish, Tunisian and Turkish coasts where all static equilibrium beaches both natural and artificial we evaluated and give good results.

$$C_0 = 1 - \beta\cot(\beta) + m\beta + q \tag{10}$$

$$C_1 = \beta\cot(\beta) - 2(m\beta + q) \tag{11}$$

$$C_2 = m\beta + q \tag{12}$$

Here m and q are two added dimensionless coefficients. For sand beaches m = −0.8460 and q = 0.2281 and for gravel beaches m = −1.0235 and q = 0.2476. All of the aforementioned modifications generate a polar curvature of which, the part between angles β and 150 to 180° is going to be fitted over the aerial photo of the coastline in hand for assessment. The generated plot would merely show the layout of the static equilibrium of the beach according to the wave diffraction point and wave direction.

After the application of the PBSE using either one of the methods explained, the stability of the beach could be examined as if the generated static equilibrium prediction is landward comparing to the currently existing shoreline periphery the beach is considered to be in dynamic equilibrium and it would degraded, in case of sediment supply interruption. On the other hand, if the predicted static equilibrium is seaward from the existing shore line the embayment is in unstable to natural reshaping mode and it would gain sediment seaward in time until the distance is filled up, and obviously if the predicted curve of static equilibrium matches the existing beach it would be in static state where change in sediment supply would not affect its periphery and layout.

3 Results and Discussion

3.1 Kanlidere (Pedieos) Stream's Delta

Considering the long period evolution of embayment had maintained its planform in a stable state. First we would choose the most viewable unsubmerged points, observable on the satellite map as tips of the two headlands (diffraction point). The result is demonstrated in Fig. 3 where the discrepancy of the plotted parabolic curves and the actual shoreline periphery is observable. As it can be seen in Fig. 3 the Tan and Chiew [24] modification is similar to the original model drawn in green color. Schiaffino et al. [25] model suggests that the northern bay beach is already in static equilibrium while the southern bay beach is in unstable situation and should be in natural nourishment state, which is not compatible with the historical data of the embayment. This difference would be either because of the miss placing the refraction point or because of the shoreline equilibrium being dynamic considering the predicted curvature is landward of the shoreline mostly. By considering the possibility of other diffraction points at the submerged rocky parts located further seaward of the chosen diffraction points, the shoreline could be divided into four different diffraction points.

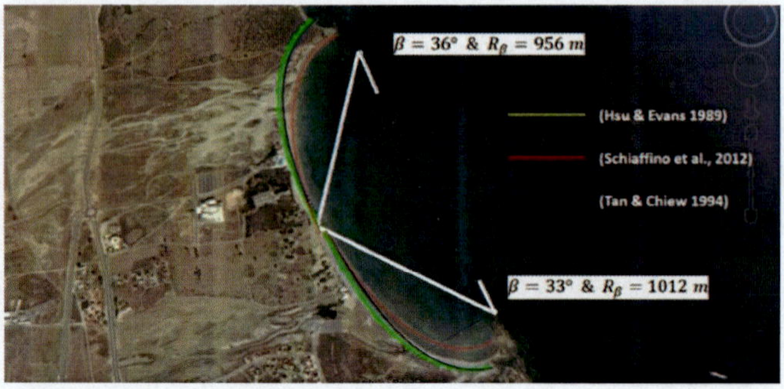

Fig. 3. The PBSE and its alternative modifications for unsubmerged headland tips to Kanlidere stream's delta (map source: Google Earth).

After investigating the beach periphery more closely, four different curves were distinguishable, two for each headland. The great agreement between the actual shoreline and the predicted planform, indicates that if the headland is in the static equilibrium, the wave diffraction point for headland is not singular but it is more likely that the submerged rocks are effecting the head land formation alongside at the tip of the headlands. The results shows that the shoreline is not in a desirable situation. For the first case, as the both sides seems to be in dynamic situation, in the absence of sediment supply, they are to be eroded. In the second approach, the unavailability of a certain physical diffraction point can be relied on long term beach protection. Hence, construction of jetties to stretch both headlands to a position where the predicted static landform, at least matches the existing shoreline, is very reasonable in order to protect this shoreline. The result of minimum distance that the diffraction point has to be moved to satisfy the matching position of the static equilibrium is estimated as 50 m in line with northern headland and 25 m in line with southern headland.

3.2 Bafra Beach

At Bafra Beach, the unsubmerged headland tips were chosen as diffraction points and the results shows that the northern bay is approximately in static equilibrium according to the original equation (the green curve), while in case of southern landform, the predicted static equilibrium curve is significantly landward comparing to the actual landform (Fig. 4(a)) indicating that the southern bay is still in dynamic equilibrium. Despite the fact that it is completely eroded, the coastal morphology is covered with rocks. Since the shorelines historical and geographical are also aligned with the result of the parabolic equation analysis, there is no need for further assessment to evaluate other possibilities for the position of diffraction points.

The result of the modified equation illustrated in Fig. 4 are mostly in great agreement with the original form, except the result of the Schiaffino et al. [25] modification for the northern bay beach. Again this modification fails to match the situation rationally for the south to north curved beach while it can be considered applicable for the south to north curved bay. Following is the protective construction suggestions that could be derived from the conducted analysis.

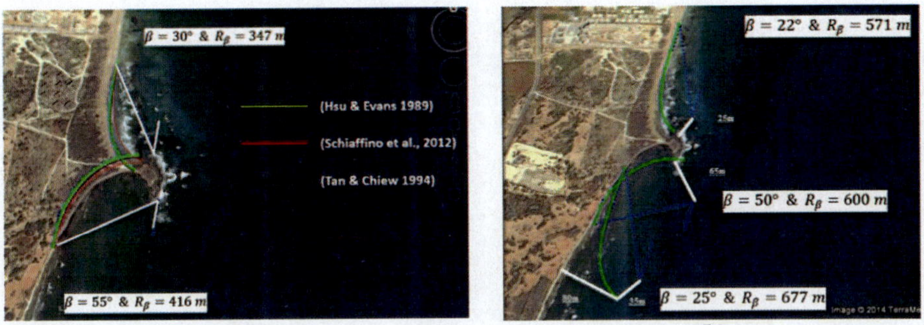

(a) (b)

Fig. 4. (a) First application of the PBSE and its alternative modifications for unsubmerged headland tips at Bafra Beach. (b) Proposed length and direction of the headlands to reach the satisfactory diffraction point for an acceptable headland bay static equilibrium (map source: Google Earth)

The three structures in Fig. 4(b) are jetty suggestions which could transform the shoreline equilibrium into the unstable state and natural nourishment would take place in time until the actual shoreline reaches the green lines. The main purpose is to protect the coastal area from erosion without going through an expensive process of artificial nourishments and other beach development methods.

Very close to the discussed region at Bafra Beach, there is another location which is protected by jetties and breakwaters. Hence, the assessment for this area is done by focusing on the effects of these artificial structures on the morphology of the shore. The results show a relatively acceptable coincidence between the shoreline and the predicted results. However while the original equation (the green curve) is closely matched with the shoreline periphery in general, there is still a little gap to be filled with sand particles extending shoreline in seaward direction (Fig. 5).

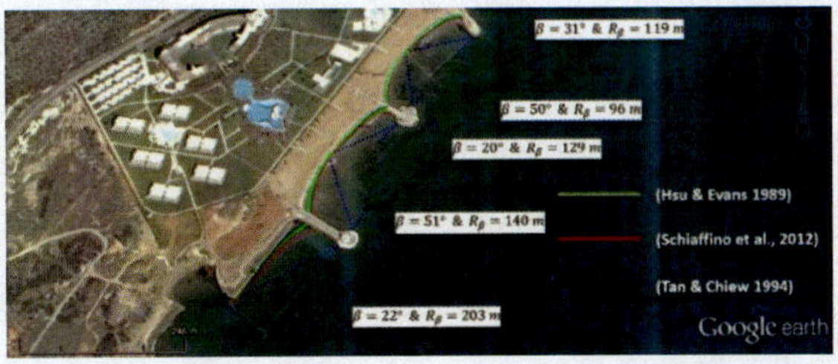

Fig. 5. Application of the PBSE and its alternative modifications for unsubmerged headland tips at Bafra Beach (map source: Google Earth).

3.3 Golden Beach

As shown in Fig. 6 the headland bay at the left is in static equilibrium according to the original Hsu and Evans (1989) equation and also the static equilibrium prediction, modified version by Schiaffino et al. [25], have close agreement with the shoreline. This is while the other PBSE expression by Tan and Chiew [24] shows that this headland bay is in dynamic equilibrium and it is going to need constant sediment supply to maintain this landform. For the right headland bay beach, all three plotted graphs are landward from the shoreline, suggesting that the shoreline is in dynamic equilibrium, which means it is being supplied with sediment supply continuously to sustain the present periphery. However, the satellite images from previous years show that the beach is stable. This can be interpreted in two ways; (1) there is another diffraction point submerged in an unknown location; (2) There is an excess longshore and cross-shore sediment transport at the region supplying sediments to the region.

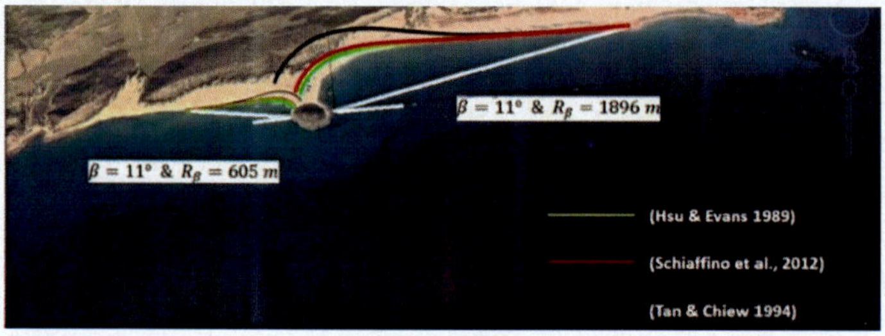

Fig. 6. Application of three alternative PBSE modifications to the Golden beach embayment, picking the unsubmerged points on the headland as diffraction point (map source: Google Earth).

An alternative submerged diffraction point as a right side headland bay is selected to find an appropriate diffraction point (Fig. 7(a)). At this new location both [16] and [25] models are approximately covering the shoreline landform. Therefore, the assumption that this is the real diffraction point is more reasonable. However, creating a physical tangible diffraction point for headland bays is more reliable than a natural diffraction point which is always associated with lots of uncertainties; hence, modifying the diffraction position by a simple coastal structure would be more logical. Figure 7(b) is the suggestible structure in according to Hsu and Evans [16] equation applied by MEPBAY.

(a)

(b)

Fig. 7. (a) Result of choosing an alternative submerged diffraction point for the right headland of the Golden beach; (b) construction of a jetty based on PBSE to protect the Golden beach shoreline (map source: Google Earth).

4 Conclusions

The results have shown that different modifications of the PBSE can show different headland bays' stability and equilibrium conditions. Therefore, the correct evaluation of the PBSE requires more details and elimination of uncertainties in terms of initial morphology of coasts under consideration. By looking at the results of this study, it can be seen that most of the sandy shorelines that have maintained around the east coast of Cyprus are due to the natural headlands. These natural headlands have just provided an appropriate diffraction position which protected the considered coast from erosion.

Engineering developments added artificial beaches as common environments as natural beaches all around the world, achieving to manage coastal zones, strengthening their stability and protecting their hinterland. The software used in this study (MEPBAY), as an integrated software was successful on modeling the headland bay beaches and help to understand the static and dynamic equilibrium conditions of coastal environment. With the correct use of the software, it would be advantageous to re-examine the coastal regions where beach erosion is evident and even propose how much nourishment can be implemented for the stability of the coastal zones.

The analysis in this study were performed for a Mediterranean Island in which the fetch length of considered sites were small and under the effect of developing sea conditions. Even though the effectiveness of parabolic beach profile model were valid for oceanic beaches, herein, it is also depicted that the model works even in closed sea conditions. Therefore, an important property of the model is its ability to provide correct results independent of the size of the material that built up the coast (sand, gravel etc.) and independent of the energy of the waves that disturbs the shoreline.

References

1. McCarroll, R.J., et al.: High-efficiency gravel longshore sediment transport and headland bypassing over an extreme wave event. Earth Surf. Process Landf. **44**, 2720–2727 (2019). https://doi.org/10.1002/esp.4692
2. Noujas, V., Kankara, R.S., Rasheed, K.: (2018) Estimation of longshore sediment transport rate for a typical pocket beach along west coast of India. Mar. Geodesy **41**, 201–216 (2018). https://doi.org/10.1080/01490419.2017.1422818
3. Bayram, A., Larson, M., Hanson, H.: A new formula for the total longshore sediment transport rate. Coast. Eng. **54**(9), 700–710 (2007). https://doi.org/10.1016/j.coastaleng.2007.04.001
4. Ding, Y., Styles, R., Kim, S.C., Permenter, R.L., Frey, A.E.: Cross-shore sediment transport for modeling long-term shoreline evolution. J. Waterw. Port Coast. Ocean Eng. **147**(4), 04021014 (2021). https://doi.org/10.1061/(ASCE)WW.1943-5460.0000644
5. Aragonés, L., Pagán, J.I., López, M.P., Serra, J.C.: Cross-shore sediment transport quantification on depth of closure calculation from profile surveys. Coast. Eng. **151**, 64–77 (2019). https://doi.org/10.1016/j.coastaleng.2019.04.002
6. Türker, U., Kabdaşlı, M.S.: Verification of sediment transport rate parameter on cross-shore sediment transport analysis. Ocean Eng. **34**(8–9), 1096–1103 (2007). https://doi.org/10.1016/j.oceaneng.2006.08.002
7. Türker, U., Yagci, O., Kabdasli, M.S.: Impact of nearshore vegetation on coastal dune erosion: assessment through laboratory experiments. Environ. Earth Sci. **78**(19), 1–14 (2019). https://doi.org/10.1007/s12665-019-8602-8

8. Pranzini, E.: Shore protection in Italy: from hard to soft engineering and back. Ocean Coast. Manag. **156**, 43–57 (2018). https://doi.org/10.1016/j.ocecoaman.2017.04.018
9. Morris, R.L., Konlechner, T.M., Ghisalberti, M., Swearer, S.E.: From grey to green: efficacy of eco-engineering solutions for nature-based coastal defense. Glob. Chang. Biol. **24**(5), 1827–1842 (2018). https://doi.org/10.1111/gcb.14063
10. Różyński, G.: Unexpected property of dean-type equilibrium beach profiles. J. Waterw. Port Coast. Ocean Eng. **147**(5), 06021001 (2021). https://doi.org/10.1061/(ASCE)WW.1943-5460.0000664
11. Mieras, R.S., Puleo, J.A., Anderson, D., Hsu, T.J., Cox, D.T., Calantoni, J.: Relative contributions of bed load and suspended load to sediment transport under skewed-asymmetric waves on a sandbar crest. J. Geophys. Res. Oceans **124**(2), 1294–1321 (2019). https://doi.org/10.1029/2018JC014564
12. Riazi, A., Türker, U.: Equilibrium beach profiles: erosion and accretion balanced approach. Water Environ. J. **31**(3), 317–323 (2017). https://doi.org/10.1111/wej.12245
13. Li, Z.L., Li, W.Q., Chen, Z.S., Zhu, Y.M.: Influencing factors and classifications of arc-shaped coasts in South China. Acta Geogr. Sin. **69**(5), 595–606 (2014)
14. Yu, J.T., Chen, Z.S.: Study on headland-bay sandy coast stability in South China coasts. China Ocean Eng. **25**(1), 1 (2011). https://doi.org/10.1007/s13344-011-0001-1
15. Yasso, W.E.: Plan geometry of headland bay beaches. Geology **73**, 702–714 (1965). https://doi.org/10.1086/627111
16. Hsu, J., Evans, C.: Parabolic bay shapes and applications. Proc. Inst. Civ. Eng. **87**(4), 557–570 (1989). https://doi.org/10.1680/iicep.1989.3778
17. Moreno, L.J., Kraus, N.C.: Equilibrium shape of headland-bay beaches for engineering design. In: Proceedings of the Coastal Sediments 1999, pp 860–875. American Society of Civil Engineers, New York (1999)
18. Li, B., Zhuang, Z., Cao, L., Du, F.: Application of the static headland-bay beach concept to a sandy beach: a new elliptical model. J. Ocean Univ. China **19**(1), 81–89 (2020). https://doi.org/10.1007/s11802-020-3899-1
19. Benedet, L., Klein, A.H., Hsu, J.: Practical insights and applicability of empirical bay shape equations. In: Coastal Engineering 2004, pp. 2181–2193. ICCE, Portugal (2004). https://doi.org/10.1142/9789812701916_0175
20. Kayan, G., Riazi, A., Erten, E., Türker, U.: Peak unit discharge estimation based on ungauged watershed parameters. Environ. Earth Sci. **80**(1), 1 (2021). https://doi.org/10.1007/s12665-020-09317-4
21. Raabe, A.L.A., Klein, A.H., González, M., Medina, R.: MEPBAY and SMC: software tools to support different operational levels of headland-bay beach in coastal engineering projects. Coast. Eng. **57**(2), 213–226 (2010). https://doi.org/10.1016/j.coastaleng.2009.10.008
22. Klein, A.H., Vargas, A., Raabe, A.L.S.A., Hsu, J.R.: Visual assessment of bayed beach stability with computer software. Comput. Geosci. **29**(10), 1249–1257 (2003). https://doi.org/10.1016/j.cageo.2003.08.002
23. Silvester, R., Hsu, J.R.: Coastal Stabilization. Advanced Series on Ocean Engineering, vol. 14. World Science, Singapore (1997)
24. Tan, S.K., Chiew, Y.M.: Analysis of bayed beaches in static equilibrium. J. Waterw. Port Coast. Ocean Eng. **120**(2), 145–153 (1994). https://doi.org/10.1061/(ASCE)0733-950X(1994)120:2(145)
25. Schiaffino, C.F., Brignone, M., Ferrari, M.: Application of the parabolic bay shape equation to sand and gravel beaches on Mediterranean coasts. Coast. Eng. **59**(1), 57–63 (2012). https://doi.org/10.1016/j.coastaleng.2011.07.007

Enhanced Saturated Seepage Analysis Using Fractal Hydraulic Conductivity (Case Study: Gale Chay Dam)

Abdollah Ojaghi[1](✉), Vahid Nourani[1,2], and Elnaz Sharghi[1]

[1] Center of Excellence in Hydroinformatics and Faculty of Civil Engineering, University of Tabriz, Tabriz, Iran
Abdollah.ojaghi@yahoo.com

[2] Faculty of Civil and Environmental Engineering, Near East University, North Cyprus, via Mersin 10, 99138 Nicosia, Turkey

Abstract. The seepage analysis of an earth-fill dam is of prime importance in its stability and design. The accurate estimation of the permeability and the hydraulic conductivity of the material used in the dam core is highly influential on the estimated seepage. In this research, the Gale chay dam was chosen for the seepage analysis, and the dam is located in the northwest of Iran in East Azerbaijan province. The fractal theory was employed to study the porous structure of the core material and approximate the corresponding permeability hydraulic conductivity. For this purpose, the dam core material was gathered and analyzed using the Scanning Electron Microscope (SEM) and the image processing technique. The steady-state seepage analyses were conducted by the Finite Element Model (FEM) using the estimated fractal hydraulic conductivity, as well as the minimum and maximum hydraulic conductivities within the measured range indicated by the laboratory experiment. Providing a relative error of 12%, the estimated fractal hydraulic conductivity enhanced the seepage analysis by almost 70% compared to the result of maximum hydraulic conductivity. Also, the implication of fractal hydraulic conductivity resulted in a better seepage analysis in terms of safety factors compared to the estimated seepage by the minimum hydraulic conductivity. The mentioned results signify the influence of a more precise observation of the soil structure using the fractal theory.

Keywords: Seepage analysis · Finite Element Model (FEM) · Saturated hydraulic conductivity · Scanning Electron Microscope (SEM) · Fractal · Gale chay dam

1 Introduction

The necessity to control the available water supplies makes dam construction a viable choice for managing the water resources. Different types of dams were constructed to store and transfer water from which earth-fill dams were widely used [1]. However, this type is highly exposed to the piping phenomenon as a potent failure mode [2]. Zhang

H. Gökçekuş and Y. Kassem (Eds.): NRSEM 2021, EESCI, pp. 183–192, 2022.
https://doi.org/10.1007/978-3-031-04375-8_21

et al. [3] indicated that 64% of dam failures were caused by piping which was directly related to the amount of seepage through the dam. Hence, an accurate seepage estimation is imperative in the designing of an earth-fill dam.

Several studies were carried out for estimating the seepage through earth-fill dams. The finite element model (FEM) and the artificial neural network (ANN) was employed for the seepage analysis of the Jeziorsko dam, and it was declared that the ANN model could perform an acceptable simulation [4]. Nourani et al. [1] used single and integrated ANN models for the seepage estimation of the Sattarkhan dam, and it was concluded that the ANN models could perform appropriate seepage estimations. Seepage analysis was carried out using Casagrande, numerical, and physical models by Al-Janabi et al. [5], and it was observed that the numerical approach was highly affected by soil compaction, and the Casagrande method could not provide precise seepage estimations.

A detailed study of hydraulic conductivity has an indispensable impact on accurate seepage estimation [6]. Several functions were proposed for estimating the hydraulic conductivity [7]. The hydraulic conductivity was observed to be directly related to the permeability of the soil structure, which is an innate characteristic of porous media [8]. Consequently, to estimate the hydraulic conductivity more accurately, it is necessary to estimate the permeability of the soil precisely.

It was declared that the porous soil structure could be considered as fractal objects of nature [9]. Mandelbrot [10] presented fractals as a type of geometry to describe the fragmented or irregular patterns that Euclidean geometry considered shapeless forms. Xu [9] mentioned that the porous structure complexities (i.e., pore or particle size distribution) could be defined by the concept of fractal dimension. Several methods were proposed to estimate the fractal dimension of a fractal object from which the scanning electron microscope (SEM) images and mercury intrusion porosimetry (MIP) were widely used in practice [11, 12].

Using the fractal assumption, numerous saturated permeability models were proposed. Yu and Cheng [13] developed a fractal permeability model for bi-dispersed porous media, and it was concluded that the model could give plausible estimations. An analytical model was developed for the permeability of the porous environment by Xu and Yu [14], which was related to the maximum pore size, porosity, and fractal dimension. Shen et al. [15] introduced an improved permeability model by the concept of fractals in which a new probability density function was developed, broadening the scope of practicality.

Regarding the beforementioned researches related to the estimation of permeability, it was indicated that the fractal models could be considered as practical alternatives. Also, to the best of the author's knowledge, it seems there is not any study regarding the application of fractal permeability models in the seepage estimation of an earth-fill dam. This study aims to perform the saturated seepage analysis of the Gale-chay dam using FEM and a fractal permeability model.

2 Materials and Methods

2.1 Fractal Permeability Model of the Porous Media

The porous structure of a soil sample contains several complexities, which must be studied accurately to understand the water flow through that media precisely. The introduction of fractal geometry led to the development of enhanced permeability and hydraulic conductivity models [15]. Sierpinski carpet is one of the widely utilized geometric patterns to model a porous environment. The pattern can be developed by simple steps illustrated in Fig. 1. First, the main square is split into nine even squares, and then the middle one becomes removed. Subsequently, the same procedure is applied to the remaining squares.

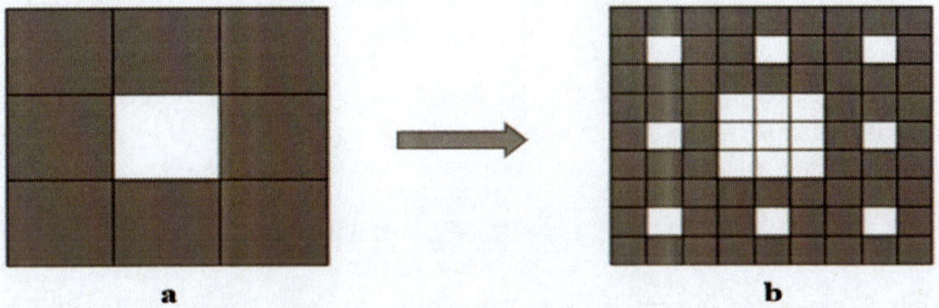

Fig. 1. Sierpinski carpet with the side length of L: a) r = L/3, and b) r = L/3^2

The number of remaining squares can be calculated as:

$$N(r) = C \times r^{-D} \tag{1}$$

where $N(r)$ is the number of remaining gray squares, r indicates the length of small squares, C is a constant, and D is the fractal dimension [16]. The white and gray segments of the Sierpinski carpet were considered to be the voids and matrix, respectively [15]. Hence the pore area was defined as:

$$A(\geq r) = A_f\left[1 - (\frac{r}{L})^{2-D}\right] \tag{2}$$

In which A defines the pore area, $A_f = L^2$ declares the area of the view field, and r indicates pore diameter. Considering the Hagen-Poiseain relation and Darcy's law, a fractal permeability model was proposed by Shen et al. [15] as:

$$K = \frac{\pi}{128}\frac{L_0^{1-D_T}}{A_f}\frac{2-D}{3+D_T-D}3r_{\max}^{3+D_T}\frac{1-(\frac{r_{min}}{r_{\max}})^{3+D_T-D}}{1-(\frac{r_{min}}{r_{\max}})^{3}} \tag{3}$$

where L_0 shows the side length of the square field of view, r_{max} and r_{min} indicate the maximum and minimum pore diameter, respectively, and D_T is the tortuosity fractal dimension. The porous media was considered to be composed of capillary tubes where the tortuosity fractal dimension can be described by

$$D_T = 1 + \frac{Ln\tau_{av}}{Ln\frac{L}{r_{av}}} \tag{4}$$

$$\tau_{av} = \frac{1}{2}\left(1+\frac{\sqrt{1-\varphi}}{2}+\frac{\sqrt{(1-\sqrt{1-\varphi})^2+\frac{\sqrt{1-\varphi}}{4}}}{1-\sqrt{1-\varphi}}\right) \tag{5}$$

where r_{av} indicates the average pore diameter, τ_{av} is the average tortuosity, φ is the porosity [17]. Regarding the porosity fractal model and the unrestricted probability density function developed by Shen et al. [15], the porosity and the average pore diameter were defined as:

$$\varphi = \varphi(\geq r_{min})= 1-\left[1-\frac{A(r_{max})}{A_f}\right](\frac{r}{r_{max}})^{2-D} \tag{6}$$

$$r_{av} = \frac{Dr_{min}}{D-1}\left[1-(\frac{r_{min}}{r_{max}})^{D-1}\right] \tag{7}$$

The saturated hydraulic conductivity was observed to be directly related to the permeability of porous media as:

$$k_s = \frac{K\rho g}{\mu} \tag{8}$$

In which K specifies the intrinsic permeability of the porous media, which can be estimated using the fractal theory via Eq. 3. Furthermore, μ and ρ are the viscosity and the fluid density, and g indicates the gravitational constant.

2.2 Governing Equation and FEM

The water flow phenomenon through porous media such as soil was considered to be described by the classical Richards equation for the saturated and unsaturated conditions [18]. The governing differential equation for the isotropic cases in the steady-state condition can be written as:

$$\frac{\partial}{\partial X}\left(K_u\frac{\partial H}{\partial X}\right)+\frac{\partial}{\partial Y}\left(K_u\frac{\partial H}{\partial Y}\right)+Q = 0 \tag{9}$$

where H is the total water head, Q indicates flux, and K_u specifies the unsaturated hydraulic conductivity, which can be defined as $K_u= k_r k_s$, in which k_s and k_r are the saturated and relative hydraulic conductivity, respectively. Considering the water flow occurring through the saturated zone, the relative hydraulic conductivity equals one [18–20]. A set of proper boundary conditions must be specified as either a hydraulic head or a flow rate on the boundary nodal points [20]. The defined hydraulic heads at the boundary nodes are referred to as the Dirichlet boundary conditions. Also, the specified flow rates across the boundary are called Neumann boundary conditions [19].

One of the most suitable and widely applied numerical methods for estimating the Eq. 9 on the spatial domain is the FEM, broadly employed for two-dimensional problems [1]. It is important to mention that in this study, Seep/w software was used for conducting the seepage analysis employing the FEM.

2.3 Case Study

In the current study, the Gale chay dam was chosen for the seepage estimation using the fractal hydraulic conductivity model. Considering the geographic coordinate of 37°31′35.9" North and 46°07′55.6" East, the dam is located on Gale chay river in Ajabshir county, East Azerbaijan, Iran. As the dam was built for various purposes such as fish farming, irrigation, and hydropower generation, it is pivotal to estimate the seepage accurately. The saturated hydraulic conductivity of the core material was measured via laboratory experiments, and it was observed that the value is within the range of 1×10^{-8} - 1.2×10^{-7} (m/sec). Also, the amount of seepage through the dam was observed to be $2.5 \times 10^{-5}\ m^3/s$ at its maximum water level.

3 Results and Discussion

3.1 Estimating the Saturated Hydraulic Conductivity

The clay material of the dam core was collected from three different spots of the dam construction site. To decrease the moisture content and prevent the shrinkage of the specimen, the freeze-cute drying method was implemented. In this way, the samples were frozen immediately at −196 °C by liquid nitrogen and cut with a sharp knife. Subsequently, they were dried for 48 h in a freeze drier. The SEM images of the samples were captured using a scanning electron microscope. Figure 2 illustrates the captured images of three samples. The proposed methodology of Rabbani and Salehi [21] was employed to analyze the microscale images.

Fig. 2. Grayscale image of clay sample

The identification of the porous space in the provided SEM images was conducted using multi-level thresholding of Otsu [22]. The watershed segmentation, which can be utilized as a potent means to detect the overlaying shapes, was employed to separate the pores of the indicated porous space. Figure 3 depicts the detected porous space and segmented pores.

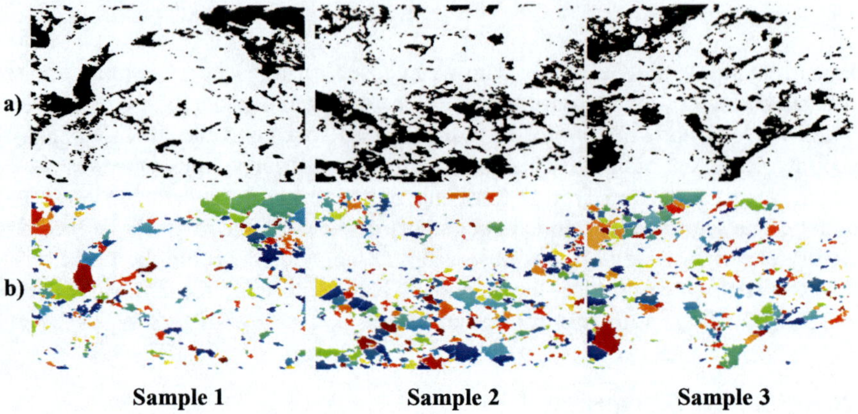

Fig. 3. Image processing of SEM images **a)** Detecting the porous space **b)** Segmentation of the porous space

The SEM images were analyzed, and the area (A), diameter (r), and number (N) of pores were calculated via MATLAB software which can be seen in Table 1. It was observed that the total pore size could be described by 7, 8, and 9 groups which were increased from zero to maximum diameter by 0.5 intervals.

Table 1. The obtained statistical data from SEM image analysis.

Pore grades	Sample 1			Sample 2			Sample 3		
	$A(\mu m^2)$	$r(\mu m)$	N	$A(\mu m^2)$	$r(\mu m)$	N	$A(\mu m^2)$	$r(\mu m)$	N
1	0.379	0.695	118	0.346	0.664	149	0.348	0.666	126
2	1.835	1.528	19	1.619	1.436	18	1.635	1.443	17
3	4.383	2.362	6	3.829	2.208	14	3.87	2.22	6
4	8.023	3.196	2	6.976	2.98	4	7.052	2.997	2
5	12.755	4.03	1	16.079	4.525	3	11.183	3.773	3
6	42.607	7.365	1	28.928	6.069	2	16.261	4.55	1
7	76.463	9.867	1	45.524	7.613	1	29.262	6.104	1
8				65.866	9.158	1	46.054	7.658	1
9							66.639	9.211	1

The fractal dimension of each sample was derived using Eq. (2) and the provided statistical data in Table 1. Considering Eq. (2), the scatter plots of three Samples were drawn, where $ln[1 - A(\geq r)/A_f]$ and $ln(r/L)$ indicate the vertical and horizontal ordinates, respectively. Hence, the fractal dimension $D = 2 - k$ was calculated from the given expression of $ln\left[1 - A(\geq r)/A_f\right] = (2 - D)ln(r/L)$, where k is the slope of the fitted line. It is noteworthy to mention that pores in a sample exhibit better fractal behavior when their scattered data points are analogous to straight lines. Using the SEM image analysis of each sample, Fig. 4 shows the fitted lines to the scattered data points.

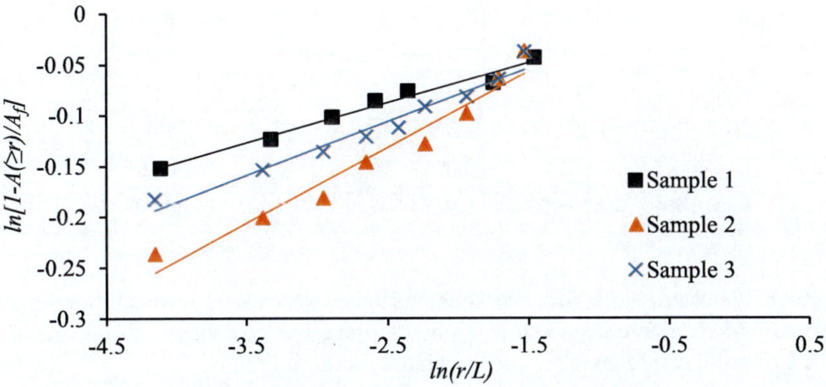

Fig. 4. Linear regression analysis of three samples

Based on the slope of the lines, the fractal dimensions of three samples were calculated, which are presented in Table 2. Considering the high correlation coefficient (CC) and low root mean square error (RMSE) of the fitted lines (provided in Table 2), good fractal behavior was observed across all samples, which could be a result of the naturally formed soil structure.

Table 2. Estimated fractal dimensions and evaluation criteria

Sample	*D*	*RMSE*	*CC*
1	1.9606	0.0062	0.9756
2	1.9252	0.0160	0.9350
3	1.9471	0.0096	0.9607

Using Eqs. (3) and (8), the permeability and hydraulic conductivity of the soil samples were approximated. The results are presented in Table 3.

In order to reach acceptable estimates of the introduced properties, the average values of fractal dimension, porosity, permeability, and hydraulic conductivity were calculated via the generated data of three samples. It can be seen that the increase of tortuosity fractal dimension D_T causes a decrease in permeability and hydraulic conductivity. This was a promising result because higher tortuosity fractal dimension D_T causes more tortuosity in the capillaries and high resistance of flow through the soil environment. The average value of the estimated saturated hydraulic conductivity via fractal theory was observed to be in the measured range ($1 \times 10^{-8} - 1.2 \times 10^{-7}$ (m/sec)), which was

Table 3. Permeability and saturated hydraulic conductivity for three samples

Parameter	Sample 1	Sample 2	Sample 3	Average
D	1.9606	1.9252	1.9471	1.9443
φ	0.1876	0.2982	0.1918	0.2258
τ_{av}	4.1464	2.9197	3.6038	3.5566
r_{av}	1.3067	1.2592	1.255	1.2736
D_T	1.4087	1.3047	1.3642	1.3592
$K(\mu m^2)$	0.0034	0.0057	0.0036	0.0042
$k_s(m/sec)$	3.29e-08	5.43e-08	3.53e-08	4.08e-08

indicated by laboratory experiments. Thus, the developed fractal permeability model of Shen et al. [15] can be considered as a potent alternative to estimate the permeability and the hydraulic conductivity of porous media more accurately.

3.2 Saturated Seepage Analysis

The FEM was carried out for the saturated seepage analysis of the Gale chay dam using See/w software. For this purpose, the previously estimated average saturated hydraulic conductivity with the fractal theory (i.e., 4.08×10^{-8} (m/sec)) was utilized. To evaluate the accuracy of the estimated fractal hydraulic conductivity, two other hydraulic conductivities were considered within the measured range for the saturated seepage estimation, namely maximum and minimum with the values of 1.2×10^{-7} and 10^{-8} (m/sec), respectively.

Table 4. Estimated seepage using saturated hydraulic conductivities

Hydraulic conductivity type	Seepage (m^3/s)	Relative error (%)
Minimum	2.21×10^{-5}	−11.6
Maximum	3.63×10^{-5}	45.2
Fractal	2.8×10^{-5}	12

Table 4 indicates the estimated seepage using three hydraulic conductivities and the corresponding relative errors compared to the observed seepage. It can be seen that the Maximum hydraulic conductivity yielded a relative error of 45.2%, which results in an uneconomical design with a high safety factor. On the other hand, the minimum hydraulic conductivity resulted in a relative error of −11.6% in seepage estimation, which indicates the estimated seepage was lower than the observed amount. Hence, a lower safety factor was gained, and it provides a hazardous dam design. Considering the estimated fractal hydraulic conductivity, a relative error of 12% was obtained. Hence, as the estimated seepage was higher than the observed value, the implementation of the fractal hydraulic conductivity provided a better seepage analysis than the minimum hydraulic conductivity in terms of the safety factor. Comparing the result of the maximum hydraulic conductivity, the seepage estimation via fractal hydraulic conductivity was almost 70% more accurate. Therefore, it provides a more economical seepage estimation.

Overall, it should be noted that considering the distribution of complexities (pores in this case) in a porous media using fractal dimension, the application of fractal theory provides an in-depth view of the soil structure resulting in a more precise permeability and hydraulic conductivity estimation as well as a more accurate seepage analysis.

4 Conclusion

In the current study, the saturated seepage analysis of the Gale chay dam was conducted using FEM. The fractal theory was employed as a new approach for estimating the saturated permeability and hydraulic conductivity of the clay material used in the dam core. For this purpose, the clay samples were gathered and analyzed using SEM images. An improved permeability model was utilized in which the restrictions in terms of pore radius were eliminated. The estimated permeabilities were then converted to the hydraulic conductivities, and the average amount was calculated. It was observed that the hydraulic conductivities were in the measured range. Subsequently, the seepage analysis was conducted with three different hydraulic conductivities of fractal, minimum, and maximum. With a relative error of 11% compared to the observed seepage, the results declared that the estimated fractal hydraulic conductivity could provide a better seepage estimation both safely and economically. As a result, it was concluded that the presented fractal model could be employed as a practical alternative for estimating the hydraulic conductivity of a soil sample.

For future studies, the unsaturated seepage analysis can be employed using the fractal and conventional hydraulic conductivity models. Also, to gain a more precise understanding of soil texture via fractal notion, the MIP technique and computed tomography (CT) images can be utilized.

References

1. Nourani, V., Sharghi, E., Aminfar, M.H.: Integrated ANN model for earthfill dams seepage analysis: sattarkhan Dam in Iran. Artif. Intell. Res. **1**(2), 22–37 (2012). https://doi.org/10.5430/air.v1n2p22.doi:10.5430/air.v1n2p22
2. Fell, R., MacGregor, P., Stapledon, D., Bell, G., Foster, M.: Geotechnical Engineering of Dams, 2nd edn. Taylor & Francis (2014)
3. Zhang, L., Xu, Y., Jia, J.: Analysis of earth dam failures: a database approach. Georisk. **3**(3), 184–189 (2009). https://doi.org/10.1080/17499510902831759
4. Tayfur, G., Swiatek, D., Wita, A., Singh, V.P.: Case study: finite element method and artificial neural network models for flow through Jeziorsko earthfill dam in Poland. J. Hydraul. Eng. **131**(6), 431–440 (2005). https://doi.org/10.1061/(ASCE)0733-9429(2005)131:6(431)
5. Al-Janabi, A.M.S., Ghazali, A.H., Ghazaw, Y.M., Afan, H.A., Al-Ansari, N., Yaseen, Z.M.: Experimental and numerical analysis for earth-fill dam seepage. Sustainability. **12**(6), 2490 (2020). https://doi.org/10.3390/su12062490

6. Zhang, Y., Schaap, M.G.: Estimation of saturated hydraulic conductivity with pedotransfer functions: a review. J. Hydrol. **575**, 1011–1030 (2019). https://doi.org/10.1016/j.jhydrol.2019.05.058
7. Dai, Y., Shangguan, W., Duan, Q., Liu, B., Fu, S., Niu, G.: Development of a China dataset of soil hydraulic parameters using pedotransfer functions for land surface modeling. J. Hydrometeorol. **14**(3), 869–887 (2013). https://doi.org/10.1175/jhm-d-12-0149.1
8. Grant, S.A.: Hydraulic properties, temperature effects. Encyclopedia of Soils in the Environment, pp. 207–211 (2005). https://doi.org/10.1016/B0-12-348530-4/00379-9
9. Xu, P.: A discussion on fractal models for transport physics of porous media. Fractals **23**(03), 1530001 (2015). https://doi.org/10.1142/S0218348X15300019
10. Mandelbrot, B.B.: The fractal geometry of nature, vol. 2. WH freeman New York (1982)
11. Song, W., et al.: Multiscale image-based fractal characteristic of shale pore structure with implication to accurate prediction of gas permeability. Fuel **241**, 522–532 (2019). https://doi.org/10.1016/j.fuel.2018.12.062
12. Zarnaghi, V.N., Fouroghi-Asl, A., Nourani, V., Ma, H.: On the pore structures of lightweight self-compacting concrete containing silica fume. Constr. Build. Mater. **193**, 557–564 (2018). https://doi.org/10.1016/j.conbuildmat.2018.09.080
13. Yu, B., Cheng, P.: A fractal permeability model for bi-dispersed porous media. Int. J. Heat Mass Transf. **45**(14), 2983–2993 (2002). https://doi.org/10.1016/S0017-9310(02)00014-5
14. Xu, P., Yu, B.: Developing a new form of permeability and Kozeny-Carman constant for homogeneous porous media by means of fractal geometry. Adv. Water Resour. **31**(1), 74–81 (2008). https://doi.org/10.1016/j.advwatres.2007.06.003
15. Shen, X., Li, L., Cui, W., Feng, Y.: Improvement of fractal model for porosity and permeability in porous materials. Int. J. Heat Mass Transf. **121**, 1307–1315 (2018). https://doi.org/10.1016/j.ijheatmasstransfer.2018.01.084
16. Turcotte, D.L.: Fractals in geology and geophysics. Pure Appl. Geophys. **131**(1), 171–196 (1989). https://doi.org/10.1007/BF00874486
17. Bo-Ming, Y.: Fractal character for tortuous streamtubes in porous media. Chin. Phys. Lett. **22**(1), 158 (2005). https://doi.org/10.1088/0256-307X/22/1/045
18. Townsend, T.G., Powell, J., Jain, P., Xu, Q., Tolaymat, T., Reinhart, D.: Sustainable practices for landfill design and operation. Springer(2015)
19. Fredlund, D.G.: Unsaturated soil mechanics in engineering practice. J. Geotech. Geoenvironmental Eng. **132**(3), 286–321 (2006). https://doi.org/10.1061/(ASCE)1090-0241(2006)132:3(286)
20. Fu, J.-F., Sheng, J.: A study on unsteady seepage flow through dam. Journal of Hydrodynamics, Ser. B. **21**(4), 499–504 (2009). https://doi.org/10.1016/S1001-6058(08)60176-6
21. Rabbani, A., Salehi, S.: Dynamic modeling of the formation damage and mud cake deposition using filtration theories coupled with SEM image processing. J. Natural Gas Sci. Eng. **42**, 157–168 (2017). https://doi.org/10.1016/j.jngse.2017.02.047
22. Otsu, N.: A threshold selection method from gray-level histograms. IEEE Trans. Syst. Man Cybern. **9**(1), 62–66 (1979). https://doi.org/10.1109/TSMC.1979.4310076

Environmental Issues and Sustainable Development in North Cyprus

İme Akanyeti[1,2,3](✉) and Sedef Çakır[2,4]

[1] Chamber of Environmental Engineers, Nicosia, North Cyprus, Mersin 10, Turkey
[2] Environmental Engineering Department, Cyprus International University, Nicosia, North Cyprus, Mersin 10, Turkey
{iakanyeti,scakir}@ciu.edu.tr
[3] Environmental Research Center, Cyprus International University, Nicosia, North Cyprus, Mersin 10, Turkey
[4] Chamber of Meteorological Engineers, Nicosia, North Cyprus, Mersin 10, Turkey

Abstract. Cyprus is an island in Mediterranean Sea and expected to be influenced by the climate change to a great extent. In the northern part of Cyprus, three important issues; domestic wastewater management, municipal solid waste management and air pollution control need to be addressed in the context of resource recovery and pollution prevention. In this study, for these three topics, some lacking data and analysis results were provided for the first time for North Cyprus contributing to the literature. Based on the results, 42% of the population was still not connected to sewage system and only 40% was connected to at least secondary wastewater treatment. Much more investment on sewage systems and wastewater treatment is required to catch up with the developed countries. Considering the large percentage of recoverable materials and resources in the composition of municipal solid waste, source segregation should be urgently put into practice. Among the criteria pollutants analyzed, PM_{10} and O_3 concentrations exceeded the threshold values. Source sampling was limited only to one of the power plants and not conducted for industrial sources. In conclusion, the implementation of the related policies and monitoring mechanisms including sufficient and reliable data collection were considered as the key priorities to address the environmental issues within the context of sustainable development.

Keywords: North Cyprus · Wastewater management · Solid waste management · Air pollution · Climate change · Sustainable development

1 Introduction

Cyprus is the third largest island in Mediterranean Sea and has two regions; where Turkish and Greek Cypriots live in the north and south, respectively. In Cyprus, sustainable management of all the natural resources is vital considering the stress caused by the population growth and climate change [1, 2]. Hence, all environmental issues need to be addressed sustainably to minimize the environmental pollution and maximize the resource recovery. Considering the United Nations Sustainable Development Goals [3],

H. Gökçekuş and Y. Kassem (Eds.): NRSEM 2021, EESCI, pp. 193–203, 2022.
https://doi.org/10.1007/978-3-031-04375-8_22

in this study, three important topics; domestic wastewater management, municipal solid waste management and air pollution control are discussed for North Cyprus (NC).

The available fresh water resources in NC is diminishing due to the less precipitation and increased evaporation caused by climate change and over-abstraction resulting in salinization of the groundwater [1, 2, 4]. The recently activated water supply project from Turkey to NC is expected to enable the growth of all the sectors and increase the quantity and quality of the groundwater [4]. To maximize the sustainable use of water resources, collection, re-use and recovery of the treated wastewaters should be enhanced with the right policies and investments [1, 2]. Solid waste pollution is another major problem that requires implementation of an integrated solid waste management plan [5, 6]. The most sustainable management plan for NC can be developed only after conducting a comprehensive and detailed characterization of the waste [7]. Implementation of such a plan is important not only to reduce the greenhouse gases emitted from the biodegradable fraction of the waste but also to recover the materials and resources contributing to the circular economy [8]. Last but not least, monitoring and controlling the air pollutants, emitted especially from industrial, transportation and agricultural activities, are required in NC as a part of the Clean Air for Europe Programme [9].

To develop the right strategies for sustainable management, the present situation for all three topics should be evaluated in detail to provide a scientific basis. Currently, collection and dissemination of environmental data by the authorities is poor [7, 10]. This study aims to provide some lacking data and results of the analysis on the specific three topics to the literature in order to help the strategy developers.

2 Domestic Wastewater Management

In the 21st century, with the diminishing resources, domestic wastewater is considered as an important source for water, energy and nutrient recovery [11]. Treatment to the desired level for re-use is possible only if the wastewater can be collected by a sewage system. To the best of the authors' knowledge, no published data is available for the population connected to sewage or to different levels of wastewater treatment. In this study, it is aimed to provide some estimates on these issues. Population of NC was 286,257 on the last census in 2011. For the analysis conducted in this study, projected population for 2020 as 329,494 [12] was used. In addition, information on the population connected to sewage were gathered from various sources [2, 13–16].

The percentage population connected to sewage is presented in Fig. 1.A for each town and overall for NC in comparison to mean OECD [17] and EU [18] values. The population connected to sewage in NC was estimated to be 58%, 9% less than the mean EU and 34% less than the mean OECD values. The domestic wastewater of the households not connected to sewage is either collected in septic tanks, treated with package treatment systems or directly discharged to surface waters [1, 2]. The results showed that much more investment is required to connect all the households to the sewage to minimize the environmental pollution while increasing the capacity of water re-use and recovery.

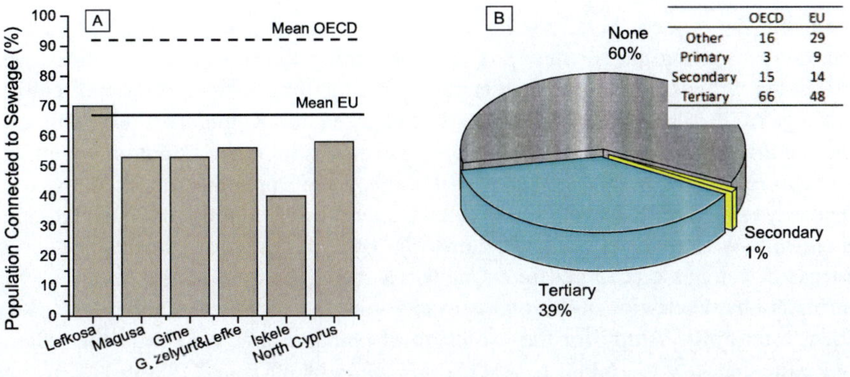

Fig. 1. Percentage of population connected to (A) sewage in each town and NC, (B) different levels of treatment in comparison to 2018 mean values of OECD [17] and EU [18] countries.

In accordance with Environmental Law (19/3) in NC, the municipalities with a population equivalent (p.e) of 2000 and higher are obliged to construct a wastewater treatment plant (WWTP). Although there are 23 municipalities with a p.e. higher than 2000, only six domestic WWTPs are in operation. The capacity and the characteristics of the plants are provided in Table 1. Information on each plant were obtained from various sources [19–22] and detailed below.

Table 1. Characteristics of domestic WWTPs located in NC (* only for NC) [19–22].

WWTP	Design Population	Design Capacity (m^3/d)	Max Capacity used (m^3/d)	Level of Treatment
New Haspolat WWTP	270,000 (81,000 north)	30,000 (9,000 north)	9,000	Tertiary
Famagusta WWTP	23,500	4,100	7,000	Tertiary
Morphou WWTP	10,750	1,350	1,000	Tertiary
Kyrenia WWTP	10,000	1,500–1,800	5,000	No treatment
Bafra WWTP	16,000	6,000	1,000	Secondary
Lapta WWTP	3,000	500	100	Secondary
TOTAL*		22,750	23,100	

New Haspolat WWTP in North Nicosia is the only bi-communal plant which was jointly funded by the Sewerage Board of Nicosia (70%), and the European Union (EU) (30%) on behalf of the Greek and Turkish Cypriots, respectively. It is considered that more than 30% of the wastewater received by the plant is from the North Nicosia, however there is no confirmed data on this matter. Due to the membrane bioreactor system involved in the plant, the effluent quality is high, well below the discharge standards [22, 23]. Discussions are on-going for new projects concerning the re-use of the treated water for irrigation with an agreement between the Greek and Turkish Cypriot communities. An average of 29,000 m^3/d treated water is discharged into Kanlıdere river. Farmers are extracting the water from the river to irrigate their lands in Mesaoria Plain, which lacks water resources. Lack of storage and distribution systems prevents the direct use of the high quality treated water and causes additional energy consumption for water extraction. *Famagusta* WWTP receives wastewater up to 7000 m^3/d, 71% above the design capacity. The treated water is discharged into Çanakkale pond. Despite of the excessive flow, the effluent characteristics of the plant comply with the discharge standards. Recently in 2021, a feasibility study for the extension of Famagusta WWTP was conducted to increase the capacity to 16,000 m^3/d [15]. *Morphou* WWTP in Yayla village receives wastewater up to 1000 m^3/d, about 22% lower than the design capacity. The effluent water has a quality meeting the discharge standards and is collected in a reservoir with a capacity of 100,000 ton. The water is distributed with a 3 km long distribution system and is used by the farmers to irrigate about 2000 acre land. There is an intention to double the WWTP capacity to 2694 m^3/d, however there is no schedule for this capacity enhancement. The WWTP located in the center of *Kyrenia* receives at least three times higher wastewater flow rate than the design flow rate. This overflow results in very poor effluent characteristics not complying with the discharge standards. Deep sea discharge of the effluent water causes a serious public health concern especially considering that Kyrenia is a touristic city and contains many bathing waters close to the discharge points. *Bafra* WWTP was designed to treat the wastewater generated in the hotels in Bafra Tourism Region. Depending on the season and the number of tourists, the wastewater flow rate varies up to 1000 m^3/d, about 17% of the total design capacity. The effluent water usually meets the discharge standards and is mostly re-used for irrigation of the green areas in the hotels. *Lapta* WWTP is another small plant treating the wastewater of some touristic places and households located in the Lapta coastal region. The treated water is discharged in the river close to the plant.

As shown in Table 1, three WWTPs, New Haspolat, Famagusta and Morphou, provide tertiary treatment and effluent characteristics with a high potential to be re-used for irrigation or recycled for consumption. Due to the overflow rates and insufficient capacity of the Kyrenia WWTP, the wastewater connected to this plant is considered as not treated for treatment level estimations. Two small scale WWTPs, Lapta and Bafra are the ones providing secondary treatment. Considering the population connected to each treatment plant, the percentage of the population without treatment, with secondary and tertiary treatment were estimated separately. The results are provided in Fig. 1.B in comparison to mean values of OECD and EU countries in 2018. EU directives [24] enforced the secondary treatment for all domestic sewage before disposal into receiving water bodies. Notably, the total percentage of the population not connected to at least secondary treatment was found to be 60% in NC, much higher than the percentages of 38% for EU and 19% for OECD countries. Hence, more investments are required to construct new tertiary WWTPs and to expand the capacity of the existing ones to increase the water re-use and recovery potential of the island.

3 Municipal Solid Waste Management

Only a decade ago, in 2012, the first sanitary landfill (Güngör) in NC was constructed by EU funds. Although Güngör landfill site receives waste from the municipalities of Nicosia, Kyrenia and recently Morphou, there are about 50 different wild dumping sites used by the other municipalities due to the long distances and lack of transfer stations [12]. Environmental Law (6/1) in accordance with EU, adopted the sustainable waste management hierarchy. However, re-use, recovery and recycling activities are still lacking; hence, Güngör Landfill is being filled up at a much faster rate than expected based on the design values. Therefore, implementation of an integrated waste management plan has a vital importance for the sustainable environmental development. Determination of the waste composition and characterization is one of the most important steps for the preparation of an integrated waste management plan [7]. The waste generation, composition and some characterization data for NC were reported in the master plan prepared by EU in 2007 [25]. However, the authors stated that reported figures contained uncertainties. Another research was conducted by Cyprus Turkish Investment Development Agency with a limited number of samples [26]. In this study, the results of the most recent and comprehensive two studies on municipal waste composition in NC were presented as shown in Fig. 2.

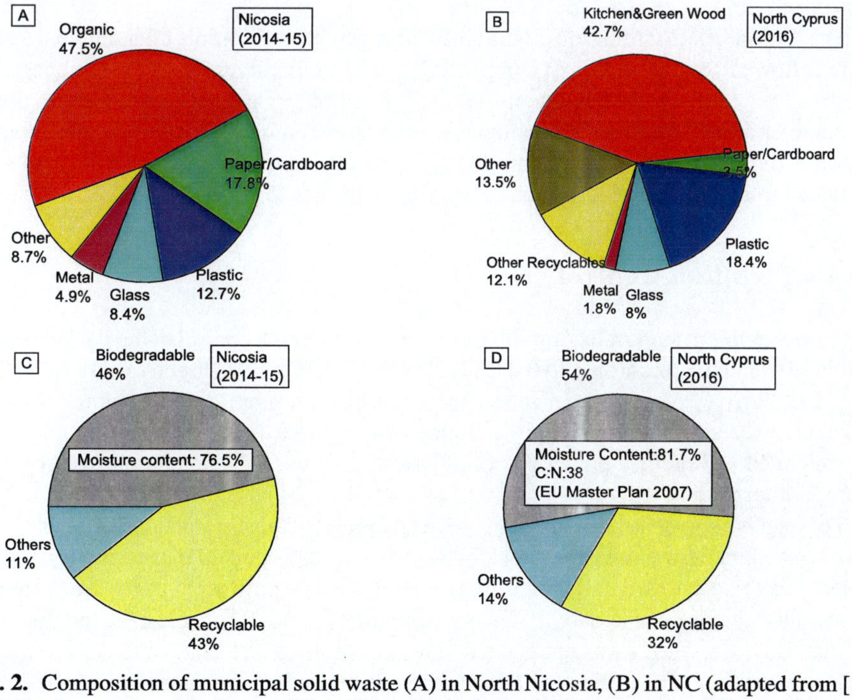

Fig. 2. Composition of municipal solid waste (A) in North Nicosia, (B) in NC (adapted from [12, 25]), Recovery potential of waste (C) in North Nicosia, (D) in NC

The first study was conducted for North Nicosia by Environmental Research Center of Cyprus International University (Fig. 2.A and 2.C). Solid waste sampling was carried out according to the random sampling method. In total 44 samples, each weighing 45–50 kg were collected in Güngör sanitary landfill, 22 in winter (2014) and 22 in spring (2015) seasons. The samples were taken from the trucks which collected waste only within the borders of Nicosia Turkish Municipality. For the moisture content analysis, 0.5 kg of wet biodegradable waste samples were dried in an oven at 105 °C for 24 h.

The second study (Fig. 2.B and 2.D) was conducted by EU funds collecting a wider range of samples for summer and spring in 2016, methodology of which was not detailed [12]. Results in Fig. 2.A and 2.B showed similar percentages for the organic waste with 47.5% and 42.7%, respectively. For paper and cardboard, there was a large difference between the data obtained for North Nicosia (Fig. 2.A) and NC (Fig. 2.B). The additional categorization of other recyclables in Fig. 2.B possibly contained a large percentage of paper and cardboards which could explain this difference. The results in Fig. 2.C and Fig. 2.D displayed that the biodegradable waste (Moisture content: 76–80%, C:N: 38) had the largest percentage in the waste composition. It is known that biodegradable fraction is the major source of greenhouse gases contributing to climate change [8]. Although gas collection system was part of the design of Güngör landfill, the system was never installed and operated resulting in undesired emissions. In both Fig. 2.C and Fig. 2.D, large percentages of recyclable materials showed the high potential for material recovery.

The European Commission highlighted that source separation of municipal solid waste followed by recycling and composting or anaerobic digestion for the biodegradable fraction results in the lowest emission of greenhouse gases in comparison to other treatment options [27]. In NC, planning and investment on infrastructure for source separation activities are urgently required [12]. In parallel, public has to be informed and motivated to participate in re-use and recycling activities [5].

4 Air Pollution Control

There are 9 ambient air monitoring stations operated under the control of the Air Emission Unit in EPD in NC. According to the EU's 2008/50/EC (CAFÉ) Directives [9], SO_2, O_3, NO_2, CO, PM_{10}, $PM_{2.5}$ and benzene concentrations are monitored to control if these pollutants exceed the limits in a given period or not. Stack (source) sampling device is only operated in Teknecik power plant [28], the data of which is not disseminated to the best of authors' knowledge.

The analyses presented in this study were conducted using the data of PM_{10}, SO_2, O_3 and NO_2 obtained from Air Emission Unit for years of 2012 and 2016. Nicosia is a traffic station and there are two rural stations around Teknecik power plant. Figure 3 shows the monthly average of the daily PM_{10} concentrations in 2012 and 2016 for the selected stations. However, PM_{10} was not presented for Teknecik in 2012 due to the lack of data. Main sources of PM_{10} in Eastern Mediterranean region can be listed as vehicles, sea salt sprays, dust re-suspension, transportation from industrialized European countries or dust storms from North Africa and Arabian Peninsula [29, 30]. During the winter season, the highest concentrations were observed in the most populated city of Nicosia where the values decreased in spring and summer. Results are in consistent with the study of Achilleos et al. [30] and winter month's higher concentrations can be explained with the lowest average temperature and wind speeds that are favorable for accumulation of pollutants.

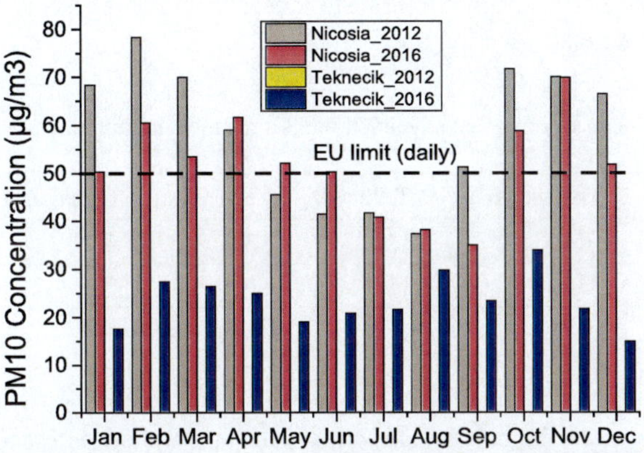

Fig. 3. Monthly mean daily concentrations of PM_{10} in 2012 and 2016.

Monthly averages of daily SO_2 concentrations of the Nicosia and Teknecik stations are presented in Fig. 4.A. Results showed that values exceeded the 125 $\mu g/m^3$ critical value only one day in Teknecik and this station had the highest concentrations during the year probably because the station is located close to the power plant. Cyprus Turkish Electricity Authority has almost 400 MG total power capacity [31, 32]. Low quality fuel with 3.5% sulphur content by weight had been used until 2016. Since 2016, 1% sulphur content fuel oil has been used in Teknecik and Kalecik power plants. The increased fuel quality clearly resulted in reduction in SO_2 emissions as observed in the data. However, there was a high monthly mean concentration observed in November 2016, which could not be explained.

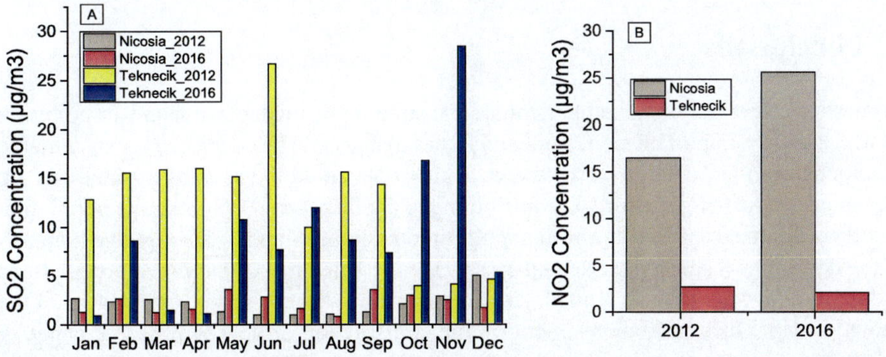

Fig. 4. (A) Monthly mean daily concentrations of SO_2, (B) Annual mean of daily NO_2 concentrations, in Nicosia and Teknecik.

The concentrations of NO_2, mainly emitted from the vehicles, did not exceed the hourly limit of 200 $\mu g/m^3$ during 2012 and 2016. Figure 4.B shows the annual mean of daily NO_2 concentrations and values showed an increasing trend from 2012 to 2016 with the higher values in Nicosia which has more population and vehicular activities.

Table 2 shows the number of days that exceeded the 8 h mean Ozone concentration threshold value of 120 $\mu g/m^3$, in different stations, except Famagusta and Kyrenia where concentrations never exceeded the limit.

Table 2. Number of days that exceeded the 8 h mean standard concentration of O_3

Year	Nicosia	Alevkayası	Teknecik	Kalecik	Famagusta	Kyrenia
2012	0	**140**	12	**36**	0	0
2016	0	**146**	11	0	0	0

As a forested rural area, Alevkayası exceeded the limit more than 25 days (the allowed number of days for exceedance). Ozone forms in the atmosphere as a secondary pollutant and the photochemical reactions of precursors control the level of O_3 in the troposphere. In the presence of sunlight, these precursors, volatile organic compounds (VOCs) and NO_x, from natural or anthropogenic sources interact to form O_3 [33–35]. Several studies showed that rural ozone concentrations can be higher than those in urban areas [33, 34, 36, 37] which can be explained by the mechanisms playing role for ozone destruction. Photochemical reactions, thermal decomposition are some of these mechanisms and even NO can contribute to ozone destruction [38, 39] which are favored in urban areas. In rural areas, amount of ozone transported especially from the urban areas does not change due to the lack of these destruction mechanisms.

The brick industry is one of the other SO_2 emission sources in NC due to coal combustion as an energy source. In addition, alternative heat sources such as rubber and olive pulp are used contributing to air emissions. On the other hand, asphalt industry contributes to the emission of greenhouse gasses, CO_2, N_2O and CH_4 [40], currently not monitored in NC.

5 Conclusions

Domestic wastewater management, municipal solid waste management and air pollution control are three important environmental issues that need to be addressed for sustainable development in NC. In this study, some lacking data and the results of the analysis are presented for the first time for NC contributing to the literature. In NC, a large population is still not connected to sewage system polluting the environment. Although the majority of the domestic wastewater collected by sewage is treated to a tertiary level, regulations and required infrastructures are still lacking to enable effective re-use or recovery of the treated water. Much more investment on the construction of new WWTPs or capacity enhancement of the existing ones is required to increase the population connected to at least secondary treatment in order to comply with the EU directives. Effective strategies should also include the recovery of industrial wastewaters and minimization of the water losses in water distribution systems. The integrated solid waste management plan prepared for NC indicates that feasibility studies are required for selecting the most suitable management methods. For these feasibility studies, more detailed characterization of the municipal solid waste is required. Initial results presented in this study show that there is a high potential for resource and material recovery indicating an urgent need for source segregation activities. The analysis of the available data from the air stations show that PM_{10} and O_3 are the two criteria pollutants which exceeded the regulated threshold values. Source sampling is limited to Teknecik power plant for which no published data is available. Considering the air quality control, for all the sectors that have emissions, it should be mandatory to measure pollutants in the chimney and strict environmental regulations should be applied to protect the environment and populace. The quality and continuity of the data from the stations are very important to correctly analyze the current conditions and to make estimations for future. Emission inventories should be urgently calculated and compared with the neighboring countries. In conclusion, developing right strategies, implementation of the related policies and monitoring mechanisms including sufficient and reliable data collection are considered as the key priorities to address the environmental issues within the context of sustainable development.

References

1. Gökçekuş, H., Orhon, D., Oğuz, G., Yücel, A.B., Sözen, S.: 2020 Integrated water and wastewater management strategy in North Cyprus - basis for an action plan. Desalin. Water Treat. **177**, 384–392 (2020). https://doi.org/10.5004/dwt.2020.25019
2. Elkiran, G., Aslanova, F., Hiziroglu, H.: Effluent water reuse possibilities in Northern Cyprus. Water **11**, 191 (2019). https://doi.org/10.3390/w11020191
3. United Nations 2021, Progress towards the sustainable development goals
4. Şenol, C.: Kuzey Kıbrıs Türk Cumhuriyeti'nin Hidrografik Yapısı. Su Sorunu ve Çözüm Önerileri, Kıbrıs Araştırmaları Dergisi. **21**(45), 77–98 (2020)
5. Akanyeti, I., Kazimoglu, C., Kanyemba, T.: Perceived versus objective knowledge towards a sustainable solid waste management in Northern Nicosia. J. Mater. Cycles Waste Manage. **22**(6), 1943–1952 (2020). https://doi.org/10.1007/s10163-020-01078-3
6. Barissever, C.: Optimization of the treatment and disposal of solid wastes in Northern Cyprus. Dissertation, Middle East Technical University, Northern Cyprus Campus (2016)
7. Günsel, C.: Environmental Problems in Northern Cyprus: A Descriptive Statistical Analysis and Forecasting of Air Pollution and Solid Waste, Dissertation, Middle East Technical University, Northern Cyprus Campus (2016)
8. Ackerman, F.: Waste Management and Climate Change. Local Environ. **5**(2), 223–229 (2000). https://doi.org/10.1080/13549830050009373
9. Clean Air for Europe Programme (CAFE), Summary Report on the Second Meeting of the CAFE Steering Group held on 8–9th October 2001, European Commission, Brussels (2001)
10. Mason, M., Bryant, R.: Water Technology and Sustainability in North Cyprus: Climate Change and the Turkey-North Cyprus Water Pipeline, PRIO Cyprus Centre Report, 1. Nicosia: PRIO Cyprus Centre (2017)
11. Metcalf & Eddy Inc.: Wastewater Engineering: Treatment and Resource Recovery. 5th ed., McGraw-Hill Professional,(2013)
12. European Union (EU): TRNC Integrated Waste Management Plan (2020)

13. Çeliker, T.E.: Head of Sewage Unit, Nicosia Turkish Municipality, Personal communication, 3 November 2021
14. Yengin, E.: Güzelyurt Municipality, Personal communication, 5 November 2021
15. UNDP, Meeting presentation on Feasibility Study for Extension of Famagusta Wastewater Treatment Plant 2.2.2021
16. Özgit, H.: Kyrenia Municipality, Personal communication, 10 November 2021
17. Organisation for Economic Co-operation and Development (OECD). Data extracted on 10 Nov 2021 12:30 UTC (GMT) from OECD.Stat
18. European Union (EU): Data extracted on 10 Nov 2021 12:30 UTC (GMT) from Eurostat (2018)
19. Alkan, İ.: Wastewater Unit, Environmental Protection Department, Personal communication, 4 November 2021
20. Environmental Protection Department in North Cyprus official website. http://www.cevrekorumadairesi.org/
21. Zorba, M.: Project Manager, Famagusta and Morphou WWTP, Personal communication, 5 November 2021
22. Önet, S.: Nicosia Turkish Municipality, Personal communication, 3 November 2021
23. Environmental Law (1991,9/19), Regulation on Water and Soil Pollution, Air Quality Control in North Cyprus
24. Angelakis, A.N.: Bontoux, L: Wastewater reclamation and reuse in Eureau countries. Water Policy **3**(1), 47–59 (2001). https://doi.org/10.1016/S1366-7017(00)00028-3
25. European Union (EU): Master plan on solid waste management in the Turkish Cypriot Community (2007)
26. Kıbrıs Türk Yatırım Geliştirme Ajansı (YAGA): Katı atık karakterizsyon çalışması sonuç raporu (2014)
27. European Commission, Executive Summary, Final Report. https://ec.europa.eu/environment/pdf/waste/studies/climate_change_xsum.pdf. Accessed 9 Nov 2021
28. Environmental Protection Department. http://www.cevrekorumadairesi.org/air/tr-pages.php?no=68
29. Querol, X., et al.: African dust contributions to mean ambient PM10 mass-levels across the Mediterranean Basin. Atmos. Environ. **43**, 4266–4277 (2009)
30. Achilleos, S., Evans, J.S., Yiallouros, P.K., Kleanthous, S., Schwartz, J., Koutrakis, P.: PM10 concentration levels at an urban and background site in Cyprus: the impact of urban sources and dust storms. J. Air Waste Manag. Assoc. **64**(12), 1352–1360 (2014)
31. Cyprus Turkish Electricity Authority (KIB-TEK). https://www.kibtek.com/uretim/
32. Erciyas, O.: Sustainability Assessment of Photovoltaic Power Plants in North Cyprus. Master thesis, Eastern Mediterranean University, North Cyprus (2014)
33. Kalabokas, P.D., Viras, L.G., Bartzis, J.G., Repapis, C.C.: Mediterranean rural ozone characteristics around the urban area of Athens. Atmos. Environ. **34**, 5199–5208 (2000). https://doi.org/10.1016/S1352-2310(00)00298-3
34. Kleanthous S., Vrekoussis, M., Mihalopoulos, N,, Kalabokas, P., Lelieveld, J.: On the temporal and spatial variation of ozone in Cyprus. Sci. Total. Environ. **476–477**, 677–87 (2014). https://doi.org/10.1016/j.scitotenv.2013.12.101
35. Cakir, S., Sita, M.: Evaluating the performance of ANN in predicting the concentrations of ambient air pollutants in Nicosia. Atmos. Pollut. Res. **11**(12), 2327–2334 (2020). https://doi.org/10.1016/j.apr.2020.06.011
36. Tangwu, Y.: Comparison of near-Ground ozone concentrations between urban and rural forests. Acta Ecol. Sin. **34**(19), 5670–5678 (2014)
37. Ripperton, L.A., Vukovich, F.M.: Gas phase destruction of tropospheric ozone. J. Geo. Res. **76**, 7328–7333 (1971). https://doi.org/10.1029/JC076i030p07328

38. Staehelin, J.: Ozone Measurements and Trends (Troposphere), Encyclopedia of Physical Science and Technology (Third Edition) (2003)
39. Liu, N., Ren, W., Li, X., Ma, X., Zhang, Y., Li, B.: Distribution and urban–suburban differences in ground-level ozone and its precursors over Shenyang, China. Meteorol. Atmos. Phys. **131**(3), 669–679 (2018). https://doi.org/10.1007/s00703-018-0598-1
40. Ceylan, H.: Asfalt Endüstrisi Çevre Kanunu İlişkisi. Erciyes Üniversitesi Fen Bilimleri Enstitüsü Dergisi **25**(1–2), 71–98 (2009)

Evaluation of Alternative Source of Rare Earth Elements Current Situation (Technological and Economic Aspects)

Şükrü Şafak[1] (✉) and Taha Altıparmak[2]

[1] General Directorate of Mining and Petroleum, Ankara, Turkey
sukru.safak@mapeg.gov.tr
[2] Ministry of Energy and Natural Resources, Ankara, Turkey
bektastaha.altiparmak@enerji.gov.tr

Abstract. Rare earth elements are increasingly used in renewable energy, catalyst and lightening applications. Since 2010, quotas applied by China create supply volatility. This canalized sector to find reliable alternative ways to obtain rare earth elements. Firstly, exploration new resources out of China increased and new mines opened (primary resources). Secondly, observed risks related to exploitation of primary resources (dust, land disturbance and toxics) promotes obtaining rare earth elements from recycling scrap materials, coal ore or/and coal combustion products and byproducts or reuse of tailings. This paper aims to evaluate current situation in secondary source of rare earth elements and make some suggestions to maintain sustainability of rare earth industry.

Keywords: Rare earth elements · Critical raw materials · Recycling · Resource policy

1 Introduction

Rare earth elements (REEs) are special name of 17 elements. These are lanthanides (15 elements), scandium and yttrium. They are not rare as given in their name. REEs are found in bastnazite, monazite, xenotime, apatite, cerrite and zircon minerals deposits. [1]. RREs are divided in two groups. These are heavy rare earth elements and light rare earth elements. This classification depends on REEs' atomic numbers and abundancy of them in the earth crust. LREE are more abundant than HREE [2].

Their excellent mechanic, magnetic and optical properties make them indispensable for modern technologies. According to European Commission's Raw Materials report LREEs and HREEs were determined as critical and strategic materials [3]. Moreover, US Department of energy listed several REEs (Yttrium, Europium, Terbium, Neodymium, and Dysprosium) as critical elements for clean energy applications [4].

The largest REEs' reserves found in China (80%). Even in oil sector, there is not such a dependency at this rate. In addition, China uses REEs reserves as a weapon. In 1992 Former Chinese President Deng Xiaoping said *"The Middle East has oil. China has rare earths."* this phrase express China's approach absolutely [5].

H. Gökçekuş and Y. Kassem (Eds.): NRSEM 2021, EESCI, pp. 204–207, 2022.
https://doi.org/10.1007/978-3-031-04375-8_23

This paper mainly focused on use of REEs and alternative and environmentally friendly ways of RREs production method by using recent studies conducted about this topic.

1.1 Use of REEs

NdFeB Permanent Magnets

NdFeB permanent magnets have wide range of use. These are wind turbine, electric and hybrid vehicles and high-tech products (computers, mobile phones, etc.). This is mainly due to magnets increasing magnetic properties with increasing use of REEs. Increased magnetic properties gave chance to produce high capacity hard discs with small dimensions. Therefore, 76% of total Nd production and all of the produced Dy used in production of permanent magnets [6].

Catalysts

Catalysts' are utilized in fluid catalytic cracking, automotive and flue gas emission abatement systems, and in hydrogen production with steam reforming. Catalysts commonly made up of CeO_2 and La_2O_3 [7].

Lightning

Low energy consumption and durability maintained by using REEs. La, Ce and Eu are utilized to produce phosphors for lightening applications.

Ni-MH Batteries

Ni-MH batteries are commonly used in electric vehicles. Use of these batteries has crucial importance for low carbon green future. Batteries contain nickel, cobalt and REEs (La, Ce, Nd, Pr) [8].

2 Recycling REEs

Modern technologies dependency of REEs increases day by day. In addition, export quotas applied by China and limited REEs deposits corresponding price hikes 2009–2011[9].

Recycling was determined one of the strategies to reduce supply volatility of REEs between 2009–2011. Recycling applications of REEs are limited with recycling scrap fluorescents and permanent magnets.

Moreover, there are some papers which imply that recycling REEs in current economic conditions was not feasible. These studies claim that price of REEs are not high as 2009–2011 period, lifetime of REEs products' is long, recycling rate of these product is not adequate and advances in substitution technologies [9].

However, there are also some studies shows recycling scrap REEs products is feasible in case of some cases. According to these studies, higher capacity recycling plants and higher recycling rate could make recycling scrap REEs products feasible [9].

2.1 Recycling Process

Recycling scrap NdFeB magnets conducted by using carbonization/hydrogenation - hydrolysis process.

Biochar and scrap NdFeB magnets mixed with each other and carbonization and hydrogenation conducted (at 1450° for 90 min.) and NdFeB-C/H alloy formed. Obtained material crushed below 45μ and passed through the magnetic separator to remove Fe-based metals. NdFeB-C/H powders poured into deionized water and during hydrolysis REOH and combustible gases formed. REOHs were oxidized to obtain REOs at 600 °C for 2 h [10].

Use of expensive chemical extractants (NH_4Cl, $MgCl_2$-KCI and graphite carbon) was eliminated with the help of this method. Therefore, recycling cost decreases without change in recovery efficiency (REOHs' recovery efficiency was 99,7%. REEs' recovery efficiency was 93%) [10].

Other study conducted by Güncan [11] scrap florescent lamps recycled to obtain REEs. After crushing scrap lamps REEs in phosphorous powders were extracted by using $(NH_4)_2SO_4$ and H_2SO_4 solutions. Eu was the most yielded REE in $(NH_4)_2SO_4$ solution (3,17%) while Ce was the most yielded REE in H_2SO_4 solution (6,224%) [11].

3 Coal Combustion Product

Obtaining REEs from Coal combustion products offer some advantages when compared with primary resources. These are use of available waste, eliminating cost of processing ore and decreasing environmental degradation (elimination of stockpile of CCPs and pollutants emitting from primary ore processing). REEs are embeded with glassy alumina-silicate minerals [12]. This causes differences in REEs extraction procedure. CCPs are not soluble in acid or alkali under ambient temperature. Therefore, high temperature/pressure roasting/ leaching applied to break alumina-silicate structure [12]. REEs extracted from CCPs by using ionic liquids and molten salts (contains organic and inorganic ions). In addition, CCPs have higher REEs content than primary sources (approximately 300–445 ppm) [13–15].

4 Conclusions and Suggestions

- Incentives should be applied to recycling of REEs by Governments (For example, recycling rate of REEs could be increased by using effective policies
- CCPs' REEs potential should be evaluated.
- Countries should organize corporation to decrease and balance Chinese dominancy (Mineral diplomacy)
- In order to decrease recycling cost research should be focused on elimination of chemical solvents (use of biochar and organic solvents were good choices)
- Funds should be allocated to research and development(R&D) work to find substitute materials of REEs. For example, in US, critical Material Institute funded by department of Energy working for divirsing supply sources of REEs [1]

References

1. USGS 2016 Minerals Year Book rare Earths [Advance Release]
2. Croat, J.: Rapidly Solidified Neodymium-Iron-Boron Permanent Magnets (2018)
3. European Commision Study on the review of the list of critical materials. Criticality assesment (2017)
4. Bauer, D., David, D., Li, J., Sandalovw, D., Telleen, P., Wanner, B.: US Department of energy Critical Materials Strategy (2011)
5. Percowski, J.: Behind China's Rare Earth Controversy (2012). https://www.forbes.com/sites/jackperkowski/2012/06/21/behind-chinas-rare-earth-controversy/#3a0186b64074 (2020)
6. Trench, A., Sykes, J.P.: Rare Earth Permanent Magnets and Their Place in the Future Economy. Engineering (2019)
7. Charalampides, G., Vatalis, K.I., Apostoplos, B., Ploutarch-Nikolas, B.: Rare earth elements: industrial applications and economic dependency of Europe 2015. Procedia Econ. Finance **24**, 126–135 (2015)
8. Korkmaz, K., Alemrajabi, M., Rasmuson, A.C., Forsberg, K.M.: Separation of valuable elements from NiMH battery leach liquor viaantisolvent precipitation 2020. Separation and Purification Technology 234 (2020)
9. Qui, Y., Suh, S.: Economic feasibility of recycling rare earth oxides from end-of-life lighting technologies Resources, Conservation & Recycling 150 (2019)
10. Lui, B., Zhu, N., Li, Y., Wu, P., Dang, Z., Ke, Y.: Efficient recovery of rare earth elements from discarded NdFeB magnets 2019. Process Saf. Environ. Prot. **124**, 317–325 (2019)
11. Güncan, A.: Extraction rare earth elements from waste florocent laps. M.Sc. thesis. Süleyman Demirel University (2015)
12. Mondal, S., et al.: Recovery of rare earth elements from coal fly ash using TEHDGA impregnated resin. Hydrometallurgy **185**, 9993–10101 (2019)
13. Wang, Z., Dai, S., Zou, J., French, D., Graham, I.T.: Rare Earth elements and Yttrium in coal ash from the Luzhou power plant in Sichuan, Southwest China: Concentration, characterisation and optimized extraction 2019. Int. J. Coal Geol. **203**, 1–14 (2019)
14. Huang, C., Wang, Y., Huang, B., Dong, Y., Su, X.: the recovery of rare earth elements from coal combustion products by ionic liquids 2019. Miner. Eng. **130**, 142–147 (2019)
15. Lanzerstorfer, C.: Pre-processing of coal combustion fly ash by classification for enrichment of rare earth elements 2018. Energy Rep. **4**, 660–663 (2018)

Evaluation of Streamflow Drought Index in Aegean Region, Turkey

Ayşe Gulmez[1](✉), Denizhan Mersin[2], Babak Vaheddoost[1], and Mir Jafar Sadegh Safari[3]

[1] Department of Civil Engineering, Bursa Technical University, Bursa, Turkey
19262821008@ogrenci.btu.edu.tr, babak.vaheddoost@btu.edu.tr
[2] Department Civil Engineering, Izmir Institute of Technology, Izimr, Turkey
denizhanmersin@iyte.edu.tr
[3] Department of Civil Engineering, Yasar University, Izimr, Turkey
jafar.safari@yasar.edu.tr

Abstract. Water is an invaluable substance of which ensures the life cycle and hydrological events across the world. In this respect, water deficit also known as drought is a natural disaster related to water scarcity in time and space. Although there is no solid definition for the phenomenon, the outcome of repeated wet and dry spells cause in economic, social, and political problems at regional, country-wide, and world-wide scale. In this study, drought associated with the streamflow in the Aegean region, which has an important economic, historical and socio-cultural role in the western Turkey, is investigated through the well-known streamflow drought index (SDI). Therefore, average discharge in the Çiçekli-Nif, Beşdeğirmenler-Dandalas, Bebekler-Rahmanlar and Koçarlı-Köprübaşı station respectively related to on Gediz, Büyük Menderes and Küçük Menderes basins were used. Then SDI with 1, 3, 6,12 months moving average are acquired to express the drought severity associated with the streamflow in the basins. Results showed that the SDI values in all of stations together with the 1, 3, 6, and 12-month moving averages depicts similar results and no abnormal situation exist during the study period.

Keywords: Aegean region · Drought · Streamflow drought index · Hydrology

1 Introduction

Water is the most vital commodity that ensures the life and hydrological cycle across the globe. For this, water deficit also known as drought can be distinguished as a natural disaster related to the long lasting dry spells across a region through time and space. To this end, there is no universally acceptable definition for the drought phenomenon mostly due to the diverse aspects and the outcome [1–3]. This disaster influences natural habitats, ecosystems, and various economic and social sectors, urban water supply, the modern multiple manufactories and from the agriculture to transportation [4]. Types of droughts have been classified under five categories respectively as meteorological,

H. Gökçekuş and Y. Kassem (Eds.): NRSEM 2021, EESCI, pp. 208–213, 2022.
https://doi.org/10.1007/978-3-031-04375-8_24

hydrological, agricultural, socioeconomic, and ecological. One of the most important types of drought is hydrological drought, as it affects many activities such as hydropower generation, industrial and urban water supply on surface water resources [5]. Several drought indices were developed to characterize hydrological droughts. Due to the index value indicates the level of drought severity, drought conditions is to use drought indices, because they provide a quantitative method for determining the onset and end of a drought catastrophe [4, 6].

In this study, drought associated with the streamflow in the Aegean region, which has an important economic, historical and socio-cultural role in the western Turkey, is investigated through the well-known streamflow drought index (SDI). Therefore, average discharge in the Çiçekli-Nif, Beşdeğirmenler-Dandalas, Bebekler-Rahmanlar and Koçarlı-Köprübaşı station respectively related to on Gediz, Büyük Menderes and Küçük Menderes basins were used. Then SDI with 1, 3, 6, 12 months moving average are acquired to express the drought severity associated with the streamflow in the basins.

2 Study Area

To this end, the Büyük Menderes, Küçük Menderes and Gediz Basin located in the west of Turkey are selected as the study area for further discussion (Fig. 1). Then, monthly streamflow time series in 4 hydrology stations during the 1973–2015 period are used, while the data were recorded at Beşdeğirmenler-Dandalas, Koçarlı-Köprübaşı, Bebekler-Rahmanlar and Çiçekli-Nif which are located Büyük Menderes, Küçük Menderes and Gediz basins respectively. Table 1, also details the location of the selected stations, position of the basins, names of the sub-basin and the study period.

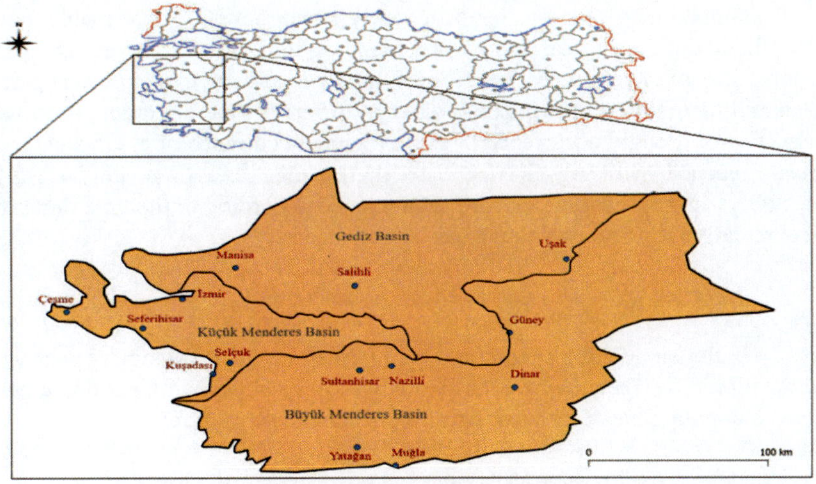

Fig. 1. Study area and location of the hydrology stations

Table 1. Observation times and geographical information of the selected stations

Station	Basin	Subbasin	Location
Çiçekli-Nif	Gediz	KemalPaşa-Nif Çayı	38°29′41"N 27°17′54"E
Beşdeğirmenler-Dandalas	B. Menderes	Akçay Havzası	37°48′18"N 28°34′48"E
Bebekler-Rahmanlar	K. Menderes	K.Menderes Havzasi	38°17′18"N 27°55′41"E
Koçarlı-Köprübaşı	B. Menderes	Aydın-Söke Havzasi	37°18′36"N 27°42′47"E

3 Quality of the Data

The streamflow time series data were obtained from the State Water Works in the daily total streamflow format. Afterward, consistency, randomness, and trends in the data were examined respectively with double mass curve, run test, linear trend test methods. Therefore, a set of complete and qualified time series are obtained to be used as the inputs to the drought analysis detailed below.

4 Streamflow Drought Index (SDI)

The SDI method is used to estimate the frequency and severity of the hydrological droughts based on the drought impact on the river basin. In this method, to compute SDI, it is assumed that a time series of monthly streamflow $Q_{i,j}$ is available where i stand for the hydrological year and j is the month within the same hydrological year. Cumulative streamflow volumes from varied time periods are used to investigate the variation and distribution of drought severity at different time durations, from which the frequency of drought occurrence of study region, occurrence of circle and drought severity are reached. In this way, it is estimated that the time series of monthly streamflow volumes ($Q_{i,j}$) are consecutive, that are accumulated according to the time duration of k. The cumulative streamflow volume can be obtained as

$$V_{i,k} = \sum\nolimits_{i}^{3k} Q_{I,J} i = 1, 2, \ldots, 12, k = 1, 2, 3, 4 \tag{1}$$

where $V_{i,k}$ is the cumulative streamflow volume for the ith hydrological year with a period duration of k. Then the SDI for the ith hydrological year with period duration k is defined computing the cumulative streamflow volumes $V_{i,k}$ as follows:

$$SDI_{i,k} = \frac{V_{i,k} - \overline{V}_k}{S_k} i = 1, 2, \ldots, 12, k = 1, 2, 3, 4 \tag{2}$$

where $\overline{V}_k$ refers to long term mean of cumulative streamflow volumes and S_k refers to standard deviation of cumulative streamflow volumes.

When evaluating an SDI value, the data for analysis must follow a normal or log-normal distribution. Nonetheless, in small basins, streamflow may follow a skewed probability distribution, whose distribution pattern is alike that of the Gamma distribution. Thus, in this study, SDI values were calculated using gamma distribution with the help of DrinC software. DrinC software was improved at the Centre for the Assessment of Natural Hazards and Proactive Planning and the Laboratory of Reclamation Works and Water Resources Management of the National Technical University of Athens [7].

For each year the computed SDI values are categorized based upon the range for which varied drought severities are described. In this study, drought severity classification was used as a reference to the different drought severity developed by Al-Faraj et al. [8]. After which the SDI values are computed, results could easily be interpreted by means of threshold values given in Table 2.

Table 2. SDI values for drought severity classification

Index value	Description
$SDI \geq 2$	Extremely Wet
$2 \geq SDI \geq 1.5$	Very Wet
$1.5 \geq SDI \geq 1$	Moderately Wet
$1 \geq SDI \geq -1$	Near Normal
$-1 \geq SDI \geq -1.5$	Moderately Dry
$-1.5 \geq SDI \geq -2$	Severe Dry
$SDI \leq -2$	Extremely Dry

5 Result and Discussion

The consistency in the time series are checked using double mass curve and then *z*-values related to the run test are checked for the 95% confidence ($-1.96 \leq Z \leq +1.96$). Since the *z*-score associated with the selected time series are outside of the mentioned range, the time series considered not to be random (Table 3).

Table 3. Z-Scores of the Run test for streamflow data

Station name	Streamflow
Çiçekli-Nif	−65.32
Bebekler-Rahmanlar	−60.36
Beşdeğirmenler-Dandalas	−57.15
Koçarlı-Köprübaşı	−58.99

The results of trend analysis also showed that, the Bebekler-Rahmanlar Station demonstrates negative direction among all stations, meaning that streamflow values decrease during the study period. However, no significant trend at the remaining 3 stations, are observed that means average streamflow values at these stations are constant.

Streamflow drought index values for each month were also calculated and evaluated by considering the average values of each month. SDI values results from time period of 1981–2015 belonging to observation stations of Çiçekli-Nif, Beşdeğirmenler-Dandalas, Bebekler-Rahmanlar and Koçarlı-Köprübaşı are obtained as shown in the Fig. 2. In this context, the highest and the lowest SDI-1 were respectively seen as 4.23 in Beşdeğirmenler-Dandalas and −3 in Bebekler-Rahmanlar stations. Similarly, the wettest and driest SDI-3 were respectively observed as 4.30 in Beşdeğirmenler-Dandalas station and −3.31 in Bebekler-Rahmanlar stations. The highest SDI-6 value was recorded as 4.36 at Bebekler-Rahmanlar station and the lowest SDI-6 value was -3.31 at Çiçekli-Nif station. Finally, the highest SDI-12 value was 4 at Bebekler-Rahmanlar station and the lowest SDI-12 value is −2.75 at Çiçekli-Nif station. Therefore, for all moving average values of all stations (1-, 3-, 6-, 12-month) based on general averages and for 1-month MA, no drought was detected, at 1981–2015.

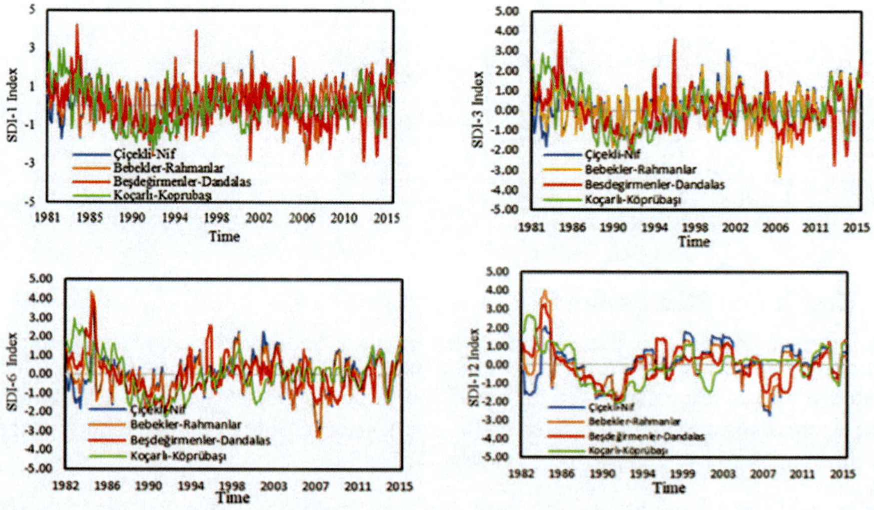

Fig. 2. SDI-1, SDI-3, SDI-6 and SDI-12 Graphs

6 Conclusion

The study area is a region where metropolitan cities such as Izmir and Aydın are located and whose population is constantly increasing. Agricultural activity is also very common. 34 years of streamflow records associated with 4 stations located in Büyük Menderes, Küçük Menderes, and Gediz basins were used to calculate the SDI drought indices on 1-, 3-, 6- and 12-month time scales. Initially, it was understood that the data was not random and consistent because of the tests applied. In addition, it was determined that there was no trend in streamflow with 3 stations and there was negative direction trend in just one station. Then, the spatial average of the historical SDI patterns was obtained separately to determine the drought events in the basins according to each index.

This study is limited to historical hydrological drought indices. Trends in drought indices could be extended to future periods based on anticipated outputs of global climate models to make an educated choice for sustainable watershed planning and management and to optimize the operating rules of current water resources. If future studies take into consideration meteorology, agricultural, and socioeconomic droughts, the findings will be more relevant.

References

1. Esfahanian, E., et al.: Development and evaluation of a compre-hensive drought index. J. Environ. Manage. **185**, 31–43 (2017)
2. Hayes, M., Svoboda, M., Wall, N., Widhalm, M.: The lincoln declaration on drought indices: universal meteorological drought index recommended. Bull. Am. Meteor. Soc. **92**(4), 485–488 (2011)
3. Lloyd-Hughes, B.: The impracticality of a universal drought definition. Theoret. Appl. Climatol. **117**(3–4), 607–611 (2013). https://doi.org/10.1007/s00704-013-1025-7
4. Heim, R.R., Jr.: A review of twentieth-century drought indices used in the United States. Bull. Am. Meteor. Soc. **83**(8), 1149–1166 (2002)
5. Vasiliades, L., Loukas, A., Liberis, N.: A water balance derived drought index for pinios river basin, Greece. Water Resour. Manage. **25**(4), 1087–1101 (2011)
6. Smakhtin, V.U.: Low flow hydrology: a review. J. Hydrol. **240**(3–4), 147–186 (2001)
7. Tigkas, D., Vangelis, H., Tsakiris, G.: The drought indices calculator (DrinC). In: Proceedings of the 8th International Conference of …, pp. 1333–1342, June 2013
8. Al-Faraj, F.A.M., Scholz, M., Tigkas, D.: Sensitivity of surface runoffto drought and climate change: application for shared river basins. Water (Switzerland) **6**(10), 3033–3048 (2014)

Integration of Electrocoagulation, Electro-Fenton Processes for Treatment of High Concentration Dye Solutions

Doğukan Yümün(✉), Eda Ceylan, Gizem B. Dindaş, Nihal Bektaş, and H. Cengiz Yatmaz

Department of Environmental Engineering, Gebze Technical University, 41400 Gebze, Kocaeli, Turkey
dogukan.yumun2016@gtu.edu.tr

Abstract. Discharge of synthetic dyes from industrial effluents to the environment is very important issue for getting good quality water and healthy ecosystems. Therefore, there is a need to apply appropriate novel treatment solutions prior to discharging them into the environment. Also, the use of well treated wastewater in the textile industry will be a sustainable answer to large demand of water consumptions in this industry. In electrocoagulation (EC), current passes through electrolyte solution between electrodes and sacrificial anode were started to dissolve resulting high metal ion concentrations and can create hydroxide ion species depending of pH value. These hydroxide species can serve as coagulants and remove dye molecules from the wastewater. In electro-Fenton process (EF), Fe electrode was used to get in situ of Fe^{2+} ions and necessarily hydrogen peroxide (H_2O_2) is externally added in solution for generation of the hydroxyl radicals to degrade dye molecules in wastewater. The use sequential and/or integrated different treatment processes can maximize dye removal efficiency. Therefore, the aim of presentation is to investigate the integration of EC and EF removal process of high concentration of different dyes from aqueous solutions. Effect of the sequential use of EC and EF processes with iron plate electrodes were examined in terms of high concentration of dye removal to get complete dye removal processes.

Keywords: Electrocoagulation · Electro-Fenton · High concentrated · Dye solutions

1 Introduction

The amount of wastewater generated by the textile industry are increasing with growing degree of industrialization. If textile wastewater is discharged directly to the receiving environment without any treatment process, it can cause many problems such as damaging the food chain and ecological balance as well as aesthetic problems [1, 2]. Dyes are generally large molecule toxic organic structures that cannot be biodegraded spontaneously and wastewater containing dyestuffs cannot be treated using only conventional methods. For this reason, the treatment of textile wastewater with innovative technologies gains great importance [2, 3].

H. Gökçekuş and Y. Kassem (Eds.): NRSEM 2021, EESCI, pp. 214–219, 2022.
https://doi.org/10.1007/978-3-031-04375-8_25

Electrocoagulation (EC) is a process that provides the formation of metal hydroxide flocs in wastewater by dissolving the anode as a result of electrolysis. The principle of EC process in removing pollutants is based on one or more of the mechanisms of coagulation, adsorption, precipitation and flotation removal [4, 5]. Electro-Fenton (EF) method is an alternative advanced oxidation treatment process in which Fe2 + ions dissolved from the iron electrode under the electrical field and hydroxyl radicals formed both as a result of the reduction of cathodic oxygen and with the help of H_2O_2 that can be added are used in the treatment process [6, 7]. These radicals, which can easily decompose organic substances, show good oxidant properties. In literature, there are several studies dye removal studies using a single process of EC and EF however the single use of these processes cannot get sufficient results since resistant nature of some organic molecules found in dye wastewaters and need additional treatment phases to get complete degradation. There are a few reports sequential use of different techniques for organic removal [8–10] however, there is a need to find an effective combination of use of these techniques.

In this study, treatment of high concentration coloured aqueous solution by sequential use of electro-coagulation, electro-Fenton and photocatalytic processes was investigated. With the different advanced oxidation processes applied, the biological degradability of large molecule toxic organic structures in the dye wastewater will increase toxicity value of wastewater will be decreased, simultaneously.

2 Materials and Methods

2.1 Chemicals

Commercially available Astrazon Blue FGRL and Reactive Orange 16 obtained from Dystar were used in the experiments (Table 1). Astrazon Blue FGRL, a cationic dye is made up of two major components: C.I. basic blue 159 and C.I. basic blue 3. The mixture was prepared with the two components have a weight-to-weight ratio of around 5:1 by

Table 1. The properties of Reactive Orange and Astrazon Blue FGRL dyes.

Textile dyes solution	Reactive Orange 16 (RO16)	Astrazon Blue FGRL (AB)
Chemical structure	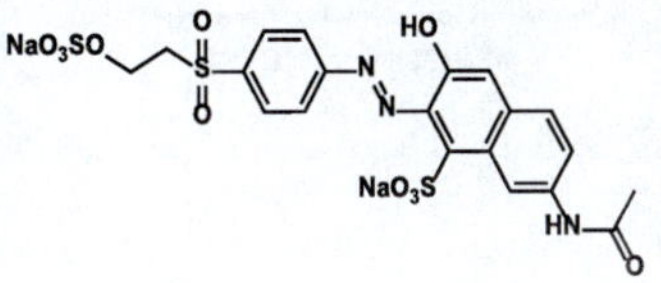 $C_{20}H_{17}N_3O_{11}S_3.2Na$	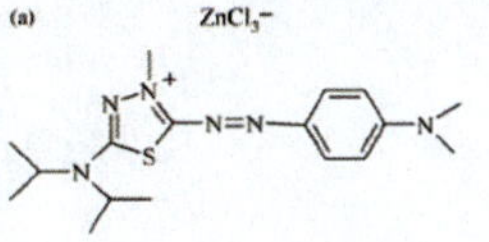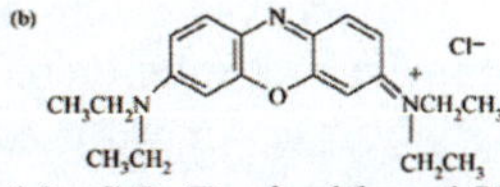 (a) C.I. Basic blue 159 (b) C.I. Basic blue 3

(*continued*)

Table 1. (*continued*)

Textile dyes solution	Reactive Orange 16 (RO16)	Astrazon Blue FGRL (AB)
The highest light absorbance	576 nm	492 nm
Zeta potential (mV), pH ~ 6	−2,39	0,664

weight, [11]. Dye concentration in aqueous solution was determined using a UV–Vis spectrophotometer (Hach Lange, DR 2800). A stock solution was used to get a diluted series of dye solutions.

2.2 Electrocoagulation (EC) and Electro-fenton (EF) Process

In EC process, 100 mg/L, 500 mL of dye solution was used to perform an electrocoagulation (EC) process in a batch electrochemical reactor. Initially, solution pH was adjusted to 7 and NaCI was added to increase conductivity. The electrochemical reactor used in the experiments has four iron electrodes (surface area of each electrode:5 cm^2; thickness: 0.3 cm; distance between electrodes: 2.0 cm). A DC power supply source (NETES 6303D) in parallel mode used to get current between the cathodes and anodes. Homogeneous mixing in reactor was maintained using a magnetic stirrer (Fig. 1).

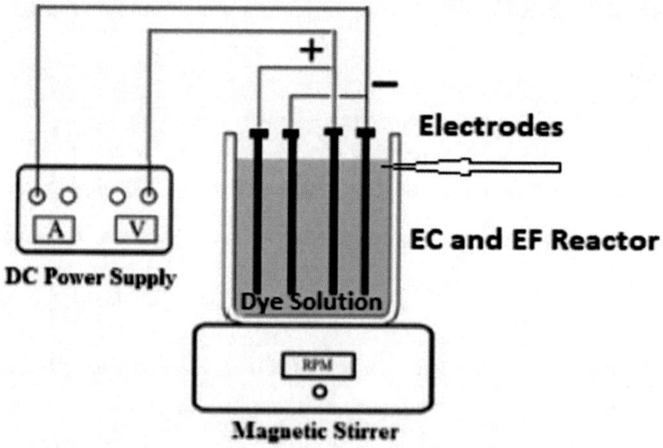

Fig. 1. Schematic diagram of experimental set-up.

In EF process, electrochemically generated iron ion and H_2O_2 entails • OH production in the bulk solution reducing cost and disadvantages many related problems. In this EF study, iron ion was used freshly produced from EC process and accordingly efficient amount of hydrogen peroxide (H_2O_2) was added to the reactor to form the Fenton reaction. After the experiment begins, dye solutions were taken and analysed at 5 th, 15 th, 30 th, 45 th and 60th minutes intervals. After the samples were kept in the centrifuge for 5 min, were analysed with a spectrophotometer.

3 Results and Discussion

In the study, electrocoagulation and electro-Fenton processes were applied to highly concentrated dye solutions, respectively. First, EC process took place using 5 mA/cm^2 current density for 10 min. After initial EC process, Fenton reagent, 0.3 mL H2O2 (35% w/w) was included to get Fe:H2O2 ratio of 1:10 at the beginning of cycles EF.

As can be seen from Fig. 2, about 50% removal rate was achieved for EC process for RO16 dye. EF process increased the removal rate up to 90%. The results proved that EC is a good a pre-treatment step for high concentration dye solutions. the similar results also shown with Astrazon Blue EC + EF process (Fig. 3). The removal of different dyes solution by sequential process provides high-efficiency treatment in wastewater containing dyestuffs in a high concentration within a short period of time.

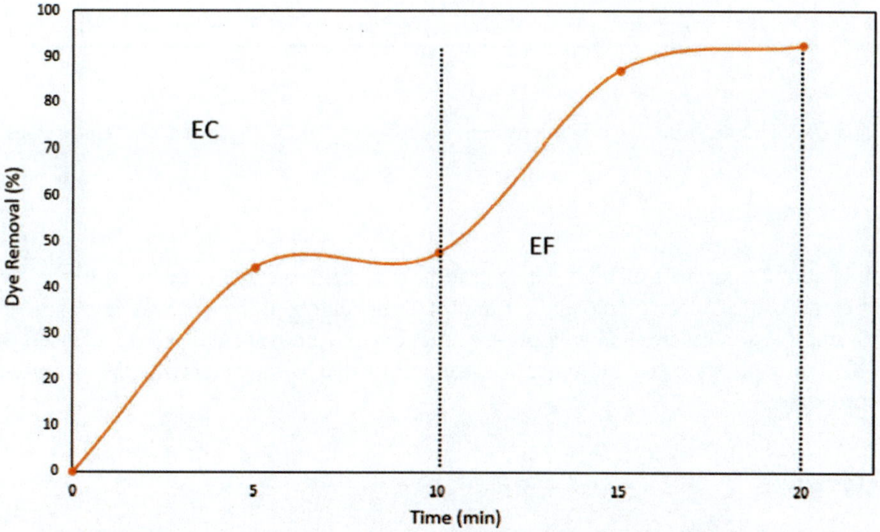

Fig. 2. The RO16 dye solutions removal efficiency of EC + EF successive processes.

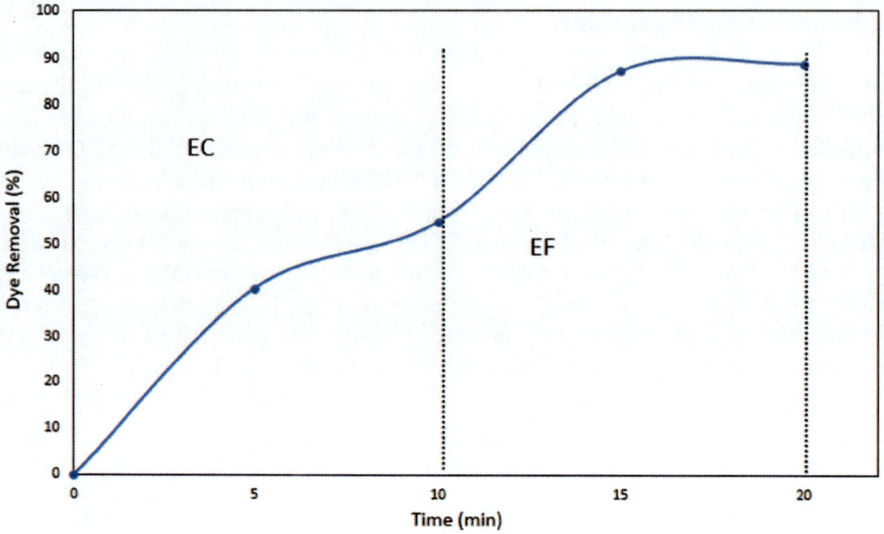

Fig. 3. The Astrazon Blue dye solutions removal efficiency of EC + EF successive processes.

4 Conclusion

In this study, treatment of high concentration two different dye aqueous solutions by sequential use of electro-coagulation and electro-Fenton processes was investigated. With the EC process applied as a pre-treatment step, removal rates were increased up to 90% also possible the biological degradability dye wastewater will be increased, accordingly.

References

1. Patel, H., Vashi, R.T.: Characterization and treatment of textile wastewater. Elsevier (2015)
2. Holkar, C.R., Jadhav, A.J., Pinjari, D.V., Mahamuni, N.M., Pandit, A.B.: A critical review on textile wastewater treatments: possible approaches. J. Environ. Manage. **182**, 351–366 (2016)
3. Samsami, S., Mohamadizaniani, M., Sarrafzadeh, M.H., Rene, E.R., Firoozbahr, M.: Recent advances in the treatment of dye-containing wastewater from textile industries: Overview and perspectives. Process Saf. Environ. Prot. **143**, 138–163 (2020)
4. Mollah, M.Y., Morkovsky, P., Gomes, J.A., Kesmez, M., Parga, J., Cocke, D.L.: Fundamentals, present and future perspectives of electrocoagulation. J. Hazard. Mater. **114**(1–3), 199–210 (2004)
5. Scott, V.K.: Electrochemical Processes for Clean Technology. The Royal Society of Chemistry, 307 (1995)
6. Casado, J.: Towards industrial implementation of Electro-Fenton and derived technologies for wastewater treatment: a review. J. Environ. Chem. Eng. **7**(1), 102823 (2019)
7. Ismail, S.A., Ang, W.L., Mohammad, A.W.: Electro-Fenton technology for wastewater treatment: A bibliometric analysis of current research trends, future perspectives and energy consumption analysis. J. Water Process Eng. **40**, 101952 (2021)

8. Dindas, G.B., Caliskan, Y., Celebi, E.E., Tekbas, M., Bektas, N., Yatmaz, H.C.: Sequential treatment of food industry wastewater by electro-fenton and electrocoagulation processes. Int. J. Electrochem. Sci **13**, 12349–12359 (2018)
9. Thiam, A., Zhou, M., Brillas, E., Sirés, I.: Two-step mineralization of Tartrazine solutions: study of parameters and by-products during the coupling of electrocoagulation with electrochemical advanced oxidation processes. Appl. Catal. B **150**, 116–125 (2014)
10. Lalwani, J., Sangeetha, C.J., Thatikonda, S., Challapalli, S.: Sequential treatment of crude drug effluent for the elimination of API by combined electro-assisted coagulation-photocatalytic oxidation. J. Water Process Eng. **28**, 195–202 (2019)
11. Marungrueng, K., Pavasant, P.: Removal of basic dye (Astrazon Blue FGRL) using macroalga Caulerpa lentillifera. J. Environ. Manage. **78**(3), 268–274 (2006)

Investigation of Meteorological Drought Characteristics of the Great Man-Made River Region (Libya)

Mustafa Ibrahim Mohamed Elhaj, Tülay Ekemen Keskin(✉), and Ali Jamali

Faculty of Engineering, Civil Engineering Department, University of Karabük, Karabük 78050, Turkey
{tulayekemen,alijamali}@karabuk.edu.tr

Abstract. In this study, located in the Great Man-Made River region, meteorological drought analysis were conducted for five monitoring stations in Northern Libya, the Standardized Precipitation Index (SPI) method and the Reconnaissance Drought Index (RDI) method were used to determine meteorological drought using monthly total precipitation data, and using mean monthly temperatures data and total monthly precipitation data, respectively. The drought analysis using DrinC software of the Great Man-Made River region for 1-, 3-, 6-, and 12-months SPI and RDI values were conducted and examined in detail. According to the SPI-12 month index values, the driest period was determined by 86% in Tripoli Airport and Nalut station, and the least dry period was determined at Sirt station by 39%, and as a result of the analyzes that were conducted, according to the values of the RDI-12 month index, the longest period was determined drought in Zuara station in the year 2000–2001. It was noted that the year 2000–2001 was one of the driest years of all stations, and the other years with high drought rates were 1981–1982, 1984–1985, and 1992–1993.

Keywords: Standardized Precipitation Index (SPI) · Reconnaissance Drought Index (RDI) · Meteorological drought analyses · The Great Man-Made River Region · Libya

1 Introduction

Drought is a pernicious natural hazard that involves a wide range of climatic processes and has far-reaching effects on both the environment and society. Due to the present harsh climatic occurrences, there has been an increase in interest in drought consequences and monitoring in recent years. These types of droughts have implications for many of the GEOSS (Global Earth Observation System of Systems) Societal Benefit Areas (SBAs), and this phenomenon set up an interconnection between various fields, such as agriculture sustainability, food security, ecosystem functions and services, biodiversity and carbon stocks, water resources, and wildfires. A drop in precipitation paired with rising temperatures associated with drought occurrences is predicted, particularly in the Mediterranean Basin, which would reduce water availability for natural and agricultural

H. Gökçekuş and Y. Kassem (Eds.): NRSEM 2021, EESCI, pp. 220–229, 2022.
https://doi.org/10.1007/978-3-031-04375-8_26

systems and human requirements, according to the newly released IPCC (Intergovernmental Panel on Climate Change) 5th Assessment Report [1]. The main concern is to find one of the driest years of all the stations and to find the years with high drought rates.

By making efficient use of the huge fresh water reservoirs in Libya's southern area, the GMMR project aims to raise the quality of life for its people. High salinity rates in the north and low groundwater levels have put the Great Man–Made River on the verge of collapse, where every meter of fresh water decreases and 40 m of salt rise. In light of Libya's dwindling water supply and increasing population, it is more important than ever to seek foreign aid funding. Money might be used to build hydraulic equipment required to harvest and transport water from the desert for use in agriculture and consumption in heavily populated regions [2]. For these reasons the subject was chosen to study and analysis the longest-drought period to avoid any future problems.

Drought is a natural disaster with multiple varieties. According to the purpose of the study and the data used, many methods assessed drought and its development in various factors. Therefore, according to methods, developed by Palmer (1965) [3] to translate drought into the language of mathematics and Palmer Drought Severity Index (PDSI) based on precipitation and temperature data [4], and had analyzed solely of precipitation data Standardized Precipitation Index (SPI) which had explored meteorological drought [5]. Also, the Reconnaissance Drought Index (RDI) had analyzed during the compare and completion of the results.

The main objective of this study is to determine the drought sensitivity by making meteorological drought analyzes of The Great Man-Made River zone where was carried out within the scope of a study aiming to transmit the groundwater from various aquifer systems in the Southern Libyan region to the coastal regions through pipes. Furthermore this study aims to determine the longest-drought period in the study area. For the purpose this study, meteorological drought analysis will be conducted in the region, for five monitoring stations in Northern Libya.

2 Literature Review

Drought is defined as a phenomenon that may be related to the area under investigation and should be addressed using a specific application. Drought cannot be used to measure a region's average rainfall over several years. As a result, the region faces environmental, economic, and social challenges. Drought is one of the many natural disasters that have occurred throughout the world. As a result, many definitions of the dangerous drought had been developed [6]. During a drought, a lack of moisture usually results in a severe hydrological imbalance. Because of this condition, drinking water is typically used for precipitation, which has serious consequences for both humans and the environment. The area had also experienced dry weather and long-term water scarcity as a result of water scarcity. According to Hagman (1984), drought is the most common natural disaster [7]. These are the most complex of all natural disasters that have affected man, but the nature of drought has been described as the event [8], in a specific period and specific circumstances; the decrease of water availability in the area is termed as drought. Every year, various regions of the world are affected by drought [9].

Drought is a natural occurrence with numerous consequences. Economic, environmental, and social effects are all common. With the emergence of drought as a result of a lack of rainfall, the region's water and water resources, which were the source of life, decrease. It is also possible that agricultural productivity will suffer as a result. Consider our own country as an example, where the effects of drought and semi-arid climate characteristics are felt.

In this situation, some issues may arise as a result of the agricultural and hydrological drought, which occurred as a result of the country's meteorological drought, at a time when Libya has been experiencing a dry period. Certain agricultural sectors, which are among Libya's most important essential sectors, have suffered as a result of the agricultural drought. Similarly, the products cultivated by farmers, which are considered the most basic elements of this sector, are generally climate dependent. As a result, a decrease in rainfall usually resulted in problems such as a decrease in product yield and an inability to meet the country's food needs [10].

An attempt is made in the study of Aksoy vd. (2018), for the drought analysis with the SPI method using the data of 35 stations in the Gediz Basin with at least ten years of measurement between 1960 and 2016. As a result of the study, they had determined the SPI values for the periods of 1, 3, 6, 12, 24 and 48 months. For all periods in the basin, they found that 32% of the time had a mild drought, 8.8% moderate drought, 5% severe drought, and 2.3% of severe intensive drought [11].

A study by Al-Faraj and Al-Dabbag [12] looked at the impact of a multi-year drought on the growth of the Diyala River basin, which is shared by Iraq and Iran.

Al-Qinna et al. (2011) [13], Standardized Precipitation Index and Normalized Difference Vegetation Index methods, a detailed exploration of average day to day temperature and metrological drought analysis of the Hashemite Kingdom of Jordan between 1970–2005 was conducted. According to the results of these two indices, an extraordinary extreme drought was observed in the period 1999–2000 and it was stated that the country was exposed to drought cycles for 35 years after on. In addition, with the Standardized Precipitation Index method, predictions were made until 30 years the local would remain victimized to extensive drought conditions.

Al-Timimi and AL-Jboori (2013) conducted a drought analysis in the Iraq Region (SPI) method. Among the stations they studied, Dohuk, Sulaymaniyah, Dokan and Erbil stations were more affected by drought and the drought size in the 1999–2000 season varied between (18.4 and 17.4) mm. The northeastern region of Iraq faced high drought during 1980–2012. 1983, 1998, 1999, 2000 and 2008 are the years most affected by drought [14].

According to the SPI drought index, the Euphrates-Tigris basin in the Middle East just underwent a recent drought study [15].

Moreover, Apak (2009) [16], his research study has analyzed the drought origins in stations with long-term rainfall measurement in the Aegean Region by using the Standard Precipitation Index method for two periods, 1938–1970 and 1971–2006. As a result of these analyzes, he mentioned the effectiveness of the investigation in two periods in determining the severity of the drought and observed that in the second period both the number of dry years and the drought intensity increased as compared to the first.

Arslan et al. (2016) [17], investigated the droughts that occurred in the Gediz Basin between these years by using the Standard Precipitation Index (SPI) for 1, 3, 6, 9, 12 and 60-month periods, using the monthly precipitation data of 8 meteorological precipitation stations between 1973 and 2013. As a result of the study, it was determined that the droughts experienced in recent years lasted longer. For periods of 12 and 60 months. It has been stated that the drought period has increased by 3–7 times in recent years compared to the past.

Drought assessment in Urmia Lake basin was attempted using the Standardized Precipitation Index approach. Aslı and Hezerani (2019). They found that short-term scales react fast to changes in precipitation, whereas long-term scales are more likely to be affected by drought. SPIs are said to be most affected by monthly precipitation, which has a significant impact on total quarterly precipitation. As a result, the length of droughts in long-term series was shown to be longer than in short-term series [18].

Atmaca (2011) [19], analyzed the drought of Konya's region with the L-moments approach and utilized the methods of Standard Precipitation Index method. In his study, he created 3, 6, 9 and 12-month cumulative rainfall series using the monthly precipitation; he had obtained from the stations in the province. When he divided 44 observation stations into three regions, he was able to achieve homogeneity and as a result, he observed that mild drought was common in the regions according to SPI values.

3 Material and Method

Methodology: This study, located in the Great Man-Made River region, meteorological and hydrological drought analysis will be conducted for five monitoring stations in Northern Libya, the Standardized Precipitation Index (SPI) method is used to determine meteorological drought using monthly total precipitation data, and the Reconnaissance Drought Index (RDI) method using mean monthly temperatures data and total monthly precipitation data. The drought analysis using DrinC software of the Great Man-Made River region for 1-, 3-, 6-, and 12-months SPI and RDI values is conducted and examined in detail.

Study Area: This study aimed to determine drought sensitivity and calculate dry years to illustrate the driest years in the Great Man-Made River region using the Standard Precipitation Index (SPI) and Reconnaissance Drought Index (RDI) via the DrinC program in 5 meteorological stations using data between 1980–2009. Temporal changes of drought index values were examined with drought analysis. Northern Libya is the study area as rainfall is lack near the Libyan coast. Hence the study area is located between the longitudes of 9 and 25 easts and 30 and 33 latitudes of the north. The analysis is made through monitoring data from the stations based on the atmospheric droughts using the precipitations index of droughts as shown in the following given Table 1 and Fig. 1 the rainfall monitoring stations and geographic information in northern Libya used in the Great Man-Made River.

3.1 Drought Indices Overview

Study Area: This study aimed to determine drought sensitivity and calculate dry years to illustrate the driest years in the Great Man-Made River region using the Standard

Table 1. Rainfall monitoring stations and its geographic locations.

Station number	Station name	Coordinates and altitudes
62007	Zuara	32.53 N 12.05 E 03 m
62010	Tripoli Airport	32.40 N 13.09 E 81 m
62002	Nalut	31.52 N 10.59 E 621 m
62016	Misurata	32,19 N 15.03 E 32 m
62019	Sirt	31.12 N 16.35 E 13 m

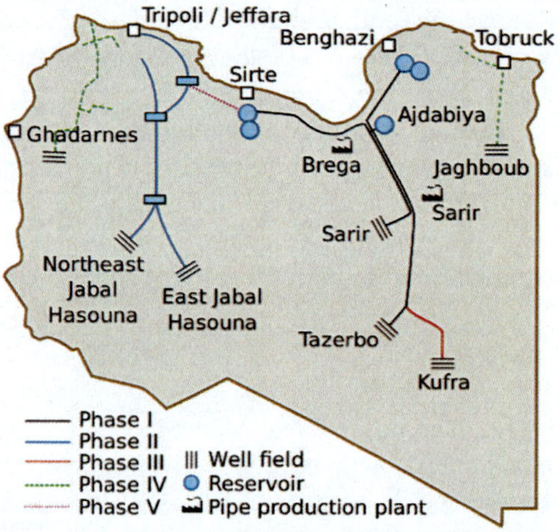

Fig. 1. Schematic view of The Great Man-Made River Area

Precipitation Index (SPI) and Reconnaissance Drought Index (RDI) via the DrinC program in 5 meteorological stations using data between 1980–2009. Temporal changes of drought index values were examined with drought analysis. Northern Libya is the study area as rainfall is lack near the Libyan coast. Hence the study area is located between the longitudes of 9 and 25 easts and 30 and 33 latitudes of the north.

The Standard Precipitation Index (SPI) is developed by McKee et al. to determine the effects of reduction in precipitation on groundwater, reservoir storage, soil moisture, snow drifts and streams. It is obtained by dividing the difference of precipitation from the mean, which is converted to normal distribution within the specified time period, by the standard deviation. In fact, SPI provides a standardized conversion of the observed precipitation probability and could be calculated for desired time periods such as 1, 3, 6, 9, 12, 24 and 48 months.

The Reconnaissance Drought Index (RDI) is developed to approach the water deficit in a more accurate way, as a sort of balance between input and output in a water system. It is based on both cumulative precipitation (P) and potential evapotranspiration (PET), which is one measured (P) and later calculated (PET) determinant.

There has been a lot of work put into developing DrinC (Drought Indices Calculator), a piece of software designed to make it easy to calculate drought indices. Reconnaissance Drought Index (RDI) and Streamflow Drought Index (SDI) may both be calculated using DrinC, as well as the Standardized Precipitation Index (SPI) and the Precipitation Deciles (PD) indices. RDI may also be calculated using a module that uses temperature-based techniques to estimate potential evapotranspiration (PET). Drought monitoring, the evaluation of drought's geographical distribution, the analysis of climate and drought scenarios, and so on, may all benefit from the software DrinC is gaining popularity as a research and operational tool for drought analysis in arid and semi-arid countries, where it has been tested extensively.

4 Research Findings and Discussion

The findings and discussion of the data that collected from 5 monitoring stations (Zuara Station (62007), Tripoli Airport Station (62010) Nalut Station (62002), Misurata Station (62016), Sirt Station (62019)) from the state meteorological station and Libya in the region of The Great Man-Made River are given below;

Within the scope of this research, the SPI and RDI values for 1-, 3-, 6- and 12-months are calculated and evaluated using the method of the Precipitation Drought Index and Reconnaissance Drought Index. The process used the values of the monthly total precipitation and monthly mean temperature of 5 monitoring stations from the state meteorological station and Libya in the region of The Great Man-Made River.

4.1 Zuara Station (62007) Reconnaissance Drought Analysis

Zuara Station computed using monthly total precipitation and mean monthly temperature continuously measured data between the periods 1980–2008, RDI values are examined during periods 1-, 3-, 6-, and 12 months. When the one-month RDI values obtained from the monthly total rainfall and mean monthly temperature data for a 29-year period for the Zuara monitoring station are examined, the presence of an exceptional drought is determined in November and December year 2000–2001 and in March year 2005–2006. The highest drought value (−3.03) was recorded in November 2000–2001.In July, all years are near normal except the year 1985–1986 exceptionally moist.

Time distributions of RDI values for 3-, 6-, and 12-months, Examination of RDI values for 3 months RDI-3 October year 1981–1982 extremely dry period, RDI-3 January year 1980- 1981 extremely dry, RDI-3 April 1998–1999 exceptionally dry period. When the RDI values are checked for 6 months, RDI-6 October year 1981–1982 and 2000–2001 is extremely dry, and RDI-6 April 1980–1981 and 1988–1989 and 1998–1999 is determined to be extremely dry. According to RDI-12 years 1981–1982 were extremely dry. The year 2000–2001 is considered one of the driest years in this station. The drought RDI value reached −2.25 and is often repeated. Drought in this station after a year or two during the study period.

4.2 Tripoli Airport Station (62010) Reconnaissance Drought Analysis

Tripoli Airport station, computed using monthly total precipitation and mean monthly temperature continuously measured data between 1980–2008, RDI values were examined during periods 1-, 3-, 6- and 12 monthly. When the one-month RDI values obtained from the monthly total rainfall and mean monthly temperature data for a 29-year period for the Tripoli Airport monitoring station are examined, the presence of an extreme drought was determined in December 1989–1990 and in March 2007–2008. The highest drought value (−2.98) was recorded in November 2000–2001. In July, most years are abnormally dry.

The time distributions of RDI values for 3-, 6- and 12-months, Examination of RDI values for 3 months RDI-3 October year 2000–2001 extremely dry period, RDI- 3 January year 2008–2009 extremely dry, RDI-3 April 1998–1999 extremely dry period. When the RDI values are checked for 6 months, RDI-6 October year 2000–2001 is extremely dry and RDI-6 April 1998–1999 is determined to be exceptionally dry. According to RDI-12, the year 2000–2001 is considered one of the driest years in this station. The drought RDI value reached −2.04.

4.3 Nalut Station (62002) Reconnaissance Drought Analysis

Nalut Station computed using monthly total Precipitation and mean monthly temperature continuously measured data between 1980–2008, RDI values are examined during periods 1-, 3-, 6-, and 12- months. When the one-month RDI values obtained from the monthly total rainfall and mean monthly temperature data for a 29-year period for the Nalut monitoring station are examined, the presence of an extremely drought is determined in December 2000–2001, and the presence of extreme drought was determined in January 2006–2007. The highest drought value (−2.56) was recorded in December 2000–2001.

Time distributions of RDI values for 3-, 6- and 12-months, Examination of RDI values for 3 months RDI-3 October year 1981–1982 exceptionally dry period, RDI-3 January year 2004- 2005 exceptionally dry, RDI-3 April 1998–1999 exceptionally dry period, RDI-3 July year 1982–1983 severely dry. When the RDI values are checked for 6 months, RDI-6 October year 1992–1993 is extremely dry and RDI-6 April 1983–1984 is determined to be exceptionally dry. According to RDI-12, the year 1992–1993 is considered one of the driest years in this station. The drought RDI value reached −2.13.

4.4 Misurata Station (62016) Reconnaissance Drought Analysis

Misurata Station computed using monthly total precipitation and mean monthly temperature continuously measured data between periods 1980–2008, RDI values are examined during periods 1-, 3-, 6-, and 12- months. When the one-month RDI values obtained from the monthly total rainfall and mean monthly temperature data for a 29-year period for the Misurata monitoring station were examined, the presence of an exceptionally dry is determined in November 2000–2001 and December 1989–1990 and in February 1985–1986. The highest drought value (−3.03) was recorded in March 1999–2000.

The time distributions of RDI values for 3-, 6- and 12-months, Examination of RDI values for 3 months RDI-3 October year 1992–1993 extremely dry period, RDI-3 January year 1984–1985 severely dry, RDI-3 April 1998–1999 extremely dry period, RDI-3 July year 1989–1990 and 1992–1993 severely dry. When the RDI values are checked for 6 months, RDI-6 October 1993–1994 is exceptionally dry and RDI-6 April 2004–2005 is extremely dry. According to RDI-12, The year 2000–2001 is considered one of the driest years in this station. The drought RDI value reached −2.02.

4.5 Sirt Station (62019) Reconnaissance Drought Analysis

Sirt Station computed using monthly total precipitation and mean monthly temperature continuously measured data between periods 1980–2008, RDI values are examined during periods 1-, 3-, 6-, and 12- months. When the one-month RDI values obtained from the monthly total rainfall and mean monthly temperature data for a 29-year period for the Sirt monitoring station are examined, the presence of an exceptionally dry was determined in November 1980–1981 and in January 2008–2009.

The highest drought value (−2.95) was recorded in February 1984–1985. The Time distributions of RDI values for 3-, 6-, and 12- months, Examination of RDI values for 3 months RDI-3 October year 2000–2001 exceptionally dry period, RDI-3 January year 2008–2009 exceptionally dry, RDI-3 April 1987–1988 exceptionally dry period. When the RDI values are checked for 6 months, RDI-6 October year 2000–2001 is exceptionally dry and RDI-6 April 1980–1981 is determined to be extremely dry. According to RDI-12, the year 2000–2001 is considered one of the driest years in this station. The drought RDI value reached −2.11.

5 Conclusion

In the SPI analysis of all stations on a monthly basis, in Zuara station it is found that August is drier than other months with 79% ratio and for Tripoli Airport station it was recorded that July was drier with 86%, at Nalut station it is discovered that August was drier than other months with 86%, and for Misurata station August turned out to be drier with 79%, at Sirte station it turned out that June is drier than other months with 76%. In calculating SPI-3 for all stations based on the monthly rainfall data in Zuara station, the highest values of drought are observed in SPI3-3 April in 1988–1989 exceptionally dry, and the highest value of humidity in SPI3-1 October in 1984–1985 exceptionally moist, at Tripoli Airport station the highest value for drought in SPI3-2 January in 2008–2009 extremely dry and the highest value for moisture in SPI3-4 July in 1996–1997 exceptionally moist, at Nalut station the highest value for drought is in SPI3-1 October in 1982–1983 exceptionally dry, and the highest value for moisture in SPI3-1 October In 1995–1996 exceptionally moist. For Misurata station, the highest value of drought is in SPI3-3 in April in 1998–1999 extremely dry, and the highest value in humidity was in SPI3-2 in January in 1994–1995 exceptionally moist, in Sirte station, the highest value of drought is in SPI3-2 in January in 2008 -2009 exceptionally dry and the highest value of moisture was in SPI3-4 July in 1985–1986 exceptionally moist.

In calculating SPI-6 for all stations in Zuara station, the highest value of dryness and humidity is observed in SPI6-1 in October, the highest value of drought in 2000–2001 extremely dry and the highest value of humidity in 84–1985 exceptionally moist; at Tripoli Airport station, the highest value of drought in SPI6-2 April in the year 1998–1999 exceptionally dry, and the highest value of humidity is in the year 1993–1994 exceptionally moist; in Nalut station, the highest value of drought was in SPI6-2 in April in the year 1983–1984 exceptionally dry, and the highest value of humidity is in SPI6-1 in October in the year 1995–1996 exceptionally moist; for the Misurata station the highest value of dryness is observed in SPI6-1 October in the year 1993–1994 and the highest value humidity in the year 1980–1981 as exceptionally moist; in Sirt station the highest value of drought was in SPI6-1 October in the year 2000–2001 exceptionally dry, and the highest value of humidity was in SPI6-2 April in the year 85-1986 exceptionally moist. In calculating SPI-12 for all stations, the maximum value of dryness and humidity in Nalut station was in relation to drought in 1992–1993 exceptionally dry (−2.25), and humidity in 1995–1996 exceptionally moist (2.96). In the monthly RDI analysis for all stations using monthly precipitation data and mean monthly temperatures, it was found that the maximum value of the Reconnaissance drought at Tripoli Airport station in year 2000–2001 is in November month (−2.98), and in the analysis of RDI-3 it is found that the highest value of drought is in Nalut station in RDI3-1 October in year 1981–1982 exceptionally dry, in the analysis of RDI-6, it is found that the highest value of drought is in Sirte station in RDI6-1 October in the year 2000–2001 exceptionally dry, in the analysis of RDI-12, the maximum value of reconnaissance drought is in Zuara station in 2000–2001 year exceptionally dry.

References

1. Marimon, C.D.: Contributions to the knowledge of the multitemporal spatial patterns of the Iberian Peninsula droughts from a Geographic Information Science perspective. Geofocus Revista Internacional de Ciencia y Tecnología de la Información Geográfica **17**, 9 (2016)
2. Government of the Libyan Arab Jamahiriya: Bankable Investment Project Profile: Great Man-Made River Distribution Facilities. Food and Agriculture Organization of the United Nations, Investment Centre Division, Libya (2006)
3. Palmer, W.C.: Meteorological drought. US Department of Commerce, Weather Bureau, USA, 20–25 (1965)
4. McKee, T.B.: Drought monitoring with multiple time scales. In: Proceedings of 9th Conference on Applied Climatology, Boston (1995)
5. Nalbantis, I.: Evaluation of a hydrological drought index. Eur. Water **23**(2), 67–77 (2008)
6. McMahon, T.A., Arenas, A.: Methods of computation of low streamflow. Imprimerie de la Manutention, Paris, 117-99 (1982)
7. Hagman, G., Beer, H., Bendz, M., Wijkman, A.: Prevention better than cure. Report on human and environmental disasters in the Third World, Washington, USA, 34–41 (1984)
8. Beran, M.A., Rodier, J.A.: Hydrological aspects of drought. UNESCO, Switzerland, pp. 67–78 (1985)
9. Hisdal, H., Tallaksen, L.M.: Estimation of regional meteorological and hydrological drought characteristics: a case study for Denmark. J. Hydrol. **281**(3), 230–247 (2003)
10. Öztürk, K.: Küresel İklim Değişikliği ve Türkiyeye Olası Etkileri. Gazi Üniversitesi Gazi Eğitim Fakültesi Dergisi **22**(1), 34–41 (2002)

11. Aksoy et al.: Drought Analysis in Gediz Basin. Scientific Congress of the Turkish National Union of Geodesy and Geophysics (TUJJB), İzmir, pp. 28–31 (2018)
12. Al-Faraj, F.A.M., Al-Dabbagh, B.N.S.: Assessment of collective impact of upstream watershed development and basin-wide successive droughts on downstream flow regime: the Lesser Zab transboundary basin. J. Hydrol. **530**, 419–430 (2015)
13. Al-Qinna, M.I., Hammouri, N.A., Obeidat, M.M., Ahmad, F.Y.: Drought analysis in Jordan under current and future climates. Clim. Change **106**(3), 421–440 (2011)
14. Al-Timimi, Y.K., Al-Jiboori, M.H.: Assessment of spatial and temporal drought in Iraq during the period 1980–2010. Int. J. Energy Environ, **4**(2), 291–302 (2013)
15. Amini, A., Zareie, S., Taheri, P., Yusof, K.B.W., ul Mustafa, M.R.: Drought analysis and water resources management inspection in Euphrates–Tigris Basin. In: River Basin Management. IntechOpen (2016)
16. Apak, E.: Drought analysis of Aegean region by standardized precipitation index (SPI). Ege University, Graduate School of Natural and Applied Sciences, Izmir, Turkey, pp. 34–45 (2009)
17. Arslan, O., Bilgil, A., Veske, O.: Standart yağiş indisi yöntemi ile kizilirmak havzasi'nin meteorolojik kuraklik analizi. Niğde Ömer Halisdemir Üniversitesi Mühendislik Bilimleri Dergisi **5**(2), 188–194 (2016)
18. Aslı, U.L.K.E., Boustani, H.A.: Analysis of Basin drought for URMIA Lake in Iran with Standardized Precipitation Index method SPI. Karaelmas Fen ve Mühendislik Dergisi **9**(2), 167–176 (2019)
19. Atmaca, D.: Standartlaştırılmış yağış indeksi (SYİ) yöntemi ile Konya ili bölgesel kuraklık analizi. Master's thesis, Gaziosmanpaşa Üniversitesi, Fen Bilimleri Enstitüsü, 50 (2011)

Investigations of Greenery Façade Approaches for the Energy Performance Improvement of Buildings and Sustainable Cities

Saeed Hussein Alhmoud(✉)

Faculty of Architecture, Department of Interior Architecture, Near East University, North Cyprus Nicosia/TRNC, Mersin 10, Turkey
Saeed.hmoud22@gmail.com

Abstract. Today, sustainability is resulting in a significant and intriguing approach to architecture and the environment. Environmental threats such as greenery façade and resulting energy shortages, negative impacts of climate change, sick building syndromes, and many building-related professional bodies have led to the recognition of the need for effective sustainable building design strategies. As a result, in recent years, there has been an increase in interest in developing efficient solutions to improve the long-term energy performance of buildings. The use of greenery on building facades is becoming more popular as a way to improve the quality of life in urban areas. The green façade has the potential to cool the building's surface. The meaning, benefits, and techniques of the green façade as a part of the urban environment's sustainability strategy will be presented in this paper. The two types of greenery façade systems are the green façade or vertical greening and the living wall. The goal of this research is to analyze the effect of green facades systems on building temperature and to address different types of green facade systems and their thermal effects as part of this quest. Finally, the study highlights emerging innovative building façade design solutions aimed at improving the building sector's energy efficiency and contributing to cities' long-term sustainability.

Keywords: Green façade · Energy efficiency · Vertical greenery · Living wall · Sustainable construction · Green wall · Passive system · Urban greening

1 Introduction

Building facades are a feature of urban environments that can be creatively used in architecture to achieve sustainable or biophilic solutions. Green facades today have the ability to draw inspiration from traditional architecture whereas using advanced materials and other technology to promote sustainable building functions is also a good idea [5].

Facade greening has several advantages, including not only making a positive contribution to the environment and Nature, but also lowering long-term operating costs. Furthermore, the greening of the facades in larger commercial areas is especially important, as the microclimate of a small settlement area has a significant impact. These green

H. Gökçekuş and Y. Kassem (Eds.): NRSEM 2021, EESCI, pp. 230–239, 2022.
https://doi.org/10.1007/978-3-031-04375-8_27

types are essential for controlling dust, humidifying, and generating cold air, thus promoting human health. As a result, they serve as a local, natural air conditioner, similar to a green roof. Because there are so many potentials and opportunities, it's worth taking a closer look at the subject in terms of long-term value [11].

The use of plants on building facades contributes significantly to the development of the built environment's sustainability. Using them as an architectural feature to improve the appearance of a building is both environmentally and aesthetically acceptable [1].

At the building and urban scales, the term "Urban Green" has been defined as a collection of man-made elements that provide multiple ecosystem services. Among these functions, building energy savings and the reduction of the urban heat effect stand out.

Furthermore, in the winter, vertical PV facades would also generate more energy than in the summer, as well as more throughout the especially in early periods of the day, whenever the sun is lower in the sky [3]. The use of climbing plants to green a city's façade can alter the interaction between the building and the environment. This interaction lowers the building's outside temperature and cools the respective interior spaces [14]. Green facades and green living walls can provide a cooling effect on a building's surface, lowering energy consumption by improving thermal performance [10]. By reducing air and surface temperature maxima and variation, increasing the amount of vegetation, or green infrastructure, in a city can help address the root cause of the problem [7].

Buildings' exterior walls and roofs have been greened. Among the reasons for doing so were improved insulation, aesthetics, indoor and outdoor climate, reduction of greenhouse gases such as carbon dioxide (CO2), carbon monoxide (CO), and nitrogen dioxide (NO2), and increased ecological values by creating habitats for birds and insects [4].

Plants on building facades play an important part in enhancing the built global environment sustainability. As an adequate architectural/interior feature that enhances facades, their use is both environmentally and aesthetically acceptable. Their use results in an energy-conscious design approach that protects the environment in densely populated urban areas.

Tall buildings also provide multiple surfaces for solar radiation absorption, which is then reradiated as heat, increasing the efficiency with which cities are energy performance [6].

Finally, the purpose of this is to analyze the green façade and roofs, to understand when façade might act as an added advantage to the rooftops, structure protection, noise reduction, climate influence, and aesthetics.

2 Materials and Methods

A literature review was undertaken to collect data on greener façade and living wall through experimental and review researches undertaken in different countries in Europe, Greece and Spain and two case studies were around the world. Fifteen original publications were reviewed (Table 1).

Table 1. Summary with basic characteristics of reviewed publication on experimental and review researches undertaken in different countries

Reference	No. of walls	Location	Research
[15]	–	Greece	Rethinking user based innovation: Assessing public and professional perception of energy efficient building facades
[14]	8	Spain	Effects of the type of façade on the energy performance of office building representative of the city
[3]	2	Portugal	The importance of facades for the solar PV of a Mediterranean city LIDAR data
[13]	–	UK	A Hedera green façade – Energy performance maritime – temperate, winter weather conditions
[12]	–	Spain	Vertical Greenery Systems (VGS) for energy saving in building: A review
[11]	–	Egypt	Green Facades as New Sustainable Approach Towards Climate Change
[2]	–	Spain	Evaluation of green walls as a passive acoustic insulation system for buildings
[9]	–	Spain	Vertical greenery systems for energy savings in building: A comparative study between green walls and green facades
[4]	2	Malaysia	A review on energy conscious designs of building façades in hot and humid climates: Lessons for (and from) Kuala Lumpur and Darwin
[7]	–	Australia	Planning for cooler cities: A framework to priorities green infrastructure to mitigate high temperatures in urban landscapes
[10]	–	Egypt	The Effect of Green Façades in Landscape Ecology
[8]	–	Egypt	Green Architecture: A Concept of Sustainability
[6]	–	Spain	Vertical Greening Systems and Sustainable Cities
[1]	–	Turkey	Façade Greening: Away to Attain Sustainable Built Environment
[5]	–	Germany	Green façades – a view back and some visions

Fig. 1. Team REBEL presenteert Kop Zuidas – Amsterdam [13].

Fig. 2. Solar Dunes Project, Nozha, New Cairo, Egypt [10].

Fig. 3. Herzog & de Meuron Designs "Horizontal Skyscraper" Above Historic Moscow Brewery [4].

Fig. 4. Vo Trong Nghia Breaks Ground on Checkerboard Building With Tree-Filled Balconies in Vietnam [7].

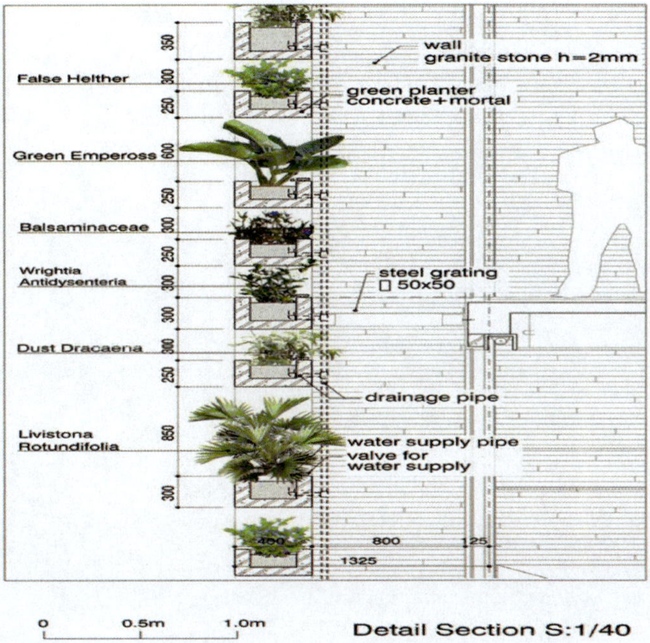

Fig. 5. Living wall diagram [14].

Fig. 6. Green façade example from Malaysia [4].

Fig. 7. Green façade example from Turkey [1].

Fig. 8. Green façade example from Germany [5].

3 Discussion

By reducing urban air as well as maximum surface temperature and contrast, rainwater management, air quality, and sustainable energy, a green façade that increases the amount of vegetation, or green infrastructure, in a city can significantly reduce the root cause of the issue. Therefore, insulation is needed to keep heat away when it is very hot outside or to keep heat inside when it is very cold outside [7]. In general, well enough and controlled vegetation can be an effective tool for passive thermal controller of buildings, saving energy. This could happen in four ways, the most common of which are thermal insulation, solar radiation interaction (shading), evaporative cooling, and changing winds on the building. This can happen in 4 ways, all of which are often related: thermal insulation, solar radiation interaction, i.e. shade, evaporative cooling, and wind variation on the structure.

Green walls have become an important part of living architecture, and they will become more common in our cities in the coming years. Architects and designers, who want to use the building envelope to achieve multiple goals, create new stand-alone design elements on the interior and exterior of structures have a variety of options with green wall building technologies. There is also a significant potential for lowering urban temperatures when the building envelope is covered with vegetation. It can be concluded that the greater the effect of vegetation on urban temperatures, the hotter and drier the climate.

Façade vegetation makes it appear as a component of interior space and layout to take into account today's architecture and urbanism, and plants have become one of the major design considerations in the development of modern buildings, due to the improved environmental influences.

Further, enlarging plants or green spaces on a building façade has been shown to improve air quality, and lower surface temperatures in the built environment. The amount of carbon dioxide and carbon monoxide in the air, as well as the temperature and relative humidity, differ significantly between areas with and without green walls [11], and enhancing the energy Performance improvement of buildings.

Table 2. Summary with basic characteristics of reviewed publication on experimental and review researches undertaken in different countries

Cubicle		TSC	East wall	South wall	West wall
	Value	Standard deviation	ΔT (°C)	ΔT (°C)	ΔT (°C)
GW	0.063	0.031	4.5	16.5	6.5
GF	0.118	0.026	-0.2	0.7	-0.3
Reference	0.099	0.023	-	-	-

Table 2 shows the thermal response of the external surface wall temperatures by facade orientation during the winter season when various vertical greenery systems

(VGS) are installed on the East, South, and West facades to determine the sensitivity of the house like-cubicles.

Greening can also aid in the mitigation of climate change and the improvement of living conditions in our rapidly expanding cities. VGSs, on the other hand, have a lot of potential to help create a better environment for sustainable cities [6]. In addition, vertical PV façades will produce more in the early and late hours of the day in the winter and less in the summer [3]. Selective glass, which provides a medium-high visible light transmission while reducing the amount of thermal radiation absorbed by the glass and transmitted to the interior of the building, could be one of the best solutions to this problem when combined with exterior shielding elements that have systems to adapt to changing external environmental conditions [6].

4 Conclusion and Recommendations

The integration of vegetation on the façades of buildings and other structures can be extremely beneficial to the urban environment, as well as a tool for passive thermal control of structures. Green façades have the ability to moderate urban temperature if they are carefully designed and maintained [1]. As a result, in enclosed areas the location of the green wall must be considered because it will affect the temperature and humidity [11]. Furthermore, many green building designers are encouraged to consider the environmental impacts of green façades as part of a sustainable design strategy that addresses our complex relationship with the natural environment, which is otherwise eroding as we live in increasingly dense urban environments [10].

There are several recommendations for incorporating green façades into design in order to develop the ambient and thermal conditions;

- Plants and vegetation must be used liberally, but with caution, on building facades in urban areas. Plants should be chosen based on their natural support mechanisms and ability to adapt to harsh environments.
- Plants and vegetation on the urban façade must be placed in such a way that they receive full sunlight for as long as possible.
- Plants implemented in the vertical direction throughout the urban environment must be considered for maintenance in order to avoid hazards. They'll need to be watered frequently and regularly trimmed.
- The thermal comfort will be compensated by the high relative humidity, particularly once the temperature is high and there is no wind to alleviate the discomfort.

References

1. Abdullahi, M.S., Alibaba, H.Z.: Facade greening: a way to attain sustainable built environment. Int. J. Environ. Monit. Anal. **4**(1), 12–20 (2016). https://doi.org/10.11648/j.ijema.20160401.13
2. Azkorra, Z., et al.: Evaluation of green walls as a passive acoustic insulation system for buildings. Appl. Acoust. **89**, 46–56 (2015). https://doi.org/10.1016/j.apacoust.2014.09.010

3. Brito, M.C., Freitas, S., Guimarães, S., Catita, C., Redweik, P.: The importance of facades for the solar PV potential of a Mediterranean city using LiDAR data. Renewable Energy **111**, 85–94 (2017). https://doi.org/10.1016/j.renene.2017.03.085
4. Halawa, E., et al.: A review on energy conscious designs of building façades in hot and humid climates: lessons for (and from) Kuala Lumpur and Darwin. Renewable Sustain. Energy Rev. **82**, 2147–2161 (2017). https://doi.org/10.1016/j.rser.2017.08.061
5. Köhler, M.: Green facades—a view back and some visions. Urban Ecosyst. **11**(4), 423–436 (2008). https://doi.org/10.1007/s11252-008-0063-x
6. Pérez-Urrestarazu, L., Fernández-Cañero, R., Franco-Salas, A., Egea, G.: Vertical greening systems and sustainable cities. J. Urban Technol. **22**(4), 65–85 (2015). https://doi.org/10.1080/10630732.2015.1073900
7. Norton, B.A., Coutts, A.M., Livesley, S.J., Harris, R.J., Hunter, A.M., Williams, N.S.: Planning for cooler cities: a framework to prioritise green infrastructure to mitigate high temperatures in urban landscapes. Landsc. Urban Plan. **134**, 127–138 (2015). https://doi.org/10.1016/j.landurbplan.2014.10.018
8. Ragheb, A., El-Shimy, H., Ragheb, G.: Green architecture: a concept of sustainability. Procedia Soc. Behav. Sci. **216**, 778–787 (2016). https://doi.org/10.1016/j.sbspro.2015.12.075
9. Coma, J., Pérez, G., de Gracia, A., Burés, S., Urrestarazu, M., Cabeza, L.F.: Vertical greenery systems for energy savings in buildings: a comparative study between green walls and green facades. Build. Environ. **111**, 228–237 (2017). https://doi.org/10.1016/j.buildenv.2016.11.014
10. Elgizawy, E.M.: The effect of green facades in landscape ecology. Procedia Environ. Sci. **34**, 119–130 (2016). https://doi.org/10.1016/j.proenv.2016.04.012
11. Sheweka, S.M., Mohamed, N.M.: Green facades as a new sustainable approach towards climate change. Energy Procedia **18**, 507–520 (2012). https://doi.org/10.1016/j.egypro.2012.05.062
12. Pérez, G., Coma, J., Martorell, I., Cabeza, L.F.: Vertical Greenery Systems (VGS) for energy saving in buildings: a review. Renew. Sustain. Energy Rev. **39**, 139–165 (2014). https://doi.org/10.1016/j.rser.2014.07.055
13. Cameron, R.W., Taylor, J., Emmett, M.: A Hedera green façade–energy performance and saving under different maritime-temperate, winter weather conditions. Build. Environ. **92**, 111–121 (2015). https://doi.org/10.1016/j.buildenv.2015.04.011
14. Planas, C., Cuerva, E., Alavedra, P.: Effects of the type of facade on the energy performance of office buildings representative of the city of Barcelona. Ain shams Eng. J. **9**(4), 3325–3334 (2018). https://doi.org/10.1016/j.asej.2017.04.009
15. Tsoka, S., Tsikaloudaki, K., Theodosiou, T., Dugue, A.: Rethinking user based innovation: assessing public and professional perceptions of energy efficient building facades in Greece, Italy and Spain. Energy Res. Soc. Sci. **38**, 165–177 (2018). https://doi.org/10.1016/j.erss.2018.02.009

Monthly Rainfall Variability and Vulnerability of Rainfed Cereal Crops in the Tellian Highlands of Algeria

Smadhi Dalila[1](✉), Zella Lakhdar[2], Amirouche Mawhoub[2], Bachir Hakim[1], and Semiani Mohamed[1]

[1] Division of Bioclimatology and Agricultural Hydraulics, National Institute for Agricultural Research, El–Harrach, 16004 Algiers, Algeria
dalsmadhi@yahoo.fr

[2] Faculty of Nature and Life Sciences, Department of Biotechnology, University of Saad Dahlab, OuledYaïch, 09000 Blida, Algeria

Abstract. Rainfed cereal production in semi-arid regions of Algeria is under the permanent influence of rainfall fluctuation and droughts. In this context, the study attempts to analyze over a period of 80 years (1940 to 2020), the average behavior of these variables, during the agricultural season, from September to August. The analysis focused on two cereal-growing wilayas, in the high plateaus of the East and West of the country. The variables analyzed on a monthly time step, showed spatio-temporal variations, with probabilities of occurrence between 40 and 80% according to a distribution of 20, 30, 50 and 80%. These occurrences have made it possible to determine rainfall deficits, which reach between 30 and 92% depending on the month of the annual series, causing drought sequences that exceed 10 days, monthly. These characteristics contribute to quantify the vulnerability of the evolution of the culture, to the climatic hazards, which affect the yield, which does not exceed 17 q/ha, in good year, as well as the water resource, in this conjuncture of climatic changes.

Keywords: Algeria · Rainfall · Drought · Production · Vulnerability

1 Introduction

Cereal cultivation in Algeria dates back to the time of the Numidians, who divided the country into three regions: the Eastern region, the Central region and the Western region [1, 2]. Cultivation of durum wheat, soft wheat barley and oats is widespread throughout the Algerian tell (from Timgad, Tipaza, Tébessa, Skikda, Annaba to Tazoult), without occupying all the soils that are suitable for it [3–6] emphasise that Algeria is a wheat country, characterised by average rainfall of between 200 and 600 mm/year. These averages, which have remained unchanged over the last few decades [7], cover relatively stable cereal-growing areas of around 2.8 million hectares and produce nearly 25 million quintals, with a yield that does not exceed 17 quintals per hectare in a good year (2017). The deficit, which regularly reaches 75%, is made up by imports. This

H. Gökçekuş and Y. Kassem (Eds.): NRSEM 2021, EESCI, pp. 240–251, 2022.
https://doi.org/10.1007/978-3-031-04375-8_28

situation is attributed to the capricious and poorly distributed rainfall, represented by annual droughts, whose loss of rainfall reaches nearly 70% in a dry year [8].

At this time scale, research [9–11] has attempted to simulate drought indices in the absence of soil moisture, runoff and evapotranspiration data. These indices, which help to describe droughts qualitatively and quantitatively, continue to be presented as major concerns, associated with rainfall trends. In this context, the main question raised is to what extent it is possible to assess and mitigate rainfall constraints for better adaptation of cropping techniques. [9] put the emphasis on the installation of drought in the short term and which can be estimated by rainfall. These authors concluded that, as a single parameter, rainfall is the best indicator of drought. [12] proposed a calculation of dry spell probabilities based on distribution quantities. In this context, [13] consider the onset of rainfall as the first 10 days with at least 20 mm total rainfall. [14] marks the date of the first rains at 20 mm or more, after 1 May and without being followed for 30 days or more by a dry period exceeding 5, 10, 15 and 20 days (d). The end date of the rains, according to the same author, occurs when there is no more rain for 20 days.

Work on cereal agroclimatology in Algeria is rare and poorly documented [15–20] its interest is all the more important at a time of climate change and rising prices for cereal products. This study attempts to highlight the evolution of monthly rainfall associated with the evolution of droughts and the number of rainy days, which can affect cereal yields in the northern part of the country. The study is based on a monthly rainfall history of the rainfed regions.

2 Material and Methods

2.1 Structuring Your Paper

The study area covers northern Algeria, located at latitudes 32° and 37° and longitudes −2° and 9°. It covers 35.9 million ha, i.e. 17% of the national territory. This area, which constitutes the natural limits of rainfed cereal cultivation, is subdivided into three regions: the Eastern Region (ER), the Central Region (CR) and the Western Region (WR), representing 36, 17 and 47% of the total respectively. The area (Fig. 1) comprises 38 territorial units, administratively limited to wilayas (or departments) of varying proportions.

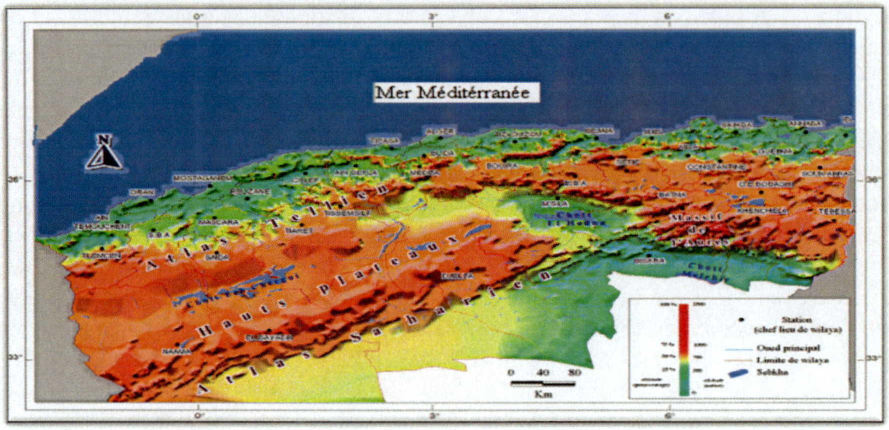

Fig. 1. Geographical location of rainfed cereal growing areas in northern Algeria.

2.2 Cereal Data

Rainfed cereal crops are dominated by durum wheat, soft wheat, barley and oats. Its evolution over time (1940–2019) is shown by data on annual sown areas, production and yields, collected from archival documents and the agricultural statistics bulletins published by the Ministry of Agriculture.

2.3 Rainfall Data

This research takes into account series of rainfall data, constructed on a monthly scale [8] over the period (1940–2019). These data, which run from September to August, represent 76 stations; spread over 38 wilayas, according to the networks of the National Meteorological Office, the National Water Resources Agency. The average density is 0.21 station per 100,000 ha (or 2 per 1 million ha). A rate of 41% is installed in the East region (ER), 22% are in the Center region (CR) and 37% in the West region (WR). The chosen time step highlights rainfall databases, representing the amounts of rain, the number of rainy days, in relation to the indices of droughts, in this conjuncture of variability and climate change in rain-fed cereal regions.

2.4 Temporal Analysis of Monthly Rainfall

The analysis of the evolution of rainfall from September to August highlights the spatio-temporal distribution of recorded rainfall amounts, numbers of rainy days and droughts during the agricultural season, with reference to the approach established by [8]. The rainfall amounts, highlighted, represent rainfall at thresholds above $R \geq 0.1$ mm, $P \geq 5$ mm, $R \geq 10$ mm, $R \geq 20$ mm and $R \geq 30$ mm. These thresholds participate in differentiating according to [21, 22] meteorological rainfall, from agronomic rainfall.

The first threshold represents the average meteorological rainfall from September to June. The second threshold characterises agricultural rainfall. The last three thresholds provide information on the water requirements of cereals from sowing to grain ripening. Monthly rainfall files (MR) from September to August, at different rainfall thresholds, as well as synthesis files relating to the number of rainy days per month (rd) and to droughts, were elaborated by Wilaya. The numerous results allowed the synthesis of a number of variables (N) for each of the parameters, equivalent to 2698 cases (71 * 38).

The monthly geo-referenced data obtained are analysed statistically, by means (My), medians (Md), standard deviations (SD), coefficients of variation (Cv), variances (V) and by correlation coefficients R2. These statistics help to characterise the average rainfall condition by taking into account the most significant dispersion indices. They remain references for comparing rainfall amounts between the 25% and 75% quartiles. They also represent the measurement of climatic variability and rainfall changes from one month to another [23] and from one region to another. The set of statistics used contribute to highlighting the rainfall trend model at monthly time step. The model is based on the third order polynomial regression, which is a simple model, taking into consideration two variables (P,Y), of which the variable to be explained is raised to increasing power. The regression Eq. (1) allows to draw the trend curves, which have as many inflection points, as there are degrees to the polynomial.

$$Y = b_0 + b_1 R + b_2 R_2 + b_3 R_3 \tag{1}$$

2.5 Spatial Analysis of Rainfall

The statistically evaluated monthly rainfall data are regionalised by the numerical kriging method [24]. According [25] this geostatistical method results in the statistical modelling of point spatial data. Its use is widely recommended for interpolating rainfall in a climatic or hydrological modelling framework [26–28]. It takes into account latitudinal and longitudinal coordinates, which are only artefacts according to [29] obscuring other parameters responsible for rainfall, such as the altimetric factor, which is a predominant element in mountainous environments. The preliminary analysis of the data distribution took into account the statistical adjustment that determines the different data values (rainfall) associated with each combination of the simulated and regionalized parameter. Statistically homogeneous areas (or surfaces) are obtained.

2.6 Drought Analysis

The statistical parameters used to quantify the spatio-temporal variability of rainfall are supported by the calculation of the [30]. This index, which takes into account the standard deviation, makes it possible to standardise the data and transform them into reduced centred anomalies (RCA), giving each value the same weight. The RCAs make it possible to clearly distinguish the intensity of the rainy months (surplus) from the less rainy months (deficit). The formula that characterises them is the following:

$$Ri = (Ri - RA)/S \tag{2}$$

with Ri: rainfall of year i; PA: average intermonthly rainfall of each month over the study period; S: standard deviation of intermonthly rainfall over the same period. A month is considered to have a surplus if the ratio is greater than 1 and a deficit if it is less than 1. This analysis is accompanied by the calculation of drought severity, based on average probabilities of 20%, 30%, 50% and 80% in sequences of 5, 10, 15 and 20 days in each month. The probabilities derived from historical data, further evaluate the monthly droughts in comparison with the consumption needs of rainfed cereals, as defined by [6] in the cereal-growing wilayas of Northern Algeria.

3 Results and Interpretations

3.1 Trends in Area Sown and Cereal Yields

The average Algerian cereal area is 2.7 million hectares per year (Fig. 2). This average represents durum wheat, which dominates at the national level, with a surface area of 1.2 million ha, i.e. 48% of the total cereal area. Barley, with 885 588 ha, comes in second place (34%) compared with common wheat, which is tending to emerge during this century, with areas of the order of 645 903 ha, i.e. 17% of the total area sown. Finally, oats occupy negligible areas of 35 591 ha or 1% of the total annual cereal sown. The minimum areas sown to cereals are equivalent to 1.2 Mha, i.e. the year (1993/1994), whereas they reached the maximum value of 3.8 Mha in (2001/2002).

Moreover, the years in which the area under cereals remained below 2.5 million ha or exceeded 3.5 million ha represent only 5 and 27% of the series respectively. It should be noted, however, that the areas harvested are most often smaller than those cultivated. Indeed, the average yield of rainfed cereals is 7 q/ha according to the same Fig. 2. However, it has increased progressively over the years. The differences recorded are estimated at −22% and -5%, evolving in the direction of their production [7]. From the 1970s onwards, average yields evolved between 7 and 10 q/ha, i.e. differences of + 5% to +38%. From one year to the next, differences in harvests are recorded in relation

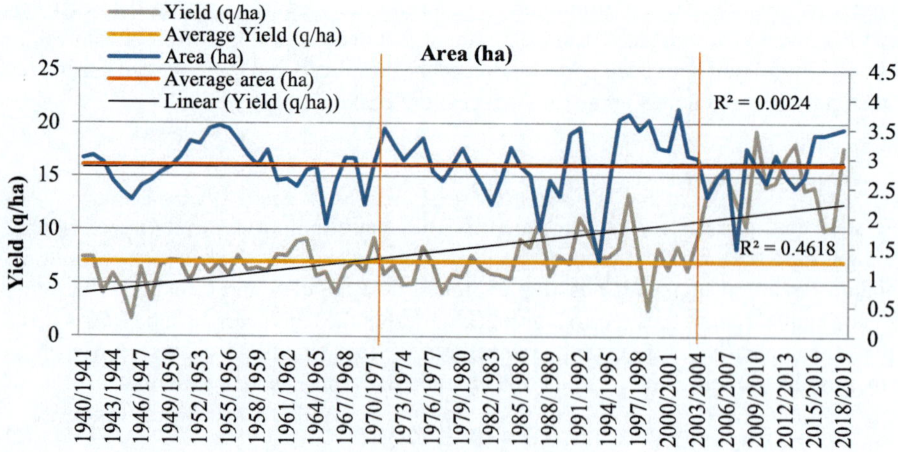

Fig. 2. Areas planted and yields of rainfed cereals in northern Algeria, over the period (1940–2020).

to the dispersed sowing of crops throughout the northern region of the country. These differences, which are rarely recorded in the statistical data, are most often attributed to climatic hazards.

3.2 Temporal Trend of Monthly Rainfall

The descriptive statistics show the average evolution of monthly rainfall quantities at the threshold of more than 0.1 mm, from September to August, over the 71-year period (Table 1). This evolution materialized by the fluctuations of the averages and medians, shows however, exceptional minimums (0 mm) and maximums which reach 140 to 576 mm representing almost 1/3 of the annual average [16]. This is the case for the years 1968/1969 for the month of September in the wilaya of Guelma in the east of the country, the years 1956/1957, 1957/1958 and 1961/1962, for the months of October, November, December and January, and the years 1937/1938, 1971/1972 in the month of March in the coastal and mountainous wilayas. The fairly high interquartile ranges (from 33.3 mm in September to 70.5 mm in December and 18.5 mm in June) indicate the monthly rainfall variability, the calculated differences of which reflect the dispersion of values around the averages. According to the coefficients of variation, the climatic variability from September to August exceeds +58% interpreting the differentiation of water inputs on a time and space scale.

Table 1. Characteristic of the statistical parameters of monthly rainfall

Months	My	Md	Min	Max	the 1900s	Wilaya	Iq	SD	Cv
S	25.5	18.7	0,0	224.7	68/69	Guelma	33.3	18.8	73.8
O	44.8	33.9	0,0	459.5	56/57	Skikda	50.0	30.5	68.0
N	58.9	44.1	0,0	521.4	57/58	Bejaia	58.7	40.1	68.0
D	76.3	59.2	0,0	551.7	61/62	Tizi	70.5	50.9	66.7
J	68.9	53.8	0,0	562.2	57/58	Tizi	64.1	43.3	62.8
F	56.5	46.2	0,0	576.8	37/38	Tizi	53.9	34.8	61.6
M	48.3	39.9	0,0	422.7	71/72	Tizi	47.1	28.5	59.0
A	39.0	31.8	0,0	385.4	52/53	Blida	45.8	22.8	58.4
M	38.0	32.3	0,0	364.6	44/45	Tlemcen	40.7	22.6	59.5
June	14.2	7.9	0,0	141.5	68/69	Tiaret	18.5	12.9	90.4
Jly	4.8	0.9	0,0	147.8			5.6	5.7	120.0
Aug	7.7	3.1	0,0	192.5			10.8	8.0	103.6

My: mean; Md: median; min: minimum; Max: maximum; Iq: interquartile range; SD: standard deviation; Cv: coefficient of variation.

3.3 Temporal Trend of Monthly Rainfall

At this time scale and at the threshold of R $\geq$ 0.1 mm, the rainfall variation from September to August is illuminated by the smoothing of rainfall with moving averages, the shape of which highlights the levels of breaks that occur during the agricultural cycles (Fig. 3a). These breaks are indicated by the arrows in the graph. According to the curve, from September onwards, with an average of 25.5 mm, rainfall at the 0.1 mm threshold begins to set in, marking the beginning of the autumn season. The average rainfall value, however, masks the minimum recorded in the wet Eastern Region (24 mm) and the dry Western Region (18 mm) and the maximum in the wet Central Region. These rainfall amounts increase rapidly in October (O) and November (N), with 45 and 60 mm. These months, which coincide with soil preparation and cereal sowing, accumulate an average of 28% during the autumn season. This percentage differs from an accumulation of 33% in the Eastern and Central Regions and nearly 19% in the Western Region. The months of December (D) and January (J) experience relatively higher rainfall. Average rainfall amounts range from 76 (D) to 69 mm (J), fluctuating between 107 and 73 mm (D), 87 and 60 mm (J) in the Eastern (ER) and Central Regions (CR), and between 46 (D) and 47 mm (J) in the dry Western Region. Rainfall decreases rapidly in February with amounts varying between 87, 60 and 42 mm from the Eastern to the Western Region, with the average in the North of the country not exceeding 57 mm. In the dry continental wilayas, February often has a relatively low total rainfall compared to the months that surround it, as in the case of the wilayas of Sétif and M'sila in the east, Djelfa in the centre and Mascara in the west. Its duration of 10% less than that of January and March would result in a lower monthly total for equal rainfall. Despite this distribution, winter rainfall, which allows tillering, records the maximum amount of water with an average percentage of 43%, i.e. 67% in the Eastern Region, 46% in the Central Region (CR) and 29% in the Western Region (WR). The months of March, April and May show a rapid decrease in rainfall amounts. The rainfall in March did not exceed 49 mm, i.e. 63 mm in RE, 53 mm in CR and 38 mm in WR. Also, like February, April is often relatively dry with 41 (ER), 43 (CR) and 34 (WR) mm, i.e. an overall average of 39 mm; whereas in May, the first heat storms sometimes provide significant rainfall, higher (45 mm in ER) or close (42 mm in CR) to that of April (41 mm: ER and 43 mm: CR). These spring rains, which coincide with fruiting and ripening of cereal grains, do not exceed 27% of the annual total, i.e. 32 (ER), 29 (CR) and 22% (WR). From the summer months of June, July and August onwards, average monthly rainfall is at its lowest in all regions of the country, marking the end of the agricultural cycle. They do not exceed 4, 7 and 5%, or an average of 2%.

The results obtained are in line with the annual rainfall [31] granting the ER, months with more rainfall than the CR followed by the OR, with reference to the average rainfall at the 0.1 mm threshold. However, calculations of rainfall over 5, 10, 20 and 30 mm over the study period, reveal genuine decreases from month to month and from region to region (Fig. 3 b, c, d, e). At the threshold of R $\geq$ 5 mm, rainfall amounts to only 18 mm, or 67% of the monthly total, whereas in March it does not exceed 23 mm, or 46% of the total. At the R $\geq$ 10 mm threshold, these percentages are 36% (September) and 23% (March). They are lowest at thresholds of R $\geq$ 20 mm and R $\geq$ 30 mm. At these thresholds, the average rainfall decreases, exceed 87% and 96%. These figures

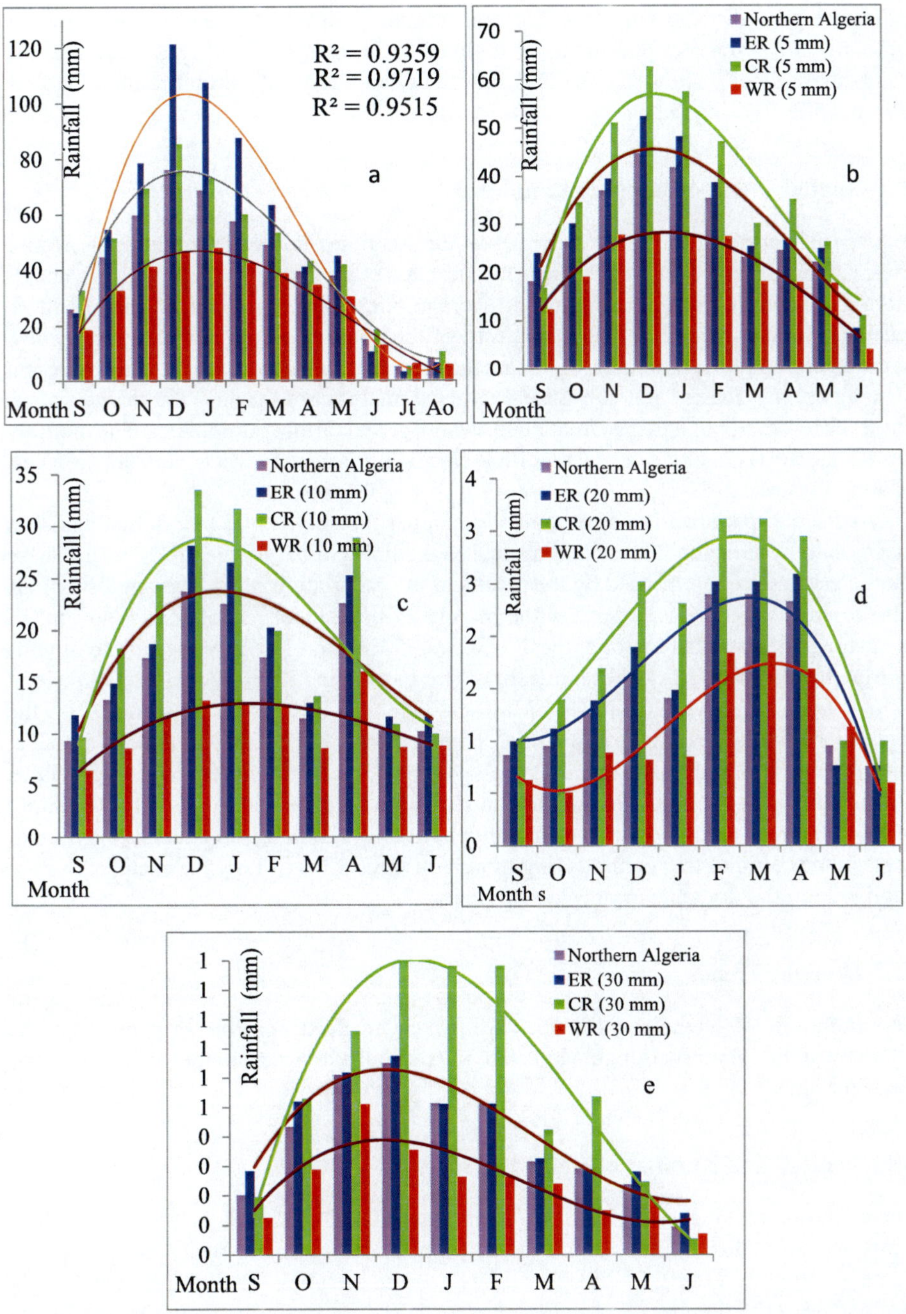

Fig. 3. Evolution of monthly rainfall above 0.1 (a), 5 (b), 10 (c), 20 (d) and 30 (e) mm in rainfed cereal growing regions

indicate a poor distribution of agricultural rainfall, followed by severe water stresses affecting cereal water requirements on a global scale. Despite these characteristics, the trend curves clearly show that the CR is more favorable to agricultural rainfall than the ER and WR.

3.4 Spatial Trend of Monthly Rainfall

The spatialization of the rains from September to June shows the impluviums whose irregular lines evolve between pluviometric classes of variable amplitudes. It shows more, the evolution of the quantities of average rains, which the rainfed cereal culture receives, monthly, from the ER to the WR, of the country. These results which evolve in the direction of the annual rains, locate the values of rains, the highest, they represent once again, the coastal and sub littoral wilayas mountainous; while the lowest values cover the lower slopes and the Great Plains where cereal farming dominates. The monthly rainfall situations indicate spatial specificities ranging from very wet to very dry surfaces, relating to cereal areas.

As for the spatialization of rainfall at the upper thresholds of 5 mm, it distinguishes the monthly characteristics of the agricultural impluviums, of the study area. These characteristics are reproduced by the results of the calculation of the average deviations with reference to $R \geq 0.1$ mm. For the months of September, October and November, the average deviations recorded, reach −38%, −50% and −35%; while for the months of March, April and May, these deviations are between −52% < March < −57%; −32% < April < −45% and −45% < May < −37%. The differences illustrated show the regression of wet areas at the expense of the extension of dry areas, on a monthly scale, over the whole territory. At thresholds above 10, 20 and 30 mm, average impluviums become increasingly low from month to month, with the lowest amounts at the threshold of $R \geq 30$ mm, marking the end of monthly agricultural rainfall. In general, the rainfall areas in the autumn, winter and spring months at thresholds of $R \geq 5$ mm and above are negligible in the dry and very dry cereal plains.

3.5 Drought Trends

The variations of Nicholson's index (1988) have made it possible to summarise the intensity of the monthly droughts linked to the rainfall averages and the number of rainy days, through.

3.6 Rainfall and Number of Rainy Days

The evolution of the number of rd (rainy days) at thresholds of $R \geq 0.1$ mm and $R \geq 5$ mm in the first three months of the year, did not exceed 11 dp in November. In Dec, this figure was highest (14 d), followed by Jan (12 d) and then Feb (10 d). Finally, in March, Apr and May the number of rd does not exceed 10–7 d, from one region to another.At the thresholds of $R \geq 10$ mm, $R \geq 20$ mm and $R \geq 30$ mm, these figures evolve between 11% and 2% from Sep to Nov, they are between 15% and 2% in Dec, Jan and Feb and do not exceed 13% and 1% in Mar, Apr and May. These results show that the number of dr is relatively more homogeneous in the dry to very dry continental wilayas than in the relatively humid coastal and sub-littoral region. These variations have highlighted the relatively wet months, with rainfall differences exceeding the average by +10%, the months close to the average and those below the average by −10%. Together with these results, Fig. 4b confirms droughts with probabilities of +80% for more than 10 and 15 days during the 30 days of each month. However, they decrease from the twentieth day (20 d) onwards. The overall results confirm the risks of rainfed cereal cultivation from the east to the west of the country.

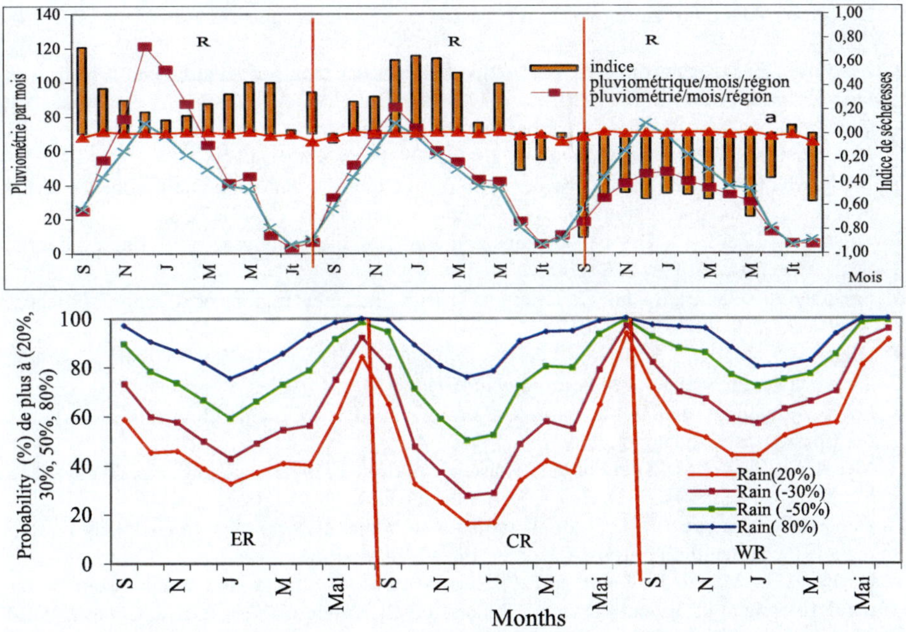

Fig. 4. Monthly drought index at thresholds greater than 0.1 mm and probabilities of dry sequences at higher thresholds of 5, 10, 20 and 30 mm, at the regional scale.

4 Conclusions

The monthly rainfall of northern Algeria over more than 70 years shows a significant aridity, manifested by progressive droughts from north to south and from east to west of the rainfed cereal-growing regions. The droughts, which are more pronounced in the western part of the country than in the eastern part, are linked to the decrease and random distribution of average agricultural rainfall (above 5 mm) and the reduced number of rainy days. These characteristics, which translate into strong water deficits ranging from +33% to +70% during the autumn and spring months, highlight the importance of considering water shortages in relation to the dates of sowing and the yield components of the cereal crop. The results of this research highlight, on the one hand, the vulnerability of cereal crops to produce adequately, in this context of climatic variability and change, over large areas. On the other hand, they show the interest in studying rainfall in association with other production factors (temperatures, sown varieties, agricultural practices, lack of know-how), for solutions to improve and sustain rainfed cereal growing in the Algerian regions.

References

1. Chalabi, A.: L'Algérie Antique (2001). www.iquebec.com
2. Lancel, S.: Algeria in Antiquity. From Massinissa to Saint Augustine. Edi. Mengès (2004). 260 p
3. Rouverou, P.: Statistics of cereal production in Algeria. Cereals of Algeria. Gouv, Gen, Alg, Direct, Agric, Colon, 2–58 (1930)
4. Gomez, G.: History of Algeria. Chronology from prehistory to independence (2005)
5. Beaumont, M.: Wheat. Ed. Que sais-je, n ° 22 109, Paris (1949). 127 p
6. Baldy, Ch.: Contribution to the frequency study of climatic conditions. Their influence on the production of the main cereal zones of Algeria. Report, ITGC (1974). 72 p
7. Smadhi, D., Zella, L.: Dry cereal crops and annual rainfall: the case of Northern Algeria. Secheresse Rev. **20**, 199–203 (2009)
8. Smadhi, D., Zella, L., Bachir, H.: Droughts in semi-arid cereal regions of Algeria. J. Fundam. Appl. Sci. **9**(2), 1063–1073 (2017)
9. Gibbs, W.J., Maher, J.V.: Rainfall deciles as drought indicators. Bulletin No. 48; Australia: Commonwealth Bureau of Meteorology (1967)
10. DRE: Mitigating the economic impacts of drought in Africa. Drought (2009). 1p http://www.nord-pas-de-calais.ecologie.gouv.fr
11. Meddi, H., Meddi, M.: Study of the persistence of drought in seven Algerian plains by using Markov chains (1930–2003). Rev. Courrier du savoir **9**, 39–48 (2009)
12. Hills, R.C., Morgan, J.H.T.: Rainfall statistic: an interactive approach to analysing rainfall records for agricultural purposes. Exp. Agric. **17**, 1–16 (1981)
13. Davey, E.G., Mattei, F., Soloman S.I.: En evaluation of climate and water resources for the development of agriculture in tne Sudano-sahelian zone of West Africa. Geneva: Wold Meteorological Organization (1976). 9 p
14. Sivakumar, M.V.K.: Duration and frequency of dry periods in West Africa. ICRISAT. Bull. inf. n°13 (1991). 181 p
15. Seltzer, P.: The climate of Algeria. Algiers (1949). 219 p
16. Smadhi, D.: Regionalization and agroclimatic modelling in rainfed cereal growing. Case of northern Algeria. These Doctorate, Agronomic Sciences. ENSA (2011). 180 p
17. Bachir, H.: Analysis and mapping of climatic parameters (rainfall, temperature) and impacts on the delimitation of rainfed cereals in the High Plateaux of Eastern Algeria. Doct. thesis, ENSA (2018). 100 p
18. Kourat, T.: Evaluation of climate change and impact on rainfed durum wheat production in the eastern high plains of Algeria. Doct. thesis, ENSA (2021). 218 p
19. Bachir, H., Kezouh, S., Ait-oubelli, M., Semar, A., Smadhi, D., Ouamer-ali, K.: Improvement of interpolation using information from rainfall stations and comparison of hydroclimate changes (1913–1938)/(1986–2016). Al-Qadisiyah J. Agric. Sci. **11**(1), 54–67 (2021)
20. Bachir, H., Semar, A., Mazari, A.: Statistical and Geostatistical analysis related to geographical parameters for spatial and temporal representation of rainfall in semi-arid environments the Case of Algeria. Arab. J. Geosci. **9**(7), 486–498 (2016). https://doi.org/10.1007/s12517-016-2505-8

21. Heathcote, R.L.: Drought perception. The Environmental, economic and social significance of drought (Lovett, J.V., ed.), Sidney, Australia: Augus and Robertson, 17–40 (1973)
22. El Hassani, T.A.: Water, wetlands and climate change. Drought preparedness and risk management in the Mediterranean region. Drought working paper. College of agriculture and natural resources. Inst. Agro. Vet. Hassan, Rabat. (2008). 3p
23. Doukpolo, B.: Procedure for processing climate observation, simulation and projection data. Final report of the institutional support project to African climate institutions ISACIP/AFRICLIMSERV. (2013). 33p
24. Krige, D.G.: A Statistical Approach to some mine valuation and allied problems on the Wit waters rand. Master thesis, University of Witwatersrand (1950)
25. Matheron, G.: Treatise on Applied Geostatistics. BRGM briefs. Ed. Tech, Paris (1962)
26. Meylan, P.: Regionalization of data marred by measurement errors by kriging. Application to rainfall. Inst. GR, Hyd. Landscaping. EPF., Lausanne (Switzerland). hydro., continent., I, 1, 25–34 (1980)
27. Enrich: Harmonization of climate prediction for mitigation of global change impact in Sudano-Sahelian West Africa Climag-west Africa a network for: EESD-ENV-99/ENRICH (European Network for Research on Global Change), Deliverable n ° 11 (2001). 22p
28. Gratton, Y.: Kriging: the optimal method of interpolation. INRS, Eau Terre Environnement, Canada (2002). 123 p
29. Marand, C., Zumstein, J.F.: Modeling of mean annual precipitation applied to the Vosges Mountains. Hyh. Cordnent ial. **5**(1), 29–39 (1990)
30. Nicholson, S.E., Kim, Hoopingarner, J.: Variability of African an dits interannual rainfall. Dept., Mateo., Tallahassee, Florida, USA (1988). 237p
31. Smadhi, D., Zella, L., Amirouche, M., Bachir, H.: Rainfall trends in semi-arid cereal regions of Algeria. Al-Qadisiyah J. Agric. Sci. **11**(1), 36–44 (2021)

Power System Reliability Assessment Considering Impacts of Climate Change

Mohammadreza Gholami[1](✉) and Parvaneh Esmaili[2]

[1] Princeton University, Princeton, NJ 08544, USA
welman@princton.edu
[2] Springer Heidelberg, Tiergartenstr. 17, 69121 Heidelberg, Germany

Abstract. Power system reliability, as one of the most significant issues in power system studies, is affected remarkably by load demand changes. On other hand, climate changes and global warming lead to increasing the electricity demand of a power system. In this paper, the impacts of global change on generation power system reliability indices have been investigated. Loss of Load Probability and Expected Energy Not Supply are considered as power system reliability indices. In addition, a Particle Swarm Optimization method is used to assess these reliability indices. IEEE_79 Reliability test system is selected as a standard test system. The results show that reliability indices are affected noticeably by temperature rising and climate change.

Keywords: Climate change · Power system reliability indices · Loss of Load Probability · Expected Energy Not Supplied · Load demand · Particle Swarm Optimization

1 Introduction

Power system reliability indices are used as the most important constraints by power system planners. As it is shown in Fig. 1, the assessment of system reliability is applied to three main hierarchical levels, termed HLI, HLII, and, HLIII. At the first level, generation system reliability, the total system generation is investigated to determine its ability to meet the total system demand requirements. At the generation system level, the transmission lines are ignored and considered as completely reliable elements with no failure rate. At the second level, composite power system or bulk system, both generation units and transmission lines are evaluated and the transmission system elements are considered completely reliable. All three parts and elements of a power system (generation units, transmission lines, and transmission system elements) are considered in HLIII studies. The three hierarchical levels are shown in Fig. 1. In this paper, generation system reliability (HLI) is assessed. The reliability of a system is evaluated using the proper indices. In this paper, LOLE (Loss of Load Probability) and EENS (Expected Energy not Supplied), as the two common reliability indices, are chosen and calculated. Obviously, reliability indices are affected significantly by changes in load demand and the LDC (Load Duration Curve) pattern of a network. Power system reliability can be

H. Gökçekuş and Y. Kassem (Eds.): NRSEM 2021, EESCI, pp. 252–261, 2022.
https://doi.org/10.1007/978-3-031-04375-8_29

improved by both enhancing the performance and efficiency of generation, transmission, and distribution elements, and applying demand-side management methods, and improving load demand patterns. In another word, reducing power consumption leads to a more reliable system without making changes in a power system.

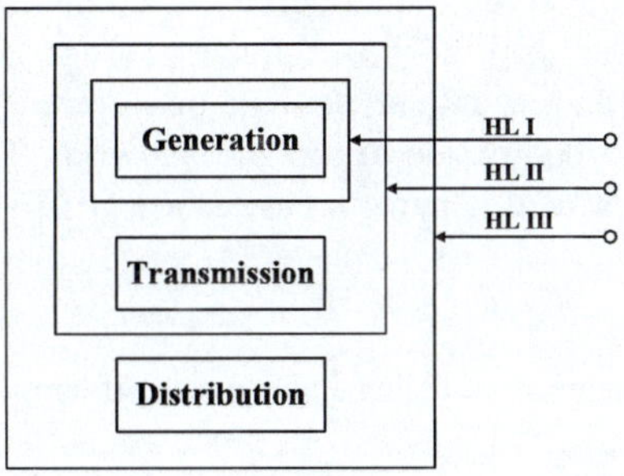

Fig. 1. Power system Hierarchical Levels (HL)

On the other hand, climate changes noticeably affect the load demand of a power system [1–4]. However, the impacts of climate change are not limited to rising ambient temperature and can lead to changes in many parameters such as wind speed, humidity, and many other factors. Although all these parameters are important in power system studies, the previous studies have shown that the temperature has more effects on load demand compared to other variables. In many researchers, the effect of rising temperatures on load demand is investigated using different methods. To determine the relationship between temperature rising and load demand, neural networks have been used widely. Neural networks are powerful techniques and are able to solve complex relations. However, they require a huge amount of information for training and most authors prefer to use simpler methods. For this, a simpler regression approach was proposed by Linder. His method uses regression models to connect demand and temperature based on daily and monthly periods. In this paper, the impact of global warming and increasing the temperature on generation power system reliability indices due to changes in load demand is investigated using the regression method.

The probabilistic adequacy reliability assessment methods can be categorized as analytical and simulation-based methods (Monte Carlo Techniques). Due to their shortcomings, many researchers have used hybrid and meta-heuristic algorithms to calculate the power system reliability indices. In [5–11] generation and composite power system, reliability indices are assessed using the Genetic Algorithm (GA) as a search tool to sample the system states. Also, [12–18] uses other optimization methods to reduce the amount of sampled system states, computational time and achieve better coverage. By using meta-heuristic methods, not only most probable failure states and contribution of each state and system elements at a given load can easily be achieved, but also, computation time is decreased and parallel computation can be done. In this paper, a particle swarm optimization algorithm is used to calculate the generation reliability indices. The algorithm is tested on the standard reliability test system RTS-79 consists of 32 generating units.

The effects and impact of climate change on electricity demand have been analyzed in many papers. The researchers have investigated these impacts on both short-term and long-term periods. Also, the investigations show that temperature has the most effective factor among other parameters such as wind speed and humidity. Authors in [27] have used the neural network as the main method to relate the temperature and load demand. Also, in [28, 29], and [2] the regression models are used instead of neural networks. In this paper, we have used the result of the reference [2] regarding the effect of rising 1 centigrade degree on peak and mean load demand on different seasons. The changes in peak load demand due to 1 centigrade degree are shown in Table 1, which is applied to the load demand amount of this paper's case study (IEEE-RTS_79 Reliability test system).

Table 1. Changes in peak demand due to 1° of temperature rise.

Demand	Winter	Summer	Monsoon
Changes on peak demand %	4.2	4.6	2.8

The previous papers have proposed algorithms for assessment of reliability of power system and investigated the changes in load demand due to climate change separately. In this paper, the impact of global warming on generation reliability indices is analyzed.

2 Power System Reliability Adequacy Assessment Methods

Generally analytical (including both discrete and continuous methods) and simulation-based (Monte Carlo Techniques) are the main two basic methods of system adequacy assessment. The main differences between these methods are related to the process of selecting system states and calculating reliability indices.

In analytical techniques, the system is represented using mathematical models and the reliability indices are calculated by solving the equations. The state space, contingency enumeration, and minimal cut set are the most commonly used analytical methods.

Simulation-based techniques, generally termed Monte Carlo Simulation (MCS) methods, solve the problems using random variables. These methods are widely used by researchers for the evaluation of power system reliability. MSC techniques are iteration based and results of all iterations are converted to a distribution function. Then the reliability indices are calculated by the achieved functions.

MCS methods are able to collect information about the both mean value and probability distribution of the reliability indices. In these methods, contingencies with higher probability are more likely to select and maybe simulated several times. Monte Carlo Simulation methods are divided into two main categories: Non-Sequential and Sequential Techniques [19–26].

In analytical methods, the system states are selected in terms of different contingency levels. This selecting process ends while a specific stop criterion is reached. The stop criteria can be a particular element outage numbers or when the probability of a selected state is less than a threshold value. Finally, indices are calculated using mathematical formulation according to the evaluation result of the selected states. Simulation-based techniques select the system states based on the random failure behavior of system elements and the states with a higher probability of occurrence are more likely to select. The stop criterion in these methods can be a specific number of simulations or other stopping rules. Finally, the reliability indices are calculated by averaging the indices obtained from each simulated state.

Both methods have their own advantageous and disadvantageous. The computational time of reliability evaluating is much less independent from the reliability level of the system. One important thing worthwhile to be considered is that in simulation-based methods, the number of selected states increases remarkably by increasing the system reliability level. Also, the outage of a component in a power system may be affected by the outage of other components and simulation methods cannot handle this in simulation process of simulation based methods. Another advantage of analytical methods is that selected states are independent of the system load curve while in simulation methods, the process should be done separately for each selected state.

On other hand, analytical methods have some shortcomings. Many simplifying assumptions are needed in using these methods while the effect of these approximations is not clearly known. Furthermore, they are proper for reliability assessment of small systems, and simulation methods have been proposed to be used in large and complex systems considering the behavior of their random components.

Because of all disadvantages mentioned above, meta-heuristic algorithms have been widely used to assess and calculate the reliability indices. They are able to calculate the reliability indices using evaluating fewer states of the system with acceptable accuracy. Also, more information of the system such as the most failure state and the most probable failure state can be determined easily using these methods. In this paper, a Particle Swarm Optimization algorithm is used to calculate the reliability indices of the test system.

3 Algorithm Approach

In this section, the reliability definitions and formulas are given first. Then it is described how to calculate the generation reliability indices by PSO in detail.

3.1 Reliability Definitions

Every generation unit is considered as an element and has its own Failure (μ) and Repair (λ) rate. The Force Outage Rate (FOR) parameter is determined for each generation unit based on Eq. 1:

$$FOR_i = \frac{\lambda_i}{\lambda_i + \mu_i} \tag{1}$$

This parameter is used to calculate the availability of a generation unit (Eq. 2). Also, each generation unit is considered to have two statuses: (1 = on and 0 = off). Then a probability value (PS) is calculated for each state of the system depending on the unit's FOR and status (Eq. 3).

$$Availibility\ of\ unit\ i = \begin{cases} FOR\,i\ if\ related\ binary\ is\ 0 \\ 1 - FOR\,i\ if\ related\ binary\ number\ is\ 1 \end{cases} \quad (2)$$

$$PS_j = \prod\nolimits_{i=1}^{ng} Availibility\,j \quad (3)$$

where:

Ng = number of generation units.

As mentioned above, a system state is shown by generation unit's status. The state of the system in which all units are in upstate is shown in Table 2.

Table 2. The sample status of the system.

Capacity MW	Status	Capacity MW	Status
115	11	400	11
	11		11
	11	50	11
	11		11
76	11		11
	11		11
	11		11
	11		11
197	11	20	
	11		11
	11		11
100	11		11
	11		11
	11		
12	11		
	11		
	11		
	11		
	11		

The total generated power of a system state is calculated by Eq. 4. The total generated power is compared to hourly load amount. If the generated amount bigger, it is considered a successful state. Otherwise, it is considered a failure state and will be used to calculate reliability indices.

$$G_i = \sum\nolimits_{j=1}^{ng} g_j * b_j \quad (4)$$

where G_i is the generated power of state i., ng is the number of generation units., g j is the capacity of unit j, b j is a binary number equals to 1 if the unit is on and equals to 0 if the unit is off.

There are 2^{ng} possible states for a system that are evaluated in analytical methods. The most two commonly used indices are Loss of Load Probability (LOLE) and Energy Expected Not Supplied (EENS). These indices are calculated by Eqs. 5–8 for the system:

$$LOLE_{system} = \sum_{i=1}^{8760} LOLE_{Loadi} \tag{5}$$

$$LOLE_{Loadi} = \sum_{i=1}^{k} PS_i \tag{6}$$

$$EENS_{system} = \sum_{i=1}^{8760} EENS_{Loadi} \tag{7}$$

$$EENS_{Loadi} = \sum_{i=1}^{k} PS_i * (Load_i - PG_i) \tag{8}$$

where 8760 is number of hours in a year (for RTS_79, 8736 h is given in a year), k is number of failure state at load I, PS_i = probability of sampled failure state i, $Load_i$ = amount of load at hour i, PG_i = power generated at failure state i.

3.2 Reliability Indices Calculation Using PSO

PSO, used frequently by researchers to solve complicated problems, is used to lower the computation burden of the system reliability calculation by searching in the power system possible states and choosing the most probable one. The number of possible states of a system with n generation units is 2^n, while PSO is able to search and find the most probable ones, save them and calculate the reliability indices by saved sampled states. In the PSO algorithm, the swarms are updated based on Eqs. 9 and 10:

$$v_i(t) = v_i(t-1) + \rho_1 \left(x_{boss} - x_{i(t)}\right) \tag{9}$$

$$x_i(t) = x_i(t-1) + v_i(t) \tag{10}$$

where, x_i (t) presents i_{th} swarm, and x_{boss} is the position of the best swarm.

Therefore, the fitness function of the algorithm is defined as PS (probability of a system state). The first collections of states are generated randomly, but the others are selected intelligently according to the PSO operations. Each system state is shown by a swarm in PSO algorithm. The algorithm ends when it reaches the stop criteria which can be a specific number of iterations. The number of proper iterations can be determined based on the size of the problem or the number of the generation units. Finally, reliability indices can be calculated based on the information of saved states' information and the LCD (Load duration Curve). The algorithm steps are given below:

1. Save the power system reliability information (the information includes the number of generation units, each unit power output in MW, and failure and repair rate of each generation unit).

2. Select PSO parameters (including population size, Iterations, C1 and C2 parameters).
3. Save Load Duration Curve information (hourly load demand for a period of one year).
4. Randomly generate the first population consisting of swarms (each swarm represents a system state).
   – While current iteration < maximum iteration do:
5. Calculations of fitness function for each swarm
6. Swarms moving toward their best historical information and the best swarm of the current generation.
7. Evaluating the states (to evaluate states as a failure a success compared to maximum load amount).
8. Saved failure states in an array.
9. Back to item 5.
10. Calculate the LOLE and EENS as generation reliability indices.

4 Case Study and Results

In this paper, The RTS-79 is chosen as a standard reliability test system. The test system includes 32 generation units. The capacities of units are from 12 to 400 MW as smallest and largest respectively. The sum of generated capacities is 3405 MW, while the maximum load is considered 2850 MW. Other necessary information such as unit sizes, number of units, forced outage rates, and hourly load demand is given. In the first step, the reliability indices of the system, based on the original load demand information and without considering the temperature rising, are calculated by both an analytical method (unit addition algorithm) and PSO as a meta-heuristic algorithm. The results are given at Table 3.

Table 3. Reliability indices of the test system without considering the temperature rising.

Reliability index	Unit addition algorithm	PSO algorithm
LOLE	9.355	9.343
EENS	1168	1164.5

In the next step, the increases in load hourly demands are considered based on the 1-degree temperature rise in different seasons. Then the reliability indices are calculated using the PSO algorithm and compared to the original values calculated by the same algorithm. The results are shown in Table 4. Finally, the reliability indices, both LOLE and EENS, are calculated again considering 2° of temperature rise. The results are given in Table 5.

Table 4. Reliability indices of the test system with considering the 1° temperature rising.

Reliability index	Original value	Considering 1° temperature rising
LOLE	9.343	19.9441
EENS	1164.5	2679.3

Table 5. Reliability indices of the test system with considering the 2° temperature rising.

Reliability index	Original value	Considering 2° temperature rising
LOLE	9.343	39.3824
EENS	1164.5	5744.8

The results show that global warming and the temperature rising have remarkably decreased the reliability of the power system. In other words, global warming causes more blackouts and energy loss in a power system. The LOLE index (showing the number of blackout hours in a year), is increased by 10.6 and 30.03 h in a year by increasing the 1 and 2° of temperature respectively. In addition, the amount of not supplied energy has been increased 1514.8 and 4580.3 MWh in a year due to increasing the 1 and 2° of temperature.

5 Conclusion and Suggested Work

In this paper, the impacts of global warming and temperature rise as one of the most important effects of climate change are investigated on the reliability of a power system. Two LOLE and EENS indices have been chosen to represent the reliability of a system. Also, and Particle Swarm Optimization was used to calculate the power system reliability indices. The standard RBTS-79 reliability test system was considered as the case study. The increase in hourly load demand was considered as the impact of global warming. The reliability indices were calculated considering the 1 and 2° of temperature rise. The results show that global warming has a noticeable negative effect on system reliability by increasing the power blackout hours and not supplying energy amounts. Furthermore, an investigation of the effects of global warming on other power system factors is suggested by the authors. The effect of temperature rise on the failure rate of power system elements such as generation units and transmission lines will be considered in future works.

References

1. Tung, N.X., Dat, N.Q., Thang, T.N., Solanki, V.K., Anh, N.T.N.: Analysis of temperature-sensitive on short-term electricity load forecasting. In: 2020 IEEE-HYDCON, pp. 1–7. IEEE (2020)
2. Parkpoom, S., Harrison, G.: Analyzing the impact of climate change on future electricity demand in Thailand. IEEE Trans. Power Syst. **23**, 1441–1448 (2008). https://doi.org/10.1109/TPWRS.2008.922254

3. Gasca, M.V., Bueno-López, M., Ibáñez, F., Pozo, D.: Ambient temperature impact on the aggregated demand response flexibility in microgrids. In: 2021 IEEE Madrid PowerTech, pp. 1–6. IEEE (2021)
4. Samaan, N., Singh, C.: Using of genetic algorithms to evaluate frequency and duration indices for generation system reliability. In: Proceedings of the ISAP 2001 Intelligent System Application to Power System Conference, pp. 251–256 (2001)
5. Samaan, N., Singh, C.: An improved genetic algorithm based method for reliability assessment of generation system. In: Proceedings of the MEPCON 2001 8th International Middle-East Power Systems Conference, pp. 235–242 (2001)
6. Samaan, N., Singh, C.: Adequacy assessment of power system generation using a modified simple genetic algorithm. IEEE Trans. Power Syst. **17**(4), 974–981 (2002)
7. Samaan, N., Singh, C.: A new method for composite system annualized reliability indices based on genetic algorithms. In: IEEE Power Engineering Society Summer Meeting, vol. 2, pp. 850–855. IEEE (2002)
8. Samaan, N., Singh, C.: Using genetic algorithms for composite system reliability indices considering chronological load curves. In: Proceedings of Intelligent Systems Application to Power Systems, ISAP (2003)
9. Samaan, N., Singh, C.: Genetic algorithms approach for the evaluation of composite generation-transmission systems reliability worth. In: 2003 IEEE PES Transmission and Distribution Conference and Exposition (IEEE Cat. No. 03CH37495), vol. 1, pp. 113–119). IEEE (2003)
10. Samaan, N., Singh, C.: Genetic algorithms approach for the assessment of composite power system reliability considering multistate components. In: 2004 International Conference on Probabilistic Methods Applied to Power Systems, pp. 64–69. IEEE (2004)
11. Green, R.C., Wang, L., Alam, M., Singh, C.: Intelligent state space pruning using multi-objective PSO for reliability analysis of composite power systems: observations, analyses, and impacts. In: 2011 IEEE Power and Energy Society General Meeting, pp. 1–8. IEEE (2011)
12. Benidris, M., Elsaiah, S., Mitra, J.: Power system reliability evaluation using a state space classification technique and particle swarm optimisation search method. IET Gener. Transm. Distrib. **9**(14), 1865–1873 (2015)
13. Green, R.C., Wang, L., Alam, M., Singh, C.: State space pruning for reliability evaluation using binary particle swarm optimization. In: 2011 IEEE/PES Power Systems Conference and Exposition, pp. 1–7. IEEE (2011)
14. Benidris, M., Mitra, J.: Composite power system reliability assessment using maximum capacity flow and directed binary particle swarm optimization. In: 2013 North American Power Symposium (NAPS), pp. 1–6. IEEE (2013)
15. Gholami, M., Sharifi, R., Radmanesh, H.: Development of composite power system effective load duration curves by using a new optimization method for assessment composite generation and transmission reliability. Int. J. Power Energy Res. **1**, 1 (2017)
16. Gholami, M.R., Hoseini, S.H., Taheri, M.M.: Assessment of power composite system annualized reliability indices based on improved particle swarm optimization and comparative study between the behaviour of GA and PSO. In: 2008 IEEE 2nd International Power and Energy Conference, pp. 1609–1612. IEEE (2008)
17. Billinton, R., Sankarakrishnan, A.: A comparison of Monte Carlo simulation techniques for composite power system reliability assessment. In: IEEE WESCANEX 1995. Communications, Power, and Computing. Conference Proceedings, vol. 1, pp. 145–150. IEEE (1995)
18. Li, W.: Reliability Assessment of Electric Power Systems Using Monte Carlo Methods. Springer, Heidelberg (2013)

19. Melo, A.C.G., Pereira, M.V.F., Da Silva, A.L.: A conditional probability approach to the calculation of frequency and duration indices in composite reliability evaluation. IEEE Trans. Power Syst. **8**(3), 1118–1125 (1993)
20. Pereira, M.V., Balu, N.J.: Composite generation/transmission reliability evaluation. Proc. IEEE **80**(4), 470–491 (1992)
21. Bhuiyan, M.R., Allan, R.N.: Modelling multistate problems in sequential simulation of power system reliability studies. IEE Proc.-Gener. Transm. Distrib. **142**(4), 343–349 (1995)
22. Sankarakrishnan, A., Billinton, A.: Effective techniques for reliability worth assessment in composite power system networks using Monte Carlo simulation. IEEE Trans. Power Syst. **11**(3), 1255–1261 (1996)
23. Billinton, R., Chen, H., Ghajar, R.: A sequential simulation technique for adequacy evaluation of generating systems including wind energy. IEEE Trans. Energy Convers. **11**(4), 728–734 (1996)
24. Jayatheertha, H.J., District, A.: Evaluation of composite electric system performance indices using sequential Monte Carlo simulation. Int. J. Adv. Eng. Res. Stud. **1**, 4–7 (2012)
25. Mello, J.C.O., Da Silva, A.L., Pereira, M.V.F.: Efficient loss-of-load cost evaluation by combined pseudo-sequential and state transition simulation. IEE Proc.-Gener. Transm. Distrib. **144**(2), 147–154 (1997)
26. Li, X., Sailor, D.J.: Electricity use sensitivity to climate and climate change. World Resour. Rev. **7**(3) (1995)
27. Linder, K.P.: National impacts of climate change on electric utilities. The Potential Effects of Global Warming on the United States, Environmental Protection Agency, Washington, DC (1990)
28. Hor, C.L., Watson, S.J., Majithia, S.: Analyzing the impact of weather variables on monthly electricity demand. IEEE Trans. Power Syst. **20**(4), 2078–2085 (2005)

Representation of Rainfall in Regions with a Low Distribution of Rain Gauging Stations

Bachir Hakim[1(✉)], Etsouri Salim[2(✉)], Smadhi Dalila[3], and Semar Ahcène[2]

[1] National Institute of Agronomic Research, El Harrach, Algeria
akm7.62@hotmail.fr
[2] Higher National School of Agronomy, Oued Smar, Algeria
Salim.etsouri@edu.ensa.dz, a.semar@ensa.dz
[3] INRAA, El Harrach, Algeria

Abstract. Random climate variables are constraints that can directly or indirectly affect economic performance. They limit the interventions of decision-makers, whether at the macroeconomic or microeconomic level, and even more so when it comes to elaborating development strategies. Classical methods of representing climatic parameters are giving way to computerised means and the tools currently available allow the processing of considerable masses of observation data and their rapid representation. In our study, we are interested in the climatic parameter "precipitation" in the North of Algeria as a case study. This region is of particular interest because it is subject to a high risk of flash floods due to a rainfall and hydrological regime characterised by a strong intermittence of rainfall and flows. To this end, we will propose an approach to represent rainfall with very low uncertainty, based on the multiple linear regression approach and taking into account other parameters such as altitude, longitude and latitude. The method is best suited for mid-mountain regions where relief plays a particular role in the occurrence of precipitation. It also allows a better estimation of precipitation in the space-time plane and the determination of the structure of the rainfall gradient in an area poorly covered by weather stations. The results obtained show that the proposed approach is a useful tool that could be easily applied to other regions with a sparse climate network. It could also be used for other climatic parameters needed for climatological, environmental, agronomic or other studies.

Keywords: Rainfall · Multiple linear regression · North of Algeria · Geographical parameters

1 Introduction

The random climatic variables therefore constitute constraints that can affect either directly or indirectly the performance of agriculture. By their very nature, they limit the interventions of decision-makers, whether at the macroeconomic or microeconomic level [1]. Climate conditions both the distribution of plant species and their behaviour in terms of growth and reproduction [2]. In this context, in order to highlight the characteristics of climate at a regional scale, especially in semi-arid areas characterised by variable climatic conditions, it is important to study climate statistically, while considering it as

H. Gökçekuş and Y. Kassem (Eds.): NRSEM 2021, EESCI, pp. 262–271, 2022.
https://doi.org/10.1007/978-3-031-04375-8_30

an intrinsic factor with defined and uniform characteristics [3]. Rainfall and temperature fluctuate in space and time. Their values and distribution are affected by many factors, including geographical factors such as longitude, latitude, altitude, distance from the sea, seasonal variations such as air humidity movement, temperature, air pressure and topography. Many studies have been conducted to correlate these variables with the above factors based on mathematical modelling (multiple regression, interpolations, etc.), as an alternative to improve the estimation using data provided by the available meteorological measuring stations [4–8].

Precipitation could be incorrectly estimated if topographical variables influencing precipitation are not taken into account [9]. However, the maximum cumulative precipitation does not necessarily coincide with the highest altitude [10]. It is often the case that the variation in a dependent variable is explained by the action of several explanatory variables, when the relationship between these variables is linear. Multiple linear regression (MLR) is a mathematical method that models the direct linear relationship between a dependent variable and one or more independent variables [11, 12]. It has been widely used in the assessment of climate variables [13]. In regions where climate factors are significantly correlated with topography [14], independent variables typically include station position and elevation. Multiple regressions are also used and applied to analyse and explain orographic rainfall) [15] taking into account topographic variables. Multiple regression can be used for several purposes: (i) to find the best linear prediction equation (model) and to evaluate its accuracy and significance; (ii) to estimate the relative contribution of two or more explanatory variables on the variation of a variable to be explained; to detect the complementary or, on the contrary, antagonistic effect between various explanatory variables; (iii) to judge the relative importance of several explanatory variables on a dependent variable in relation to a causal theory; (iv) to evaluate the relative importance of several explanatory variables on a dependent variable in relation to a causal theory. dependent variable in relation to a causal theory underlying the research.

One of the main concerns of researchers is to provide decision-makers with tools to better adapt future sustainable development and natural resource management strategies. The importance of analysing some climatic parameters in different parts of Algeria has been addressed in previous studies. These climatic parameters have been developed.

In this work, we found it interesting and useful to focus more on one climatic parameter, namely rainfall. A study was therefore undertaken in order to define the gradient as well as the factors that influence the spatial variation of monthly rainfall. In addition, we aim to analyze to what extent it is possible to interpret the selected rainfall patterns within a framework provided by classical meteorological data.

2 Material and Methods

2.1 Study Area

The study area is located in the eastern high plateau regionof Algeria, known for its predominantly semi-arid bioclimatic affiliation. It stands between 4.2° to 8.3° latitude North and 35.00° to 36.6° longitude East, extending over an area of 33,610 km^2 (Fig. 1). It is part of the Alpine orogen, which forms the backbone of the relief of the whole of northern Algeria. It is separated from the African platform by the South Atlas geological Accident. The relief offers much more open horizons, the sedimentary cover less thick and more discontinuous. Several chains of mountains naturally limit the study area. To the North, the Atlas chain is including the mountains of Constantine and Sidi-Dris with the highest points reaching1285 and 1363 m, respectively. In the North West is the Djurdjura Mounts with a culminating point reaching 2308 m (Lala Khedidja crest). To the West is the Bibans with the relatively high points (Takoucht Mont, 1900 m, and Megress Mont, 1737 m), the highest point reaching 2000 m at Babor Mont. To the east are the mountains of Tébessa (Doukhane and Bou-Roumane Monts) with the highest points reaching 2249 and 2250 m, respectively. To the South, the Saharan Atlas chain includes the Aurès Mountains, Mounts with culminating point's approaching1550 m. Moreover, the study area is dominated by plains with very low slopes (0–5%). The main plains are those of Bordj Bou Arreridj, Setif, Oued El Othmania and El Khroub at Constantine province. Furthermore, other plains occupy the area, plains of Mila in the north and plains of Touffana and Batna in thesouth [16].

2.2 Data Analysis and Approach Methodology

The data is provided directly by the weather stations that form the observation network in the study area. When only professional stations in aerodromes measure all climate parameters. The requirement to take into account the temporal and spatial variability of climate data and the limitations of available rainfall stations (Fig. 1), contributed to the choice of an approach to rainfall modelling in its spatio-temporal dimension. The approach considers the rainy field as a realization of a random process and the modelling of the fields seeks to reproduce the phenomenon, respecting the observed statistical and geometric properties.

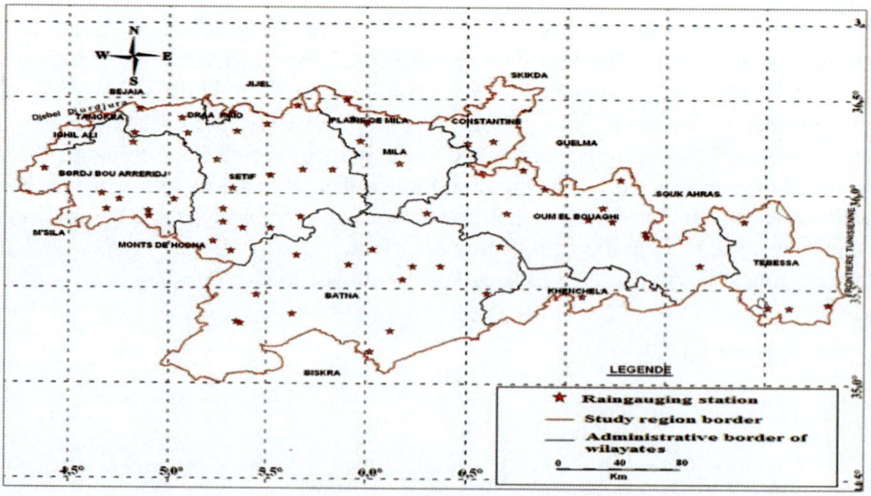

Fig. 1. Spatial location of selected rain gauging stations in the study area.

Since it should be noted that there is variability in monthly rainfall amounts, we described the rainfall data individually and assigned the same numerical codes to the stations in the study area to code, them as was done in the work of Bachir et al. 2021 [17]. This issue will allow for practical presentation of the stations in graphs and ease of interpretation.

The authors have put in place a reflection on the methodology and the tools that must be used to estimate the monthly precipitation parameter in the study area following a very specific framing for the period (1986/2016). To this end, we have adopted a methodology wich consists in expressing the rainfall variable (P) from a multiple regression equation to three (03) simple type factors.

$$P = \beta_1.X + \beta_2.Y + \beta_3.Z + \beta_0 + \varepsilon \quad (1)$$

With:

P: represents rain and is considered as a dependent variable.

β1, β2 and β3 are the coefficients of the multiple linear regression respectively of the independent variables X, Y and Z.

We also note, that X, represents the longitude, Y the latitude and Z the Altitude. The parameter 'β0' represents intercepts and errors. However, this step required the establishment of a reflection on the methodology and tools that must be used to determine the parameters of the equation. Bachir et al. 2021 [17], develop this demarche using same rain gauging stations. The author in this paper want to develop and monthly aspect of the rainfall.

3 Results and Discussion

3.1 Interpretation of the Results of the Multiple Linear Regression Application

The representations of the monthly rainfall classes obtained show the number of stations with close average rainfall. Furthermore, it is interesting to explain the behavior of monthly rainfall according to the combined effect of longitude, latitude and altitude as a function of these variables, using the different components of the multiple regression equations; coefficients, intercept and correlation coefficient (Table 1 and Fig. 2).

Table 1. Presentation of components of the multiple regression equations per months

	January	February	March	April	May	June
Longitude	3,51	1,25	1,57	−0,85	2,36	4,62
Latitude	46,42	44,04	19,14	26,53	15,48	2,27
Altitude	0,8 10^{-3}	−6,2 10^{-3}	1,6 10^{-3}	3,5 10^{-3}	17,9 10^{-3}	3,8 10^{-3}
Intercept	−1647,27	−1552,30	−666,95	−915,64	−547,82	−95,03

(continued)

Table 1. (*continued*)

	January	February	March	April	May	June
Coeff of correlation	0,681	0,767	0,625	0,706	0,601	0,551
	July	August	September	October	November	December
Longitude	0,99	2,81	0,03	0,11	2,36	0,43
Latitude	−0,42	−3,32	4,33	6,26	28,00	67,76
Altitude	$-2{,}5\ 10^{-3}$	$-0{,}6\ 10^{-3}$	$12{,}6\ 10^{-3}$	$5{,}2\ 10^{-3}$	$4\ 10^{-3}$	0
Intercept	17,92	116,09	−131,83	−202,18	−986,60	−2385,7
Coeff of correlation	0,272	0,362	0,421	0,326	0,630	0,766

According to Table 1, the coefficients of the longitude parameter vary over the year in a random fashion. However, these longitudinal coefficients have a positive sign, which is not the case for the month of April, where the coefficient takes on a negative value, i.e. −0.85. This value represents the lowest value. This value represents the lowest value. On the other hand, the highest value is for June with 4.62. It can be seen, longitude has almost no influence on the rainfall parameter during the months of September and October, with 0.03 and 0.11 respectively. The majority of the values of these coefficients are positive, which shows that longitude has weak positive effects on a monthly rainfall. This weak action of longitude is related to the close relationship between the variation of the effect of longitude and the seasonal variation of climate. Its effect correspond to global changes and astronomical phenomena influencing precipitation [18].

The coefficients of the latitude are positive except for those of July and August. The effect of the geographical factor "latitude" decreases from January until July and August when the values become negative, the values then become ascending until December when it reaches the highest value of 67.76. The effect of latitude is not very marked for the months of June to October, although the effect of this factor is significant for the remaining months, in particular for December (67.76), January (46.42), and February (44.04). In addition, the coefficients of altitude. These are very low compared to the other coefficients of longitude and latitude. These coefficients vary between-6,2 10^{-3} and 17,9 10^{-3}. Table 1, also shows the values of the intercept or constant of the MLR equation of monthly rainfall. It is composed of a series of values, most of which are very large with a negative sign. The only positive values are those for July and August with values of 17.92 and 116.09 respectively.

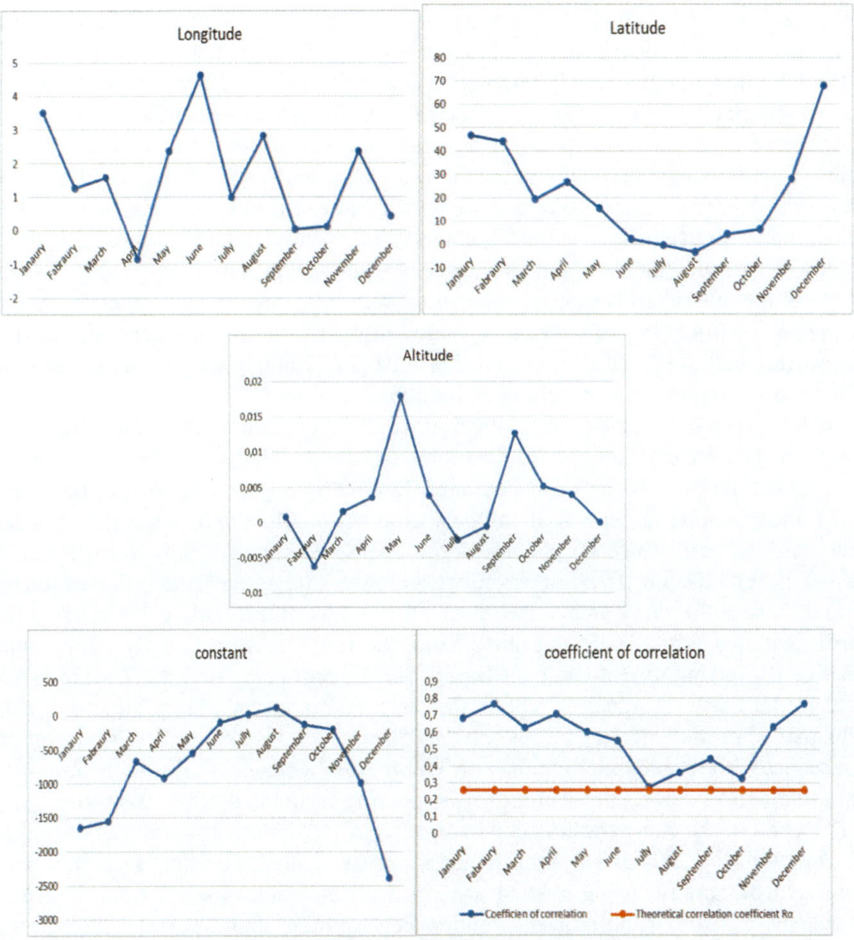

Fig. 2. MLR coefficients at the monthly scale

Figure 2, shows that the effect of longitude on precipitation is variable and develops an irregular way in different months of the year. The shape of the graph follows a zigzag form, indicating that the coefficient values of the longitude in the multiple regression equation do not follow a logical sequence. However, two important stages in the evolution of the longitude effect stand out, the first starting in January when the longitude effect decreases, too sharply to reach a negative value of −0.85 (April), the second starting in April marked by a steady increase in the longitude effect which ends in June with a maximum of 4.62. This can be explained by the fact that precipitation in winter is oceanic in nature, due mainly to large-scale westerly disturbances; it is generally of low instantaneous intensity but of long duration [19]. The effect of latitude is not very significant for the months of June to October although the effect of this factor is significant for the remaining months, in particular for the months of December (67.76), January

(46.42), and February (44.04). In the remaining months of the year the effect is medium. The results obtained show that there is a proportional relationship between the coefficient of latitude and rainfall, with December being the wettest month in the study area, coinciding with the highest value, and decreasing with decreasing rainfall. These results are supported by the fact that the study area is located at the southern limit of the North Atlantic storm tracks, the Mediterranean region is particularly sensitive to inter-annual shifts in the trajectories of mid-latitude cyclones that can lead to remarkable anomalies of precipitation. Given the seasonal characteristics of the Atlantic storm-tracks, this is particularly true in winter when the influence of mid-latitude variability is at its greatest [20]. Also in winter in arid climates, most of the rainfall comes from thunderstorms over small areas. The intensity of the rain is usually high. These rains are generally of short duration and are linked to the movement of fronts [21]. Somme topo-climatic situations facilitate the frequent development of relief thunderstorms [22].

On the other hand, it is clear from the figure that the effect of altitude on precipitation is irregular. The effect of altitude in the month of May is relatively stronger than in the other months, and it is weakest in February. In contrast, it has no effect in the winter and summer months. In general, it can be seen that the effect of altitude is very small, or even almost zero, since the values of the coefficients are close to zero. It can be said that altitude has no effect on precipitation in the study area. This is due to the low relief contrast in the study area and confirms the results of the study conducted by [23] regarding annual rainfall. On the other hand, the relief acts as a catalyst for rainfall formation. Referring to the results of Barros and Lettenmaier, 1993 [24], which states that the relief has a significant effect on the onset and intensity of rainfall and that this mechanism of rainfall increase varies between 50 and 85% for a low relief. This situation is consistent with what happens in the northern region, which is limited by an important orographic barrier that prevents clouds from passing from the north to the south.

The trend of the curve the constant or "intercept" factor monthly rainfall during the studied period (Fig. 2) shows that this factor has an inversely proportional effect, causing the amount of rainfall in the studied area to decrease each time its value increases. It should be noted that the rainfall is influenced by other environmental factors such as distance from sea, air humidity movement, land cover vegetation, temperature and atmospheric pressure [25–28]. In our case the intercept combines all other factors into one value. The intercept plays an important negative impact on the annual rainfall. The combination of the warm sea in summer and the large energy reservoir of autumn convection gives the potential for a disturbance to set in during autumn and especially winter. This type of instability requires an external mechanism to initiate the uplift. The most natural and effective mechanism is orographic uplift [29]. These two conditions (proximity to the Mediterranean Sea and presence of relief) often give rise to both deep and shallow convection events. In winter, rainfall is oceanic, due mainly to large-scale westerly disturbances; it is generally of low instantaneous intensity but long duration. There is a reduction in precipitation at high altitudes, but rainfall increases at medium altitudes, on windward massifs. In arid regions, their relative importance decreases faster than the total rainfall [30]. A study has shown that precipitation is always linked to cyclonic or frontal convergences, mesoscale convergences associated with valley winds, atmospheric instability or moisture [31].

The theoretical correlation coefficient Rα of the different explanatory equations of the precipitation parameter is considered as the critical value of the correlation coefficient, probability $\alpha = 5\%$. In our study it is R$\alpha = 0.274$ according to Snedecor's table [32]. The correlation coefficients are significant except for the month of July. We have drawn a curve of the behaviour of the monthly correlation coefficients. This graphic allows us to clearly see the months for which the equations of the multiple regression are most explanatory. The curve indicates that the majority of the months, except for July, August, September and October, the variation in precipitation is considerably explained by the topographical parameters longitude, latitude and altitude.

4 Conclusion

Rainfall spatial variability is clearly related to a north/south gradient. Considering the position of the study area, which is situated on the south side of the Tellian Atlas chain, the northern part of this area can take much advantage of orographic rains contrarily to the central and south part [33]. The intercept plays an important negative impact on the monthly rainfall due lot of parameters; High temperatures and dry air masses coming from the Sahara generate warm front that is playing the role of barrier against the cold front coming from the north. In addition, the lack of vegetation in the study area, which is dominated by annual crops like wheat and barley, contributes to reduce rainfall amounts. The large size of the surveyed area and its economic importance underline the importance of strengthening and expanding the current climate network. The complexity of estimating rainfall in the region suggests that other variables, such as temperature, air mass movement and vegetation cover, should be taken into account to better understand the influence of these parameters on rainfall.

The complexity of estimating precipitation in the region suggests that other variables, such as temperature, air mass movement and vegetation cover, should be taken into account to better understand the influence of these parameters on precipitation. In this study, the MLR provides a tool for the representation of precipitation with very low uncertainty.

References

1. Domingues Ramos, M.: Analyse de la pluviométrie sous des systèmes nuageux convectifs. Thesis, Ph.D., Uni. Grenoble (2002). 165 p
2. Kramer, K., Leinonen, I., Loustau, D.: The importance of phenology for the evaluation of impact of climate change on growth of boreal, temperate and Mediterranean forests ecosystems: an overview. Int. J. Biometeo **44**(2), 67–75 (2000)
3. Koteswaram, P.: Climat et météorologie. Edit. ISBN. U.N/E/S.C.O. Paris, pp. 29–52 (1974)
4. Marks, D., Winstral, A., Reba, M., Pomeroy, J., Kumar, M.: An evaluation of methods for determining during-storm precipitation phase and the rain/snow transition elevation at the surface in a mountain basin. Adv. Water. Resour. **55**, 98–110 (2013). https://doi.org/10.1016/j.advwatres.2012.11.012
5. Gachon, P., Dibike, Y.B.: Temperature change signals in northern Canada: convergence of statistical downscaling results using two driving GCMs. Int. J. Clim. **27**(12), 1623–1641 (2007). https://doi.org/10.1002/joc.1582

6. Dibike, Y.B., Coulibaly, P.: Temporal neural networks for downscaling climate variability and extremes. Neural Netw. **19**(2), 135–144 (2006). https://doi.org/10.1016/j.neunet.2006.01.003
7. Lloyd, C.D.: Assessing the effect of integrating elevation data into the estimation of monthly precipitation in Great Britain. J. Hydol. **308**, 128–150 (2005). https://doi.org/10.1016/j.jhydrol.2004.10.026
8. Johnson, G.L., Hanson, C.L.: Topographic and atmospheric influences on rainfall variability over a mountainous watershed. J. Appl. Meteo. **34**, 68–74 (1995)
9. Myoung-Jin, U.M., Hyeseon, Y., Chang-Sam, J., Jun-Haeng, H.: Factoranalysis and multiple regression between topography and precipitation on Jeju Island, Korea. J. Hydrol. **410**, 189–203 (2012). https://doi.org/10.1016/j.jhydrol.2011.09
10. Subyani, A.M., Al-Dakheel, A.M.: Multivariate geostatistical methods of mean annual and seasonal rainfall in southwest Saudi Arabia. Arab. J. Geosci. **2**, 19–27 (2009). https://doi.org/10.1007/s12517-008-0015-z
11. Sheather, S.J.: chapter 5: Multiple linear regression. In: A Modern Approach to Regression with R. STS, pp. 125–149. Springer, New York (2009). https://doi.org/10.1007/978-0-387-09608-7_2
12. Walpole, R.E., Myers, R.H., Myers, S.L., Ye, K.: Probability and Statistics for Engineers and Scientists, 8th edn. Pearson Prentice Hall (2007). 791p
13. Cook, E.R., Briffa, K.R., Jones, P.D.: Spatial regression methods in dendroclimatology - a review and comparison of 2 techniques. Int. J. Climatol. **14**, 379–402 (1994)
14. Hay, L., Viger, R., Mccabe, G.: Precipitation interpolation in mountainous regions using multiple linear regressions. Hydrol Water Resour. Ecol. Headwaters **248**, 33–38 (1998)
15. Naoum, S., Tsanis, I.K.: A multiple linear regression GIS module using spatial variables to model orographic rainfall. J. Hydroinf. **06**(1), 39–56 (2004)
16. Bahlouli, F., Bouzerzour, H., Benmahammed, A.: Effets de la vitesse et de la durée du remplissage du grain ainsi que de l'accumulation des assimilats de la tige dans l'élaboration du rendement du blé dur (Triticum durum Desf.) dans les conditions de culture des hautes plaines orientales d'Algérie. Biotechnol. Agron. Soc. Environ. **1**, 31–39 (2008)
17. Bachir, H., Kezouh, S., Ait-oubelli, M., Semar, A., Smadhi, D., Ouamer-ali, K.: Improvement of interpolation using information from rainfall stations and comparison of hydroclimate changes (1913–1938)/(1986–2016). Al-Qadisiyah. J. Agric. Sci. **11**(1), 54–67 (2021). https://doi.org/10.33794/qjas.2021.129350.1002
18. Barancourt, C.: Etude de l'intermittence et de la variabilité des champs de précipitation par une approche stochastique. Thesis, Ph.D., University Joseph Fourier-Grenoble I (1990). 187p
19. Baldy, C.: Comportement des blés dans les climats meditérraneans. Ecologia Mediterranea **112**(3–4), 73–88 (1986)
20. Trigo, R., et al.: Relations between variability in the Mediterranean region and mid-latitude variability. Dev. Earth Environ. Sci. (4), 179–226 (2006). https://doi.org/10.1016/S1571-9197(06)80006-6
21. Amirouche, M., Smadhi, D., Zella, L., Bachir, H.: Rainfall trends in semi-arid cereal regions of Algeria. Al-Qadisiyah. J. Agri. Sci. **11**(1), 36–44 (2021)
22. Corte-Real, J., Zhang, X., Wang, X.: Large-scale circulation regimes and surface climatic anomalies over the Mediterranean. Int. J. Climatol. **15**, 1135–1150 (1995)
23. Bachir, H., Semar, A., Mazari, A.: Statistical and Geostatistical analysis related to geographical parameters for spatial and temporal representation of rainfall in semi-arid environments the Case of Algeria. Arab. J. Geosci. **9**(7), 486–498 (2006). https://doi.org/10.1007/s12517-016-2505-8
24. Barros, A.P., Lettenmaier, D.P.: Dynamic on modelling of the spatial distribution of precipitation in remote mountainous areas. Mon. Wea. Rev. **121**, 1195–1214 (1993)
25. Subyani, A.M.: Geostatistical study of annual and seasonal mean rainfall patterns in southwest Saudi Arabia. Hydrol. Sci. J. **49**(5), 803–817 (2004)

26. Zekai, S., Zeyad, H.: Spatial precipitation assessment with elevation by using point cumulative semivariogram technique. Water. Resour. Manag. **14**, 311–325 (2000)
27. Chua, S., Bras, R.L.: Optimal estimators of mean annual rainfall in regions of orographic influences. J. Hydrol. **57**, 23–48 (1982)
28. Rosenberg, M.: Hydrologie. Choix d'un modèle régional expliquant la répartition des précipitations annuelles dans l'espace en fonction des facteurs climatiques et topographiques. CRAC Sci. **268**, 761–764 (1969)
29. Banta, R.M.: The role of montain flows in making clouds. In: Blumen, W. (eds.) Atmospheric Processes over Complex Terrain. METEOR, vol. 45, pp. 173–228. American Meteorological Society, Boston (1990). https://doi.org/10.1007/978-1-935704-25-6_9
30. Brunsdon, C., Mcclatchey, J., Unwin, D.J.: Spatial variations in theaverage rainfall–altitude relationship in Great Britain: an approachusing geographically weighted regression. Int. J. Climatol. **21**, 455–466 (2001)
31. Cosma, S., Richard, E., Miniscloux, F.: The role of small scale orographic features in the spatial distribution of precipitation. Q.J.R.M.S. **128**(579), 75–92 (2002)
32. Dagnelie, P.: Théorie et méthode statistique applications agronomiques. Ed. Presses de Gembloux, Belgique. 2, 458 (1973)
33. Miniscloux, F., Anquetin, S., Creutin, J.D.: Geostatistical analysis of orographic rainbands. J. App. Meteo. **40**(11), 1835–1854 (2001)

Check for updates

Salihli Granitoid, Menderes Massif, Western Anatolia: A Sustainable Clean Energy Source for Mitigating CO2 Emissions

Tolga Ayzit(✉), Dornadula Chandrasekharam, and Alper Baba

Izmir Institute of Technology, 35430 Urla/Izmir, Turkey
{tolgaayzit,dornadulachandra,alperbaba}@iyte.edu.tr

Abstract. Turkey has a great opportunity to promote renewable energy, which is produced from high heat-generating granitoids using EGS (Enhanced Geothermal Systems) technology. Exploiting the energy from the radiogenic granitoid will help the country save about 32211 million kg of CO_2 from gas-based electricity power plants. In addition to the hydrothermal energy sources, energy from EGS will make the country free from energy deficit and provide sustainable power, water, and food. In the present paper, we assess the power generation capacity of Salihli granitoid (SG), with an outcropping area of about 100 km^2 located within the western Anatolian plateau, and describe the technology involved in harnessing the heat from these granitoids. The Anatolian Plateau is known for extension tectonics and is explained by the westward tectonic escape and subduction rollback processes. The most prominent structures of western Anatolia are E-W and ENE-WSW trending graben and horst controlled by low and high-angle oblique to dip-slip normal faults, exposing the Menderes Massif. Magmatic activity in western Anatolia is mainly related to episodic-two stage extensional regime, where the early phase is characterized mainly by calc-alkaline Early-Middle Miocene felsic lavas and pyroclastic and the latter by late Miocene-Quaternary rift-related alkaline basaltic volcanism. The plutonic activity started during 12 to 15 Ma represented by SG. The heat generation capacity of the SG varies from 5.5 to 6.7 (μW/m^3), while the heat flow values over SG range from 68 to 107 HF (mW/m^2). These values are much higher compared to the global average crustal values.

Keywords: Renewable energy · Enhanced geothermal system (EGS) · Geothermal exploration · Radiogenic granitoid · Hot dry rock · Salihli granitoid · Menderes Massif · Western Anatolia

1 Introduction

Countries need efficient and diversified energy sources to achieve sustainable economic growth, especially renewable energy sources (RES), which costs should also be affordable for economic growth. In the last few decades, greenhouse gas (GHGs) emissions have increased by 110.4% in Turkey [1]. Moreover, Turkey´s energy demand will increase by 93% in the next decade due to 1.4% annual population growth. Turkey

H. Gökçekuş and Y. Kassem (Eds.): NRSEM 2021, EESCI, pp. 272–283, 2022.
https://doi.org/10.1007/978-3-031-04375-8_31

must provide a technology-driven sustainable energy source to meet all future energy demands and achieve the desired target of the 2 D scenario recommended by [1].

There is a lot of heat energy available below the entire surface of the Earth. It has been there for billions of years, and it will be there for a few billion years more. This heat is called geothermal energy, and it provides a limitless supply of heat and power for the entire planet with almost zero CO_2 emissions. The use of geothermal power is increasing rapidly in Turkey, from 311 MWe in 2013 to 1668 MWe by the end of 2020. Notably, Turkey has the second-highest share of geothermal electricity generation after New Zealand [1]. Turkey's proven geothermal potential reached 35 gigawatts thermal (GWt) at the end of 2018 [2]. The Turkish government considered district heating plans using geothermal energy as well as increased use of geothermal heats pumps as important tools for climate adaptation [3].

On the other hand, in 2018, the International Energy Agency (IEA) predicted geothermal energy production about 5% growth per year until 2023. However, [4] announced that geothermal energy production reached just a 3% growth per year. According to [4], geothermal energy should grow by 10% annually till 2030 within the scope of sustainable development goals (SDGs). It means that we must be finding ways to improve and increase the use of more geothermal energy. Turkey has a great opportunity to promote geothermal energy from hot radiogenic granitoid using EGS technology (Enhanced Geothermal Systems). These energy sources will help Turkey to mitigate CO_2 emissions and sustain a sound Gross Domestic Product (GDP) growth. The potential of EGS in Turkey is huge in western Anatolian, and this source is not site-specific like the hydrothermal resources. In the present paper, we assess the power generation capacity of Salihli granitoid (SG), with an outcropping area of about 100 km^2 located within the Menderes Massif of the western Anatolian plateau [5], and describe the technology involved in harnessing the heat from these granitoids.

2 Geological and Tectonic Assessment of Salihli Granitoid

The Anatolian Plate is located on the western Alpine-Himalayan orogenic belt (Fig. 1), which led to the amalgamation of two major continents, the Arabian plate in the south and the Eurasian plate in the north and intervening continental terrains [6–9]. The Anatolian Plateau consists of numerous tectonic zones which were derived from the deeply buried northern edge of the Taurides during the closure of the northern branch of the Neo-Tethys (Fig. 2A) [10–12]. The Anatolian Plateau is known for extension tectonics and is explained by the westward tectonic escape and subduction rollback processes [13–16]. This is controlled by the right-lateral North Anatolian fault and the left-lateral East Anatolian fault. The most prominent structures of western Anatolia are E-W and ENE-WSW trending graben and horst controlled by low and high-angle oblique to dip-slip normal faults, exposing the Menderes Massif [17–23]. The Menderes Massif is divided into northern (Gördes Massif), central (Ödemiş Massif) and southern (Çine Massif) sections by Gediz (also known as Alaşehir) and Büyük Menderes detachment faults which are related to post-orogenic extension [24–30]. Both faults may have run simultaneously in the Miocene to exhume the central Menderes Massif [19].

Magmatic activity in western Anatolia is mainly related to episodic-two stage extensional regime, where the early phase is characterized mainly by calc-alkaline Early-Middle Miocene felsic lavas and pyroclastic and the latter by late Miocene-Quaternary rift-related alkaline basaltic volcanism. The source of the Salihli Granitoid (SG, Fig.1B) is the subduction of the western Anatolia floor along the Hellenic trench as a volcanic arc [31, 32]. The plutonic activity started during 12 to 15 Ma represented by SG, which is located in the footwall of the Gediz detachment surface that bounds the northern edge of the central Menderes Massif metamorphic core complex (Fig. 2B) [25, 28]. SG displays a range of SiO_2 contents from 65.3 wt.% to 75.4 wt.% [33] and classifies as an S-type, calc-alkali, granodiorite in the scope of geochemistry [34].

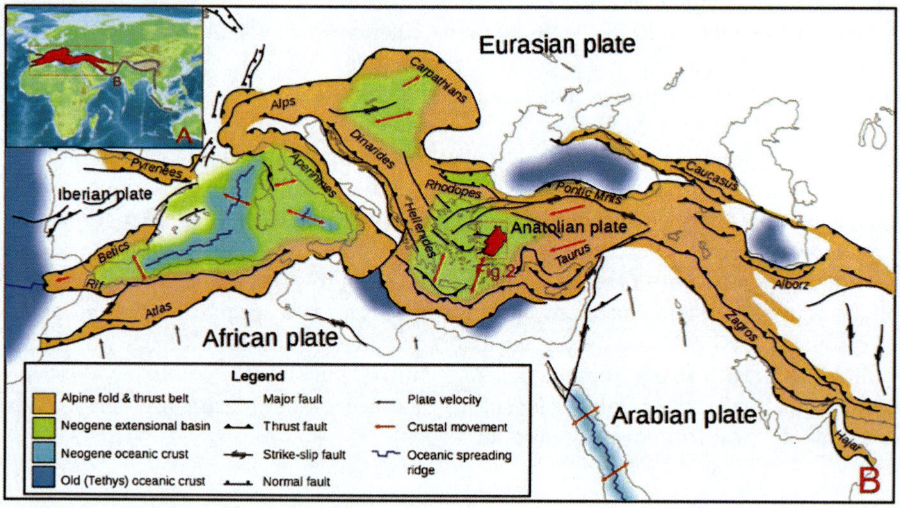

Fig. 1. **(A)** Alpine-Himalayan orogenic belt that includes an array of mountain ranges extending for more than 15000 km along the southern margin of Eurasia **(B)** Menderes Massif location in the tectonic map of southern Europe and the Middle East showing tectonic structures of the western Alpine Mountain belt [35]

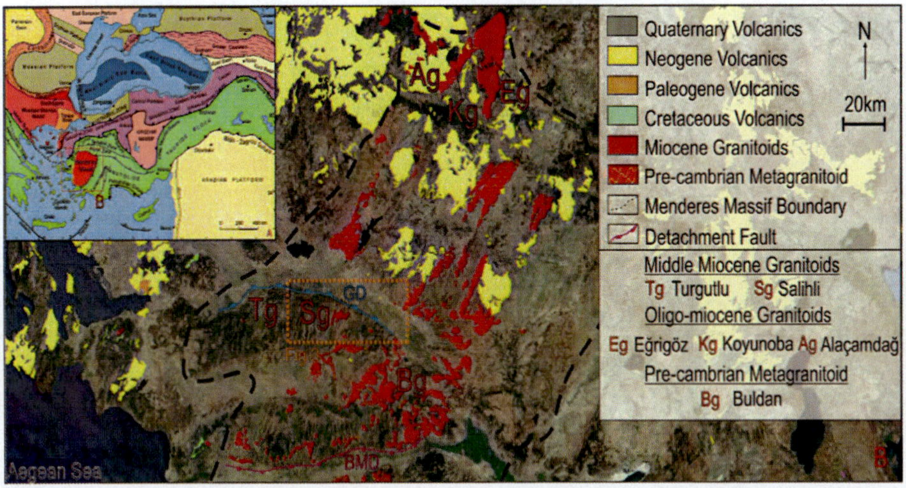

Fig. 2. (A) Menderes Massif location in the tectonic map of Turkey and the surrounding areas [9] **(B)** The spatial distribution of volcanic rocks in western Anatolia and the main structures in central Menderes Massif compiled from [26, 36–38]. Abbreviations: GD, Gediz Detachment Fault; BMD, Büyük Menderes Detachment Fault

3 Stress Regime Over the Region

The Menderes Massif is situated towards the Western Anatolia Extensional Province and restricted by the Afyon zone to the north (Fig. 1B), the Bornova flysch zone to the west, and the Lycian nappes to the south [9]. The north border of central Menderes Massif is controlled by Gediz detachment which surface dips 10–20°N, is approximately 150 km long [17, 25, 19] and is cut by more up to date high-angle normal faults [39, 40], which led to the suggestion that the structure originally opened at a high dip (48–53°) [29] and rolled back to its present-day low dip [17, 19, 25–27, 41] (Fig. 3A). Miocene intrusions present the ductile-brittle transition and deformation in the footwall of the detachment, which is related to the extension and exhumation [17, 25, 28, 42]. Most ductile kinematic indicators and brittle structures indicate NNE motion. Also, all lithologies are deformed in the same direction by asymmetric structures and folds at various scales. Slicken lines are gently plunging consistently 15 to 30°N (Fig. 3B), and kinematic indicators indicate a main dextral movement with a slight normal component. E-W trending and NNE dipping normal faults (Fig. 4), sub-vertical N-S striking strike-slip faults and NE-SW strikes, and approximately 60°N dip types of plurimetric to kilometric faults are observed in the region [43].

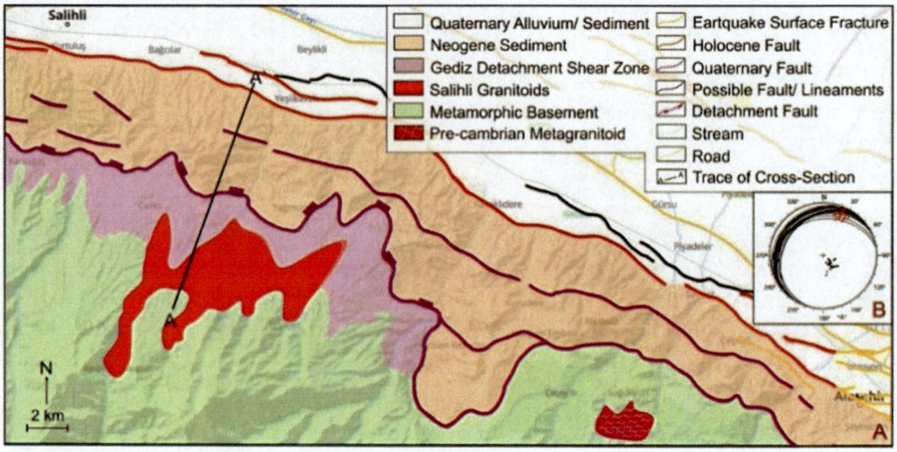

Fig. 3. **(A)** Geological and active fault map around Salihli Granitoid compiled from [36, 43-45]. Black line shows the location of the geological cross-section represented in Fig. 4. **(B)** Mylonitic foliation (solid lines), stretching lineation (red lines) measured in the Salihli granodiorite [44].

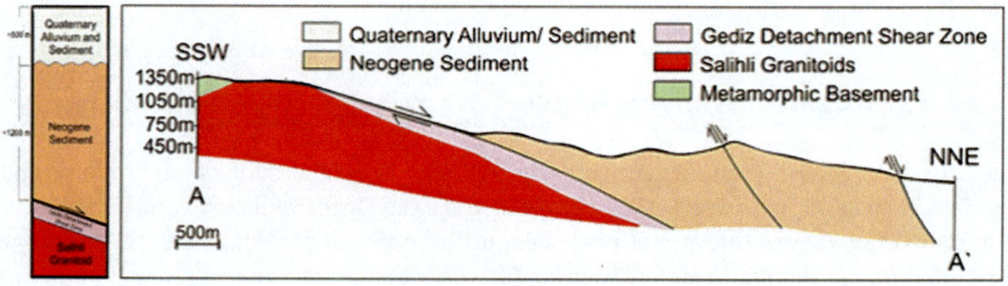

Fig. 4. Lithostratigraphic column section of Salihli area and geological cross-section refers to the longitudinal drag folds that formed on or very near the high-angle normal faults. High angle brittle faults cutting the basin sedimentary fill root into the detachment complied from [26, 43, 44].

4 Radiogenic Characteristics and EGS Potential of the Salihli Granitoid

The main hosts of the major radiogenic elements U, Th, and K in the crust are granitoids. Granitoids directly affect the surface heat flow and geothermal regime of the upper crust [46–49]. Heat production capacity and geothermal source potential are determined according to radiogenic elements content and distribution in granitoids [50]. The heat production rate (A) was obtained based on Rybach's Eq. [51]:

$$A\left(\mu W/m^3\right) = 10^{-2} * \rho * (9.52C_U + 2.56C_{Th} + 3.48C_K) \quad (1)$$

where ρ (g/cm^3) is the density of the rock (average density for granitoid is 2.7 g/cm^3), and C_U, C_{Th} and C_K are the concentrations of uranium (ppm), thorium (ppm) and potassium (%), respectively. C_K is expressed as a percentage of elemental potassium (or K_2O

multiplied by 0.83). The surface heat flow values were calculated using the equation [52].

$$Q = Q0 + D * A \quad (2)$$

where Q is the heat flow at the surface, Q_0 is an initial value of heat flow unrelated to the specific decay of radioactive element at a certain time, D is the thickness of rock over which the distribution of radioactive element is just about homogeneous, and A is the radioactive heat production. Since the thin crustal thickness (~25 km) is observed in the coastal region of the western part of the Anatolian plate [53]. Therefore, the background heat flow value 40 mW/m^2 is considered in the western part of Turkey.

Western Anatolian, Menderes Massif region hosts many intrusive that is highly radiogenic due to high concentration of minerals rich in uranium, thorium, and potassium (Table 1). The heat-generating capacity of the SG intrusive varies from 5.5 to 6.7 ($\mu W/m^3$), and this value is above the average generation value of the continental crust (5 $\mu W/m^3$). Further, the heat flow value calculated for this granite is 68 to 107 mW/m^2, and this value concurs with the value reported for this region based on curie temperature-depth deduced from the aeromagnetic investigation (110 mW/m^2).

Table 1. Radiogenic heat generation by Menderes Massif granitoids and the heat flow over the region. All granitoids in the region display high-K, calc-alkali geochemistry, shallow seated emplacement and occurred related to post-collision tectonic settings in the western Anatolian extensional regime.

Pluton Name	Eğrigöz	Koyunoba	Alaçamdağ	Salihli	Turgutlu	Buldan
Area (km^2)	400	170	30	100	25	157
Tectonic Zone	Menderes Massif (Northern)	Menderes Massif (Northern)	Menderes Massif (Northern)	Menderes Massif (Central)	Menderes Massif (Central)	Menderes Massif (Central)
Petrogenetic Rock Type	I-type granodiorite/granite	I-type granodiorite/granite	I-type granodiorite/granite	S-type granodiorite	S-type granodiorite	S-type granitic gneisses
Age	Oligo-Miocene	Oligo-Miocene	Oligo-Miocene	Middle Miocene	Middle Miocene	Pre-Cambrian
Th (ppm)	51	30,4	32	14,9	15,3	29,8
U (ppm)	9,1	4,8	10	20,6	4,5	6,4
K (%)	3,3	4	3,4	3,5	4,6	3,7
Average Curie dept (km)	10	10	8,4	6,2	8,5	10,3
Geothermal gradient ($^oC/km$)	57,8	57,8	68,7	93,4	68,7	56,3
Radioactive Heat Production ($\mu W/m^3$)	5,85	3,73	5,15	6,72	2,67	4,09
Heat flow (mW/m^2)	98,5	77,3	91,5	107,2	66,7	80,9
References	[57–60]	[58–60]	[58–60]	[5, 32, 59, 60]	[5, 32, 34, 60, 61]	[60, 61]

Radiogenic high heat-generating granitoids are considered as a future energy source due to their high heat-generating capacity and their ability to support the generation of heat and baseload electricity for societal use [54]. Each km^3 of radiogenic granitoid can generate about 79×10^6 kWh of electricity for 30 years [55]. Considering the heat flow values, heat generation values, the probable volume of the SG, the curie depth temperature of this region [56], the temperature of the granitoid at a depth of 2 km is about 90 °C. Thus, the SG intrusive is a good site for initiating EGS projects.

5 CO_2 Mitigation Using the EGS Source

Geothermal energy sources are becoming increasingly important because the energy source is clean and has a very small carbon footprint. This is especially true for countries like Turkey, which emit large volumes of CO_2. For example, Turkey's present per-capita CO_2 emissions are 4.5 tons and are expected to cross 4.5 tons soon due to 1.4% population growth annually. 500 Mt and with a population growth of 1.4% the emissions will reach beyond 500 Mt per year. The current CO_2 emissions are 400 Mt and are expected to cross 500 Mt by the year 2030 [62, 63]. This is not a good sign since the environmental consequences related to CO_2 emissions are disastrous. The CO_2 emissions are mainly from fossil fuel-based power plants. Turkey is already generating 1665 MWe of electricity from its hydrothermal power plants distributed along the western Anatolian region, and by the year 2030, this will reach 3000 MWe. Turkey's energy demand will grow to 100 Mtoe by 2040 from the current 60 Mtoe [63]. To meet the demand, the Ministry of Energy and National Resources of Turkey is encouraging to increase the share of renewable energy to 61000 MW. This is possible only when EGS sources are developed because the high radiogenic granites along western Anatolia have not been exploited for power generation [63–65]. Thus, electricity and heat from these radiogenic granites can be effectively utilized to mitigate CO_2 emissions and support sustained GDP growth [62, 63]. Thus, Turkey needs a technology-driven transformation in the geothermal energy sector by adopting a policy to develop its high radiogenic granites [63, 66].

6 Discussion

Fossil fuel-based power plants, cement industries, and heating and cooling of establishments are the major sources of GHG emissions in Turkey. The energy demand will cross 70 Mtoe in a decade resulting in CO_2 emissions crossing 500 Mt. with the increase in energy demand, and the gas imports are expected to grow by 18% [67]. However, the country has an opportunity to adopt sustainable and healthy economic growth by encouraging geothermal energy from hydrothermal and EGS sources. The hydrothermal energy source is already providing energy equivalent to 1.71 Mtoe and is expected to increase this supply by 2030. Turkey still has unexploited hydrothermal resources of 2335 MWe [68] from hydrothermal resources. By developing these unexploited hydrothermal sources, it can save 24 million kg of CO_2. In addition, by utilizing 20000 MWt for direct application (district heating and cooling), the country can save about 36% of gas imports [63]. The government has realized the importance of geothermal energy now and is expecting a growth of 61% in the coming decade supported by financial incentives [69]. EGS can play an important role in this initiative. Exploiting the EGS potential from the high radiogenic granitoids of Menderes Massif will support sustainable economic growth and provide CO_2 emissions mitigation.

There are a few successful EGS plants that are being in operation today. The Soultz-sous-Forets EGS plant in France is generating 3 MWe now by extracting heat from granites at 200 °C from a depth of 5 km. Similarly, in Cornwall, UK, the EGS project is near completion and is expected to generate 10 MWe and 55 MWt [70, 71]. Such high heat-generating granites are common around the Red Sea region, Africa, and the Himalayas [55, 72–75]. Preliminary assessment of these intrusive as a suitable candidate for initiating EGS projects is lacking. Additionally, it is highly likely that more environmentally friendly and efficient EGS projects will be developed with multidisciplinary approaches such as developing technological drilling methods or/and integrating seawater and wastewater into the system.

7 Conclusion

Western Anatolian hosts many granitoids intrusive that is highly radiogenic due to the high concentration of minerals rich in uranium, thorium, and potassium (Table 1). These high heat-generating granitoids spread over an area of almost 1000 km^2 in the Menderes Massif region. The heat-generating capacity of the SG intrusive varies from 5.5 to 6.7 ($\mu W/m^3$), and this value is above the average generation value of the continental crust (5 $\mu W/m^3$). Further, the heat flow value calculated for this granite is 68 to 107 mW/m^2, and this value concurs with the value reported for this region based on curie temperature-depth deduced from the aeromagnetic investigation (110 mW/m^2). These radiogenic granitoids have a grand potential of renewable energy sources such as EGS that will help Turkey to mitigate CO_2 emissions and sustain a sound Gross Domestic Product (GDP) growth. By extracting this heat from a merge 1 km^3 of this granite will provide minimum electricity of 79×10^6 kWh. The EGS technology is matured now due to development in drilling technology. For environmentally clean energy and for a sustainable socio-economic future, research focus should be on these high radiogenic granites. The present energy used for district heating from hydrothermal sources (1453 MWt) can be expanded, and all the districts can be brought under natural heat regime using the heat from the radiogenic granites in the future, thereby saving greater than 32211 million kg of CO_2 annually. Both hydrothermal and EGS energy sources will provide sustainable energy, food water security to the country.

Acknowledgments. This paper is part of the EGS project funded by TUBITAK (project No:120C079) through a Fellowship grant to DC.

References

1. International Energy Agency. Energy Technology Perspectives 2014 (Harnessing Electricity's Potential Explore the data behind ETP) (2014). https://doi.org/10.1787/energy_tech-2010-en

2. International Energy Agency. Review 2021 Assessing the effects of economic recoveries on global energy demand and CO 2 emissions in 2021 Global Energy (2021). www.iea.org/t&c/. Accessed: 19 Oct 2021
3. "Turkey's National Climate Change Adaptation Strategy and Action Plan," 2011
4. World Energy Outlook 2020 – Analysis - IEA (2020). https://www.iea.org/reports/world-energy-outlook-2020
5. Catlos, E.J., Baker, C.B., Sorensen, S.S., Jacob, L., Çemen, I.: Linking microcracks and mineral zoning of detachment-exhumed granites to their tectonomagmatic history: evidence from the Salihli and Turgutlu plutons in western Turkey (Menderes Massif). J. Struct. Geol. **33**(5), 951–969 (2011). https://doi.org/10.1016/j.jsg.2011.02.005
6. Jolivet, L., Brun, J.P.: Cenozoic geodynamic evolution of the Aegean. Int. J. Earth Sci. **99**(1), 109–138 (2010). https://doi.org/10.1007/s00531-008-0366-4
7. Ketin, İ.: Anadolu'nun tektonik birlikleri. Maden Tetk. ve Aram. Derg. 66(66) (1966)
8. Sengör, A.M.C., Yilmaz, Y.: Tethyan evolution of Turkey: a plate tectonic approach. Tectonophysics **75**(3–4) (1981). https://doi.org/10.1016/0040-1951(81)90275-4
9. Okay, A.I., Tüysüz, O.: Tethyan sutures of northern Turkey. Geol. Soc. Spec. Publ. **156**, 475–515 (1999). https://doi.org/10.1144/GSL.SP.1999.156.01.22
10. Okay, A.I., Harris, N.B.W., Kelley, S.P.: Exhumation of blueschists along a Tethyan suture in northwest Turkey. Tectonophysics **285**(3–4), 275–299 (1998). https://doi.org/10.1016/S0040-1951(97)00275-8
11. Okay, A.I.: Stratigraphic and metamorphic inversions in the central Menderes Massif: a new structural model. Int. J. Earth Sci. **89**(4), 709–727 (2001). https://doi.org/10.1007/s005310000098
12. Pourteau, A., Sudo, M., Candan, O., Lanari, P., Vidal, O., Oberhänsli, R.: Neotethys closure history of Anatolia: Insights from 40Ar-39Ar geochronology and P-T estimation in high-pressure metasedimentary rocks. J. Metamorph. Geol. **31**(6), 585–606 (2013). https://doi.org/10.1111/jmg.12034
13. Seyitoğlu, G., Scott, B.: Late Cenozoic crustal extension and basin formation in west Turkey. Geol. Mag. **128**(2), 155–166 (1991). https://doi.org/10.1017/S0016756800018343
14. Seyitoğlu, G., Scott, B.C., Rundle, C.C.: Timing of Cenozoic extensional tectonics in west Turkey. J. Geol. Soc. London. **149**(4), 533–538 (1992). https://doi.org/10.1144/GSJGS.149.4.0533
15. Seyitoğlu, G., Scott, B.C.: The cause of N-S extensional tectonics in western Turkey: tectonic escape vs back-arc spreading vs orogenic collapse. J. Geodyn. **22**(1–2), 145–153 (1996)
16. Jolivet, L., Goffé, B., Monié, P., Truffert-Luxey, C., Patriat, M., Bonneau, M.: Miocene detachment in Crete and exhumation P-T-t paths of high-pressure metamorphic rocks. Tectonics **15**(6), 1129–1153 (1996)
17. Hetzel, R., Ring, U., Akal, C., Troesch, M.: Miocene NNE-directed extensional unroofing in the Menderes Massif, southwestern Turkey. J. Geol. Soc. London. **152**(4), 639–654 (1995). https://doi.org/10.1144/GSJGS.152.4.0639
18. Gessner, K., Gallardo, L.A., Markwitz, V., Ring, U., Thomson, S.N.: What caused the denudation of the Menderes Massif: review of crustal evolution, lithosphere structure, and dynamic topography in southwest Turkey. Gondwana Res. **24**(1), 243–274 (2013). https://doi.org/10.1016/J.GR.2013.01.005
19. Gessner, K., Piazolo, S., Güngör, T., Ring, U., Kröner, A., Passchier, C.W.: Tectonic significance of deformation patterns in granitoid rocks of the Menderes nappes, Anatolide belt, southwest Turkey. Int. J. Earth Sci. **89**(4), 766–780 (2001). https://doi.org/10.1007/S005310000106
20. Ring, U., Johnson, C., Hetzel, R., Gessner, K.: Tectonic denudation of a Late Cretaceous-Tertiary collisional belt: regionally symmetric cooling patterns and their relation to extensional

faults in the Anatolide belt of western Turkey. Geol. Mag. **140**(4), 421–441 (2003). https://doi.org/10.1017/S0016756803007878
21. Ring, U., Collins, A.S.: U–Pb SIMS dating of synkinematic granites: timing of core-complex formation in the northern Anatolide belt of western Turkey. J. Geol. Soc. London. **162**(2), 289–298 (2005)
22. Thomson, S.N., Ring, U.: Thermochronologic evaluation of postcollision extension in the Anatolide orogen, western Turkey. Tectonics **25**(3), 3005 (2006). https://doi.org/10.1029/2005TC001833
23. Rossetti, F., Asti, R., Faccenna, C., Gerdes, A., Lucci, F., Theye, T.: Magmatism and crustal extension: constraining activation of the ductile shearing along the Gediz detachment, Menderes Massif (western Turkey). Lithos **282–283**, 145–162 (2017). https://doi.org/10.1016/J.LITHOS.2017.03.003
24. Bozkurt, E., Oberhänsli, R.: Menderes Massif (Western Turkey): structural, metamorphic and magmatic evolution - a synthesis. Int. J. Earth Sci. **89**(4), 679–708 (2001). https://doi.org/10.1007/S005310000173
25. Hetzel, R., Passchier, C.W., Ring, U., Dora, O.O.: Bivergent extension in orogenic belts: the Menderes Massif (southwestern Turkey). Geology **23**(5), 455–458 (1995). https://doi.org/10.1130/0091-7613(1995)023%3c0455:BEIOBT%3e2.3.CO;2
26. Sözbilir, H.: Extensional Tectonics and the Geometry of Related Macroscopic Structures: Field Evidence from the Gediz Detachment. Western Turkey. Turkish J. Earth Sci. **10**(2), 51–67 (2001)
27. Seyitoğlu, G., Tekeli, O., Çemen, İ, Şen, Ş, Işık, V.: The role of the flexural rotation/rolling hinge model in the tectonic evolution of the Alaşehir graben, western Turkey. Geol. Mag. **139**(1), 15–26 (2002)
28. Isik, V., Seyitoğlu, G., Çemen, I.: Ductile–brittle transition along the Alaşehir detachment fault and its structural relationship with the Simav detachment fault, Menderes massif, western Turkey. Tectonophysics **374**(1–2), 1–18 (2003). https://doi.org/10.1016/S0040-1951(03)00275-0
29. Bozkurt, E., Sözbilir, H.: Tectonic evolution of the Gediz Graben: Field evidence for an episodic, two-stage extension in western Turkey. Geol. Mag. **141**(1), 63–79 (2004). https://doi.org/10.1017/S0016756803008379
30. Buscher, J.T., et al.: Quantifying rates of detachment faulting and erosion in the central Menderes Massif (western Turkey) by thermochronology and cosmogenic 10Be. J. Geol. Soc. London. **170**(4), 669–683 (2013)
31. Cemen, I., et al.: Kinematics of post-collisional extensional tectonics and exhumation of the Menderes massif in the western Anatolia extended terrane, Turkey. AGUFM, vol. 2006, pp. T41E-01 (2006). Accessed 19 Oct 2021
32. Dilek, Y., Altunkaynak, Ş, Öner, Z.: Syn-extensional granitoids in the Menderes core complex and the late Cenozoic extensional tectonics of the Aegean province. Geol. Soc. Spec. Publ. **321**, 197–223 (2009)
33. Catlos, E.J., Baker, C.B., Çemen, I., Ozerdem, C.: Whole rock major element influences on monazite growth: examples from igneous and metamorphic rocks in the Menderes Massif, western Turkey. Mineralogia **39**(1–2), 7–30 (2008). https://doi.org/10.2478/v10002-008-0002-8
34. Catlos, E.J., Baker, C., Sorensen, S.S., Çemen, I., Hançer, M.: Geochemistry, geochronology, and cathodoluminescence imagery of the Salihli and Turgutlu granites (central Menderes Massif, western Turkey): Implications for Aegean tectonics. Tectonophysics **488**(1–4), 110–130 (2010)
35. Alpine orogeny - Wikipedia. https://en.wikipedia.org/wiki/Alpine_orogeny. Accessed 19 Oct 2021

36. Akbaş, B., et al.: Geological Map of Turkey, 1:1.250.000 scaled. General Directorate of Mineral Research and Exploration Publication, Ankara-Turkey (2011)
37. Candan, O., et al.: Late Neoproterozoic gabbro emplacement followed by early Cambrian eclogite-facies metamorphism in the Menderes Massif (W. Turkey): implications on the final assembly of Gondwana. Gondwana Res. **34**, 158–173 (2016). https://doi.org/10.1016/j.gr.2015.02.015
38. Tatar Erkül, S., Özmen, S.F., Erkül, F., Boztosun, İ.: Comparison between natural radioactivity levels and geochemistry of some granitoids in western Turkey. Turkish J. Earth Sci. **25**(3), 242–255 (2016). https://doi.org/10.3906/yer-1511-6
39. Seyitoğlu, G., Çemen, İ, Tekeli, O.: Extensional folding in the Alaşehir (Gediz) graben, western Turkey. J. Geol. Soc. London. **157**(6), 1097–1100 (2000). https://doi.org/10.1144/JGS.157.6.1097
40. Purvis, M., Robertson, A.: Sedimentation of the Neogene-Recent Alaşehir (Gediz) continental graben system used to test alternative tectonic models for western (Aegean) Turkey. Sediment. Geol. **173**(1–4), 373–408 (2005). https://doi.org/10.1016/J.SEDGEO.2003.08.005
41. Seyitoğlu, G., Koçyiğit, A., Bozkurt, E., Yusufoğlu, H.: Discussion on evidence from the Gediz Graben for episodic two-stage extension in western Turkey. J. Geol. Soc. London. **156**(6), 1240–1242 (1999). https://doi.org/10.1144/GSJGS.156.6.1240
42. Emre, T.: Gediz grabeni'nin (Salihli-Alasehir arasi) jeolojisi. no. 45 (1992)
43. Roche, V., et al.: Structural, lithological, and geodynamic controls on geothermal activity in the Menderes geothermal Province (Western Anatolia, Turkey). Int. J. Earth Sci. **108**(1), 301–328 (2018). https://doi.org/10.1007/s00531-018-1655-1
44. Asti, R., et al.: The Gediz Supradetachment system (SW Turkey): magmatism, tectonics, and sedimentation during crustal extension. Tectonics **38**(4), 1414–1440 (2019). https://doi.org/10.1029/2018TC005181
45. Emre, F., Duman, Ö., Özalp, T.Y., Elmacı, S., Olgun, H., Şaroğlu, Ş.: Active fault map of turkey with an explanatory text. 1:1,250,000 Scale. General Directorate of Mineral Research and Exploration, Special Publication Series-30, Ankara-Turkey (2013). ISBN: 978-605-5310-56-1
46. Mclaren, S., Sandiford, M., Powell, R., Neumann, N., Woodhead, J.: Palaeozoic intraplate crustal anatexis in the Mount Painter Province, South Australia: timing, thermal budgets and the role of crustal heat production. J. Petrol. **47**(12), 2281–2302 (2006). https://doi.org/10.1093/PETROLOGY/EGL044
47. Mareschal, J.C., Jaupart, C.: Radiogenic heat production, thermal regime and evolution of continental crust. Tectonophysics **609**, 524–534 (2013). https://doi.org/10.1016/J.TECTO.2012.12.001
48. Jaupart, C., Mareschal, J.C.: Post-orogenic thermal evolution of newborn Archean continents. Earth Planet. Sci. Lett. **432**, 36–45 (2015). https://doi.org/10.1016/J.EPSL.2015.09.047
49. Jaupart, C., Mareschal, J.C., Iarotsky, L.: Radiogenic heat production in the continental crust. Lithos **262**, 398–427 (2016). https://doi.org/10.1016/J.LITHOS.2016.07.017
50. Pleitavino, M., Carro Pérez, M.E., García Aráoz, E., Cioccale, M.A.: Radiogenic heat production in granitoids from the Sierras de Córdoba, Argentina. Geothermal Energy **9**(1), 1–24 (2021). https://doi.org/10.1186/s40517-021-00198-9
51. Haenel, R., Rybach, L., Stegena, L.: Fundamentals of Geothermics. Handb. Terr. Heat-Flow Density Determ., pp. 9–57 (1988). https://doi.org/10.1007/978-94-009-2847-3_2
52. Lachenbruch, A.H.: Preliminary geothermal model of the Sierra Nevada. J. Geophys. Res. **73**(22), 6977–6989 (1968). https://doi.org/10.1029/JB073I022P06977
53. Tezel, T., Shibutani, T., Kaypak, B.: Crustal thickness of Turkey determined by receiver function. J. Asian Earth Sci. **75**, 36–45 (2013). https://doi.org/10.1016/j.jseaes.2013.06.016
54. Massachusetts Institute of Technology. The future of geothermal energy (2006)

55. Chandrasekharam, D., Lashin, A., Al Arafi, N., Varun, C., Al Bassam, A.: Climate change mitigation strategy through utilization of geothermal energy resources from Western Arabian Shield, Saudi Arabia. J. Clim. Chang. **1**(1–2), 129–134 (2015). https://doi.org/10.3233/JCC-150011
56. Aydin, I., Karat, H.I., Koçak, A.: Curie-point depth map of Turkey. Geophys. J. Int. **162**(2), 633–640 (2005). https://doi.org/10.1111/j.1365-246X.2005.02617.x
57. Ozgenc, I. Ilbeyli, N.: Petrogenesis of the Late Cenozoic Egrigöz Pluton in Western Anatolia, Turkey: implications for magma genesis and crustal processes. **50**(4), 375–391 (2008). https://doi.org/10.2747/0020-6814.50.4.375
58. Akay, E.: Geology and petrology of the Simav Magmatic Complex (NW Anatolia) and its comparison with the Oligo-Miocene granitoids in NW Anatolia: implications on Tertiary tectonic evolution of the region. Int. J. Earth Sci. **98**(7), 1655–1675 (2009). https://doi.org/10.1007/s00531-008-0s325-0
59. Erkül, S.T., Erkül, F.: Magma interaction processes in syn-extensional granitoids: the tertiary Menderes metamorphic core complex, western Turkey. Lithos **142–143**, 16–33 (2012)
60. Bilim, F., Akay, T., Aydemir, A., Kosaroglu, S.: Curie point depth, heat-flow and radiogenic heat production deduced from the spectral analysis of the aeromagnetic data for geothermal investigation on the Menderes Massif and the Aegean Region, western Turkey. Geothermics **60**, 44–57 (2016). https://doi.org/10.1016/J.GEOTHERMICS.2015.12.002
61. Erkül, S.T., Erkül, F.: Magma interaction processes in syn-extensional granitoids: the Tertiary Menderes Metamorphic Core Complex, western Turkey. Lithos **142–143**, 16–33 (2012)
62. Difiglio, C., Guray, B.S., Merdan, E.: Turkey Energy Outlook (2020)
63. Chandrasekharam, D., Baba, A.: High heat generating granites of Kestanbol: future Enhanced Geothermal System (EGS) province in Western Anatolia. J Earth Sci, Spec. (in press)
64. Erdin, C., Ozkaya, G.: Turkey's 2023 energy strategies and investment opportunities for renewable energy sources: site selection based on ELECTRE. Sustainability **11**(7), 2136 (2019)
65. WECTNC. Energy report of Turkey for 2009 (2009)
66. Chandrasekharam, D., Baba, A.: Carbon dioxide emissions mitigation strategy through Enhanced Geothermal systems: western Anatolia, Turkey. Environ. Earth Sci. (under review)
67. Turkey Natural Gas: Imports, 1975 – 2021 | CEIC Data (2021). https://www.ceicdata.com/en/indicator/turkey/natural-gas-imports. Accessed 20 Oct 2021
68. Baba, A., et al.: Assessment of opportunities and interest in direct uses of geothermal energy (2021)
69. International Energy Agency. Turkey 2021: Energy Policy Review (2021)
70. Koelbel, T., Genter, A.: Enhanced geothermal systems: the Soultz-sous-Forêts project. In: Uyar, T.S. (ed.) Towards 100% Renewable Energy. SPE, pp. 243–248. Springer, Cham (2017). https://doi.org/10.1007/978-3-319-45659-1_25
71. Letcher, T.M.: Future energy: improved, sustainable and clean options for our planet (2013). Accessed 20 Oct 2021. https://www.researchgate.net/publication/27718079
72. Chandrasekharam, D., Lashin, A., Al Arifi, N.: The potential contribution of geothermal energy to electricity supply in Saudi Arabia **35**(9), 824–833 (2014). https://doi.org/10.1080/14786451.2014.950966
73. Chandrasekharam, D., Lashin, A., Al Arifi, N., Bassam, A.: Red Sea Geothermal Provinces, p. 220. CRC Press, London (2016)
74. Lashin, A., Chandrasekharam, D., Al Arifi, N., Al Bassam, A., Varun, C.: Geothermal energy resources of wadi Al-Lith, Saudi Arabia. J. African Earth Sci. **97**, 357–367 (2014). https://doi.org/10.1016/J.JAFREARSCI.2014.05.016
75. Omenda, D., Varun, P., Chandrasekharam, C.: High heat generating granites of East Africa: Possible EGS sources. In: Proceedings of the 4th African Rift Geothermal Conference Nairobi, Kenya, 2012, pp. 1–4 (2012)

Sorting Greenhouse Gases Based on Human and Environmental Impacts Using (MCDA)

Nivin Ghaboun[1(✉)], Hüseyin Gökçekuş[1], Berna Uzun[2,3], and Dilber Uzun Ozsahin[3,4,5]

[1] Civil and Environmental Engineering Faculty, Near East University, TRNC, Mersin 10, 99138 Nicosia, Turkey
{nivinahmedhassan.ghaboun,huseyin.gokcekus}@neu.edu.tr

[2] Faculty of Arts and Sciences, Department of Mathematics, Near East University, TRNC, Mersin 10, 99138 Nicosia, Turkey
berna.uzun@neu.edu.tr

[3] DESAM Research Institute, Near East University, TRNC, Mersin 10, 99138 Nicosia, Turkey

[4] College of Health Science, Medical Diagnostic Imaging Department, University of Sharjah, Sharjah, United Arab Emirates
dozsahin@sharjah.ac.ae

[5] Faculty of Engineering, Department of Biomedical Engineering, Near East University, TRNC, Mersin 10, 99138 Nicosia, Turkey

Abstract. Greenhouse gases are a global problem due to their dangerous impacts on the human health and environment. Climate change is the main result of greenhouse gases emissions which cause severe weather conditions, droughts, wildfires, disruption of eco-system, floods, pollution of air, disturbance of food supply system and extinction of animal species. The main sources of greenhouse gases are fossil fuels combustion in manufacturing such as cement manufacturing, agriculture activities and deforestation. Greenhouse gases consist of many gases; the most important greenhouse gases are Carbon Dioxide (Co2), Methane, Nitrous oxide, fluorinated gases including ChloroFluoroCarbon, HydroFlouroCarbon, Sulfur Hexafluoride and Nitrogen Trifluoride. The aim of this study is to sort greenhouse gases based on their impacts on human and environment, radiative forcing and atmospheric life time using Multi criteria Decision Analysis (MCDA). Fuzzy PROMETHEE method was used for sorting; it's a combination between PROMETHEE method and Fuzzy logic to consider ambiguous conditions. The results of the analysis sort Nitrogen Trifluoride as the first gas because it has less impacts on human and environment with less radiative forcing. Sulfur Hexafluoride, HydroFlouroCarbon and ChloroFluoroCarbon were sorted on second, third and fourth position. Carbon dioxide was sorted as the last gas because it has the worst impacts on human and environment due to high radiative forcing.

Keywords: Greenhouse gases · Fluorinated gases · Multi criteria decision analysis · PROMETHEE method · Fuzzy logic

H. Gökçekuş and Y. Kassem (Eds.): NRSEM 2021, EESCI, pp. 284–295, 2022.
https://doi.org/10.1007/978-3-031-04375-8_32

1 Introduction

Global warming is a critical problem to all inhabitants of the earth planet. Industrial activities including fossil fuels burning for manufacturing, urbanization, agriculture activities, mining and logging are the main reasons of greenhouse gases emissions. These gases are chemical compounds exist in the earth's atmosphere absorb and trap infrared radiation from the earth's surface in the lower atmosphere and reradiating it again causing the temperature of the earth's surface to increase [1, 2]. As a result of this increase in earth's temperature the ice glaciers melts causing floods, disturbance of the ecological system, ocean acidification, creating dead zones, severe storms, tornadoes, vanishing of plants and animal species, in addition to forest burning, aridity and air pollution which affect human health adversely [3]. Greenhouse gases can be classified as natural gases such as carbon dioxide, methane, water vapor and nitrous oxide, while the others are man-made. These man-made gases include chlorofluorocarbons, HydroFlouroCarbon, Sulfur Hexafluoride and Nitrogen Trifluoride [4]. Water vapor cannot be considered as a greenhouse gas because it has a short life time in the water cycle. The concentration of these gases varies from one season to another; generally these concentrations are high in warm seasons and low in cold seasons. Industrialized countries are the most one suffering from high concentration of greenhouse gases especially China, United States and India. Each gas of the greenhouse gases has different impact on earth's warming. Global warming potential (GWP) is a measurement used to compare the global warming impact of greenhouse gases through their ability to trap heat while they are in the atmosphere and how long they can stay in the atmosphere before breakdown. Many researches discussed greenhouse gases problem and try to provide solutions to reduce their impacts. Daniel H. et al., 2011, studied greenhouse gases emissions per capita in large cities and possible opportunities of reducing these emissions [11]. Darkwah W. K. et al., 2017, reported the importance of greenhouse gases in maintaining the earth's temperature warm and their effects in global warming problem [13]. Latake P.T. et al., 2015, presented the physical and biological impacts of greenhouse gases and the cost of reducing their impacts [14]. V. Ramanathan and Y. Feng, 2009, tried to analyze atmospheric brown clouds in South Asia and North Africa to understand the main reasons behind climate change and the impacts of greenhouse gases on global warming [19]. Stephen E. S., 2017, illustrated the nature and properties of greenhouse gases and their impacts on climate change and helped climate scientists in their decision [20]. Chunlin Xin et al., 2020, studied the greenhouse gases resulting from five scenarios of solid waste disposal, the results indicate that sorting and recycling waste then the remaining residue is incinerated to generate electricity can reduce greenhouse gases emissions to a great extent [22].

2 Main Greenhouse Gases

Due to the increase in population and human activities, this increases the concentration of greenhouse gases above normal levels which cause the global warming problem. So to realize the effect of greenhouse gases on the weather conditions, environment, people, plants and animals it's important to familiarize with the main sources of these gases [5,

6, 7, 8, and 9]. This section gives a brief description about the major greenhouse gases which include:

Carbon Dioxide (CO_2)

The major gas of greenhouse gases is carbon dioxide. This gas exists in the atmosphere through the carbon cycle with normal concentration, but due to the burning of fossil fuels, solid waste, trees, wood products and cement manufacturing the concentration of CO2 is increased during the last hundreds of thousands of years. The life time of this gas in the atmosphere is long, usually after 100 years 40% of the gas remains in the air, 20% remains after 1000 years and 10% remains after 10000 years [10–16].

Methane (CH_4)

The second important greenhouse gas in the atmosphere is Methane. This gas is stronger than carbon dioxide due to its high ability in absorbing heat which makes its warming impact 25 times greater than carbon dioxide over 100 years period. This gas resulted from synthetic sources such as the production of coal, natural gas and oil, decay of organic waste, landfills and raising of livestock. Natural resources of this gas include volcanoes, oxidizing bacteria, tropical and northern wet lands [18].

Nitrous Oxide (N_2O)

The global warming potential of this gas is from 265 to 300 times of the carbon dioxide over 100 year's period. The life time of this gas is more than a century (120 years). This gas resulted from industrial and agriculture activities and combustion of waste. The concentration of N2O in atmosphere is small due to the natural biological reactions in soil and water [19].

Fluorinated Gases

These gases are man-made gases resulting from many industries, the concentration of such gases in the atmosphere are very small compared with other greenhouse gases. These gases are strong, so they have high global warming potential [20, 21]. The common fluorinated gases are discussed below:

- **Hydro-Flouro Carbon (HFC):** This gas contains hydrogen, fluorine and carbon atoms. It's emitted from manufacturing activities such as air conditioning, refrigeration, fire protection and insulating foams. This gas exists in small concentration but it traps more heat than carbon dioxide. This gas is used as a substitute for ozone depleting substances.
- **Chloro-Fluoro Carbon (CFC):** This gas consists of chlorine, carbon and fluorine atoms, resulting from the manufacturing and industrial activities. The life time is 75 years. The global warming potential of this gas is very high approximately more than ten thousands of years.
- **Sulfur Hexa-Fluoride (SF6):** This gas is a colorless and odorless gas used in electrical circuit interrupters, electric piping and gaseous insulator. Sulfur hexafluoride gas has negative impacts on human health if it's being breathed, so care should be taken to avoid exposure to this gas.
- **Nitrogen Tri-Fluoride (NF3):** This gas consists of nitrogen and fluorine atoms. The main source of this gas is the manufacture of LCD panels, solar panels and semi-conductors. The life time of the gas is more than five hundred years and the global

warming potential is higher than ten thousands of years which indicate that it's stronger than carbon dioxide in trapping heat. The following table illustrates the main sources of greenhouse gases and their impacts on human health and environment.

Table 1 illustrates the properties of the sources of the main greenhouse gases.

Table 1. Sources and impacts of main greenhouse gases

Greenhouse gases	Sources of greenhouse gases	Human health effects	Environmental effects
Carbon dioxide	1-Burning of fossil fuels 2- Burning of biological materials, trees and solid waste 3-Manufacturing of cement	1- Inflammation 2- Reduction of cognitive Performance 3- Kidney and bone problems	Trap heat which cause: 1-Melting ice caps 2-Rising of ocean levels 3- Flooding
Methane	1-Production and transport of coal, natural gas and oil 2-Agricultural practices and livestock 3-Decay of organic solid waste	Respiratory deaths globally	1-Air pollution 2-Increase levels of ozone
Nitrous Oxide	1-Industrial activities 2-Combustion of fossil fuels and solid waste 3-Treatment of waste water 4-Fertilizer application	1-Dizziness, sound distortions, severe headache 2-Deficiency of vitamin B12 which cause anemia 3-Serious damage of nerves	1-Absorbs radiation and traps heat in the atmosphere 2-Depleting the ozone layer

(*continued*)

Table 1. *(continued)*

Greenhouse gases	Sources of greenhouse gases	Human health effects	Environmental effects
Hydro Flouro Carbon	1-Industrial processes, used in Refrigeration, air conditioning, insulating foams and aerosol propellants, with minor uses as solvents and for fire protection	1-Skin cancer and cataracts 2-Damage of immune system 3-Increasing health risk from insect and water-borne diseases	1- Destroy the earth's protective ozone layer, which shields the earth from harmful ultraviolet rays generated from the sun. 2- Damage to terrestrial and aquatic plant life. 3- Formation of ground-level ozone (smog). 4- Rising sea levels, extinctions of species
Chloro Fluoro Carbon	1-Manufacturing of Refrigerators	1-Skin cancer and cataracts 2- Damage of immune system 3-Increasing health risk from insect and water-borne diseases	1- Destroy the earth's protective ozone layer, which shields the earth from harmful ultraviolet rays generated from the sun, Formation of smog 2- Damage to terrestrial and aquatic plant life
Sulfur Hexa fluoride	1-Electricity transmission 2-Production of magnesium 3-Manufacture of semiconductors	1-Irritate the skin causing rash 2-Severe eye pains irritate nose, throat and lungs 3-Dizziness, headache, fainting and suffocation 4- Cancer hazard	Trap heat which cause the following: 1- Melting ice glaciers 2- Raise water levels at oceans and seas 3- Flooding
Nitrogen Tri fluoride	Manufacturing of semiconductor, LCD panels, certain types of solar panels and chemical lasers	1- Reduce the capacity of red blood cells which causes bluish discoloration 2- Headache, dizziness, weakness and confusion. 3- Changes in the liver, kidneys, spleen and heart	1-Trap heat for a long time, which increases the earth's temperature

Table 2 illustrates the life time, radiative forcing and global warming potential for main greenhouse gases.

Table 2. Main properties of greenhouse gases

Greenhouse gases (GHG)	Radiative forcing	Atmospheric lifetime (years)	Global warming potential (GWP)
Carbon dioxide CO2	2.111	Variable (300 - 1000)	1 (reference gas)
Methane (CH4)	0.52	12	28–36
Nitrous Oxide	0.21	120	265–300
Hydro-Flouro Carbon	0.041	15–29	1000–14800
Chloro-Fluoro Carbon	0.248	75	3800 - 14400
Sulfur Hexafluoride	0.002	3200	23900
Nitrogen Tri-fluoride	0.21	740	17200

3 Materials and Methods

Multi criteria decision analysis (MCDA) is a multi-step process used to organize and formalize decision making processes in a clear and specific way. MCDA is used to distinguish between acceptable and unacceptable alternatives, sort multiple options in a list and to identify the best choice between multiple choices [5, 8]. There are many methods of decision analysis; each method has its own way in calculating weights and outranking flow values. So it is necessary to choose the most effective and rational method by considering the data features for the analysis. The selection of a certain technique for multi criteria analysis depends on many factors including ease of use, transparency, logical soundness, realistic time and manpower required for the problem, ability to make edits and changes in the parameters where necessary and software availability. In this study, the Fuzzy PROMETHEE method is selected.

3.1 Fuzzy PROMETHEE

This method is a combination between PROMETHEE and fuzzy logic. PROMETHEE stands for Preference Ranking Organization Method used for evaluation of many options and fuzzy logic is used to cope with the unclear and ambiguous input data. This method is used when the parameters specified for the problem is non-numerical. The preference function used in PROMETHEE method could be linear, Gaussian, U-shape or V-shape based on the problem defined. The ranking type can be either partial or complete based on the degree of accuracy and the preference of the user. The following steps illustrate the equations used for calculating weights and outranking flow values of PROMETHEE method [24, 25]:

1- Determine the number of criterias b for each alternative where (b = 1, 2,......., n) and the set of alternatives for the studied problem.
2- Calculate the weight for each criteria, this weight illustrates the importance of the criterias specified for decision makers. Most commonly selected weights satisfy the equation below;

$$\sum_{b=1}^{n} W_b = 1 \tag{1}$$

3- Normalize the decision matrix from 0 to 1 with the help of the following equation, although not necessarily:

$$R_{ab} = \frac{[X_{ab} - \min X_{ab}]}{[\max X_{ab} - \min X_{ab}]} \tag{2}$$

where a = number of alternatives, b = number of criterias, X_{ab} = evaluation values.
4- Calculate the deviation between each pairs of comparison:

$$d_b(A, B) = g_b(A) - g_b(B) \tag{3}$$

$d_b(A, B)$ = Difference of evaluations of alternatives A and B for each criteria.

5- Determine the preference function:

$$P_b(A, B) = F_b[d_b(A, B)] \tag{4}$$

The smaller the preference values the less important to the decision maker, while the closer the preference values to 1 the more important to decision maker.
6- Calculate the preference index:

$$\pi(A, B) = \sum_{b=1}^{n} P(A, B).W_b \tag{5}$$

where $\pi(A, B)$ refers that alternative A is preferred over alternative B.
7- Find the preference order using the following, the alternative with a higher value of $\varnothing^{+}(A)$ and a lower value of $\varnothing^{-}(A)$ is the best alternative for the problem.

$$\varnothing^{+}(A) = \frac{1}{a-1}\sum \pi(A, X) \text{ And } \varnothing^{-}(A) = \frac{1}{a-1}\sum \pi(A, X) \tag{6}$$

where $\varnothing^{+}(A)$ indicates that an alternative A is the best over other alternatives.
$\varnothing^{-}(A)$ indicates that the other alternatives are better than alternative A.
a = number of alternatives.

4 Criteria of Evaluation

To sort greenhouse gases from the lowest to the highest effect on human health and environment, different criteria have been identified and explained:

Human Impacts
This criterion was used to arrange greenhouse gases in terms of diseases, disorders and health problems caused to humans due to direct exposure and breathing.

Environmental Impacts
Greenhouse gases have great influences on changing the climate and weather conditions all over the world. So it is very important to order greenhouse gases from the lowest to the highest impacted on environment.

Radiative Forcing
It's also known as the climate forcing, it refers to the difference between how much energy from sun has been absorbed by the earth and atmosphere and how much is released into the space, The positive radiative forcing indicates that the incoming energy is greater than the outgoing so causing a warming effect to the earth's surface and atmosphere, while the negative values of radiative forcing result in cooling the planet surface. The radiative forcing is expressed in watt/m^2 [23].

Atmospheric Lifetime
The time the gas stays in the atmosphere before disappearing. These lifetimes range from months, millennia and even centuries. This criterion is very important in determining the overall global warming potential of a greenhouse gas. For example the lifetime of carbon dioxide is between 300 to 1000 years which means that even if human activities were stopped to reduce gas emissions, what existed in the atmosphere will affect human future adversely. Table 3 illustrates the criteria used as an input data for analysis, while Table 4 shows the importance weights and the aim of each criterion.

In this analysis, a triangular linguistic fuzzy scale was used to determine the selected data [26]. The Yager index was used to clarify the presented data. Then, the PROMETHEE technique was applied with the Gaussian preference function.

Table 3. Analysis criterias

Greenhouse gases (GHG)	Human impacts	Environmental impacts	Radiative forcing	Atmospheric life time
Carbon dioxide	Extremely high	Extremely high	Very high	Very high
Methane	Very high	Very high	High	Very low
Nitrous Oxide	High	High	Moderate	Moderate
Hydro-Flouro Carbon	Moderate	Moderate	Low	Low
Chloro-Fluoro Carbon	Moderate	Moderate	Moderate	Moderate
Sulfur Hexa-fluoride	Low	Low	Very low	Extremely high
Nitrogen Tri-fluoride	Very low	Very low	Moderate	High

Table 4. Importance weights of the criteria

Criteria	Importance weight	Aim
Human impacts	Very high	Min
Environmental impacts	High	Min
Radioactive forcing	Moderate	Min
Atmospheric life time	Moderate	Min

5 Results and Discussion

The calculation results of Fuzzy PROMETHEE method is shown in Table 5. The net flow values which is the differences of positive and negative outranking flow values indicates that Nitrogen Tri-fluoride gas has the lowest effect on human health and environment with moderate radiative forcing. Other fluorinated gases including Sulfur Hexa-fluoride, Hydro Flouro Carbon and Chloro Fluoro Carbon were ranked as the second, third and fourth gases impacting the environment and human health. Nitrous oxide is on the fifth rank and Methane is on the sixth rank due to their high negative impacts on human health and environment. Carbon dioxide was ranked as the worst gas because it affects the human health and environment adversely in addition to a very high radiative forcing and staying in the atmosphere for a very long time.

Figure 1 illustrates the positive and negative criterias for each gas above and below the zero level. Although Nitrogen tri-fluoride can stay in the atmosphere for hundreds of years (740 years) but its impacts on human health and environment is less compared with other gases, so it's displayed at the beginning of the list of main greenhouse gases. The second gas in the list is sulfur hexafluoride because it stays in the atmosphere more

Table 5. Fuzzy PROMETHEE results

Rank	Greenhouse Gases	Positive Outranking flow	Negative Outranking flow	Net flow
1	Nitrogen Trifluoride	0,0125	0,0040	+ 0,0085
2	Sulfur Hexafluoride	0,0095	0,0060	+ 0,0035
3	HydroFlouroCarbon	0,0059	0,0029	+ 0,0031
4	ChloroFluoroCarbon	0,0036	0,0044	−0,0009
5	Nitrous Oxide	0,0039	0,0049	−0,0010
6	Methane	0,0071	0,0089	−0,0019
7	Carbon dioxide	0,0037	0,0151	−0,0113

than Nitrogen tri-fluoride do. HydroFlouro Carbon and ChloroFluoro Carbon appears on the third and fourth of the list due to their moderate impacts on human health and environment. Nitrous Oxide has two negative criterias; high impacts on human health and environment compared with two positive criterias radiative forcing and lifetime. Methane has three negative criterias against one positive criteria while Carbon dioxide has four negative criterias so it's shown at the end of main greenhouse gases list.

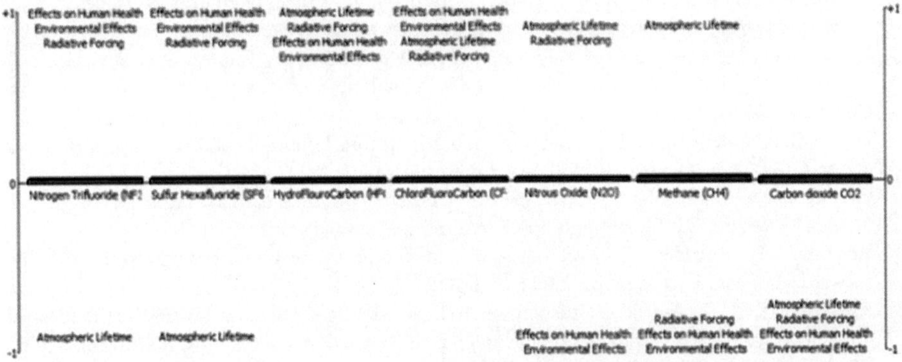

Fig. 1. Fuzzy PROMETHEE graphical chart

6 Conclusion

Greenhouse gases are the main reason of global warming and climate change all over the world. These gases have different impacts on human health and environment based on their radiative forcing, atmospheric life time and global warming potential. Sorting greenhouse gases according to human and environmental impacts is a prerequisite for decision makers to consider the synthetic sources of these gases and try to find ways to reduce the emissions of such gases. Fuzzy PROMETHEE method was used for sorting greenhouse gases based on many criterias including radiative forcing, lifetime, environmental and human impacts. Nitrogen tri-fluoride is the least gas impacts human and environment negatively, while carbon dioxide is the most harmful greenhouse gas to human and environment. The accumulation of carbon dioxide in the atmosphere increases the earth's temperature year after year which will cause serious environmental problems in the future if it is not considered now.

References

1. Arabi, S.: Global warming and its effect on the environmental human health. J. Res. Health **6**(4), 378–379 (2016)
2. Udara Willhelm Abeydeera, L.H., Wadu Mesthrige, J., Samarasinghalage, T.I.: Global research on carbon emissions: a scientometric review. Sustainability **11**(14), 3972 (2019)
3. Buha, A.: The Greenhouse Effect. Toxipedia: connecting science and people (2011). https://www.healthandenvironment.org/docs/ToxipediaGreenhouseEffectArchive.pdf
4. Britannica 2021. Greenhouse Gases: Atmospheric Science (2021). https://www.britannica.com/science/carbon-dioxide
5. Communities and Local Government.; Multi Criteria Analysis: A manual (2009). http://eprints.lse.ac.uk/12761/1/Multi-criteria_Analysis.pdf
6. Energy Information Administration (EIA): Energy and the environment explained: Greenhouse Gases (2021). https://www.eia.gov/energyexplained/energy-and-the-environment/greenhouse-gases.php
7. Environmental Protection Agency: Understanding Global Warming Potentials (2020). https://www.epa.gov/ghgemissions/understanding-global-warming-potentials
8. Jato-Espino, D., Castillo-Lopez, E., Rodriguez-Hernandez, J., Canteras-Jordana, J.C.: A review of application of multi-criteria decision making methods in construction. Autom. Constr. **45**, 151–162 (2014)
9. GHG Management Institute: What is a global Warming Potential? (2021). https://ghginstitute.org/2010/06/28/what-is-a-global-warming-potential/
10. Greenhouse Gas Protocol: Global Warming Potential Values (2013). https://www.ghgprotocol.org/sites/default/files/ghgp/Global-Warming-PotentialValues
11. Hoornweg, D., Sugar, L., Trejos Gómez, C.L.: Cities and greenhouse gas emissions: moving forward. Environ. Urbanization **23**(1), 207–227 (2011)
12. Küstermann, B., Kainz, M., Hülsbergen, K.J.: Modeling carbon cycles and estimation of greenhouse gas emissions from organic and conventional farming systems. Renewable Agric. Food Syst. **23**(1), 38–52 (2008)
13. Kweku, D.W., et al.: Greenhouse effect: greenhouse gases and their impact on global warming. J. Sci. Res. Rep. **17**, 1–9 (2017)
14. Latake, P.T., Pawar, P., Ranveer, A.C.: The greenhouse effect and its impacts on environment. Int. J. Innov. Res. Creat. Technol. **1**(3), 333–337 (2015)
15. Munda, G.: Multiple criteria decision analysis and sustainable development. In: Greco, S., Ehrgott, M., Figueira, J.R. (eds.) Multiple criteria decision analysis. ISORMS, vol. 233, pp. 1235–1267. Springer, New York (2016). https://doi.org/10.1007/978-1-4939-3094-4_27
16. National Oceanic and Atmospheric Administration (NOAA): Greenhouse Gases (2021). https://www.ncdc.noaa.gov/monitoring-references/faq/greenhouse-gases.php
17. National Science Centers: What are Global Warming Potentials and CO2 Equivalent Emissions? (2020). https://niwa.co.nz/atmosphere/faq/what-are-global-warming-potentials-and-co2-equivalent-emissions

18. Public Health England: Methane: General Information (2019). https://assets.publishing.service.gov.uk/government/uploads/system/uploads/attachment_data/file/769766/Methane_PHE_general_information__070119.pdf
19. Ramanathan, V., Feng, Y.: Air pollution, greenhouse gases and climate change: global and regional perspectives. Atmos. Environ. **43**(1), 37–50 (2009)
20. Schwartz, S.E.: The greenhouse effect and climate change: the intensified greenhouse effect. Am. Journal of Physics, 86(BNL-205797–2018-JAAM) (2018)
21. Warm Heart Worldwide 2021. Climate Change Primer (2021). https://warmheartworldwide.org/climate
22. Xin, C., Zhang, T., Tsai, S.B., Zhai, Y.M., Wang, J.: An empirical study on greenhouse gas emission calculations under different municipal solid waste management strategies. Appl. Sci. **10**(5), 1673 (2020)
23. Yu, J., Lee, S.: The impact of greenhouse gas emissions on corporate social responsibility in Korea. Sustainability **9**(7), 1135 (2017)
24. Zarghami, M., Szidarovszky, F.: Multicriteria analysis: applications to water and environment management. Springer, Cham (2011). https://doi.org/10.1007/978-3-642-17937-2
25. Zlaugotne, B., Zihare, L., Balode, L., Kalnbalkite, A., Khabdullin, A., Blumberga, D.: Multi-criteria decision analysis methods comparison. Environ. Clim. Technol. **24**(1), 454–471 (2020)
26. Yildirim, F.S., Sayan, M., Sanlidag, T., Uzun, B., Ozsahin, D.U., Ozsahin, I.: Comparative evaluation of the treatment of COVID-19 with multicriteria decision-making techniques. J. Healthcare Eng. (2021)

Sustainable Development for a Secure Future: An Overview of Challenges and Key Solutions

P. C. Kesavan[1], O. S. Glazachev[2], Yu. M. Grishaeva[3,5], I. V. Spirin[4], and O. V. Alymova[5(✉)]

[1] Jawaharlal Nehru University, New Delhi, India
pckesavan@mssrf.res.in
[2] I.M.Sechenov First Moscow State Medical University, Moscow, Russia
[3] Moscow Pedagogical State University, Moscow, Russia
[4] Scientific Research Institute of Motor Transport, Moscow, Russia
[5] Moscow Power Engineering Institute, Moscow, Russia
alymovy@yandex.ru

Abstract. The following issues discusses in the article: loss of biodiversity, poverty, climate change and environmental pollution. Currently, the world's population is growing rapidly, so food security of mankind in the XXI century remains an unsolved problem. The satisfaction of quantitative requirements in food products is the most important task. As it is solved, the following problem arises to meet the nutritional needs in terms of quality and environmental protection. Thus, the problems of ecotechnologies for eco-agriculture and eco-enterprises for sustainable food security come to the fore. Today, the Earth's biosphere is accumulating at an accelerated rate several chemical pollutants derived from anthropogenic activities in agriculture, manufacturing industry, nuclear, space, nano- and biotechnologies. Speaking about the loss of biodiversity, it should be noted that factory farming largely involves monocropping. Chemical fertilizers and pesticides (including herbicides) are used liberally. On the environmental side, chemicals and over exploitation of groundwater lead to degradation of land, aquifers and biodiversity. In a nutshell, the shift from exploitative to sustainable agriculture needs to be in the direction of using biologically-derived inputs than chemically-synthesized inputs to augment yield increase. It is important to note that for intelligent interaction with nature, it is necessary to form environmental knowledge among new generations.

Keywords: Eco-agriculture · Clean water · Food security · Eco-enterprises · Sustainable development · Ecological education

1 Introduction

More than a decade has elapsed since a Special Issue of Scientific American (September 2005) published a series of articles bearing on «crossroads for Planet Earth». The topics covered were: population peak, loss of biodiversity, poverty, energy problems, climate change and water crisis. The Intergovernmental Panel on Climate Change (IPCC) has

H. Gökçekuş and Y. Kassem (Eds.): NRSEM 2021, EESCI, pp. 296–305, 2022.
https://doi.org/10.1007/978-3-031-04375-8_33

been regularly holding conferences involving all the member nations to find ways and means to arrest the average rise in the global temperature below 1.5 °C. This now seems rather difficult despite severe warnings by climate scientists that the Planet will be hit by «tipping point» when the global average temperature reaches between 1.8 to 2.0 °C. The «tipping point» would take the Planet into a different and altogether unknown state of equilibrium which could be hostile to the welfare and even possibly the very survival of humans and several other species, mainly the mammals which are hot-blooded and have a rather narrow range of homeostatic adaptation. The IPCC conferences with all good and serious intentions have largely been far less successful in making all the member nations to substantially limit the production and release of greenhouse gases. Disappointed on this account, *Nature* editorial wrote in the editorial «The Mask Slips» that The Durban Meeting (South Africa, 2011) shows that climate policy and climate science inhabit parallel worlds. Consequently, it seems that it is inevitable that the Planet Earth will face the tipping point. Scientists from different countries are conducting studies, the results of which allow us to judge the extent of the environmental consequences of climate warming and to make forecasts of the dynamics of these processes on this basis. It is known that an increase in the temperature of the earth's surface leads to the melting of ice in the Arctic and Antarctic. Therefore, the results of monitoring the state of the surface of the polar zones can serve as an indicator of climate change on the Planet as a whole. A group of scientists conducted a study of space-time relationships between the temperature of the earth's surface and the state of its covers. Information was collected using the satellite measurements and radar surveying in 2008–2010. The Assessment of changes and prediction of the temperature balance as the earth surface is recommended during the summer. This will provide the most reliable assessment and forecast results [1].

Today, the concern is not just about climate change only. The sustainable development goals include: the eradication of hunger (goal 2), the eradication of poverty (goal 1). Sustainable resource management and ecosystem conservation (goal 15), climate change (goal 13).

2 «Hunger-Free World»: Problem Statement

In a hierarchy of human needs, food and clean drinking water are absolutely basic. The world has always had people with total access to balanced diet as well as those who had little or no access to adequate caloric diet. Fighting for a "Hunger-free world" has to tackle three kinds of hunger on the one hand and also three essential physical dimensions on the other. The three kinds of hunger are: (a) Caloric (i.e. no food at all) hunger – can be satisfied with provision of cereals. Food security of mankind in the XXI century remains an unsolved problem, which is in line with the tasks of implementing the concept of sustainable development of the UN. The importance of the development and use of food standards that should take into account the index of consumption of energy and the macronutrients (energy and macronutrient intake index – ENI). Recently, there has been an increase in fat consumption and caloric intake of food, but not enough consumption of fruits, vegetables and animal protein. "Scenario application proves that the relationship between ENI and underneath could readily be applied in future scenarios generated by integrated assessment models to provide insights into the impacts of various climate change scenarios, socio-economic development paths and alternative global trade policies on global hunger and underneath status" [2]. The satisfaction of quantitative requirements in food products is the most important task. This is the most urgent task for poor and developing countries. As it is solved, the following problem arises to meet the nutritional needs in terms of quality. It is well known that in many developed countries where the food problem has been solved, a significant number of people are overweight. This impairs the quality of life of such people. Therefore, the next step is to balance the diet of people in terms of quality. First of all, it concerns the caloric content of food.

The choice of food from the alternative list offered to people is influenced by the work of the brain. A group of 19 respondents with normal body weight was offered a choice of products that are almost identical in taste, but differ in the estimated and actual caloric content (high-low). When selecting products, respondents performed an MRI scan of their brain (MRI (fMRI)). It is established that food irritants are stronger in comparison with the choice of non-food products. The regions of the brain, forming the preferences of food products in connection with the feeling of hunger and body weight of the person. The findings can be used as markers for assessing the biological significance of food [3].

(b) Protein hunger – inadequate or no intake of pulses, poultry, fish, meat, beef, pork etc. This results in stunting of growth especially of children. The developing countries, particularly those with large vegetarian populations have many children, women and men suffering from «protein hunger». But not a purely mechanical approach to the solution of the problem of the presence of protein in the human diet. The concepts of «saturation» and «satiety» in modern studies have shown a special importance of the composition of proteins used in food. The use of foods high in protein has a significant impact not only on satisfying hunger, but also helps to maintain body weight at the recommended level of medicine. The article [4] presents the results of studies that can be used by the food industry to produce products with a rational content of proteins in them.

(c) Hidden hunger: As already discussed, this is a rather invisible type of hunger caused by micronutrients deficiencies in the diet. This category includes about 2 billion people. Despite economic success, agricultural production and health development in recent years, hidden hunger is prevalent in South Asia. In the diet of people the most significant lack of iodine (I), iron (Fe), zinc (Zn) and vitamin A. Solution to this problem is possible at present. To this end, governments need to implement programmes of action aimed at simultaneously reducing poverty, protecting the population and overcoming hidden hunger [5]. In the report [6] considered the problem of hidden hunger due to the aging of the population. The increase in life expectancy has led to an increase in the proportion of people over 65 years of age. At this age, people are at risk of chronic diseases, which include diseases of the cardiovascular system, diabetes, and infections (influenza, bronchitis, pneumonia). Older people need to follow a diet that takes into account age-related changes in their organisms. However, the production and distribution of appropriate food does not meet the needs of older persons. The report summarizes the experience of implementation of programs and strategies to address the ageing of the population. This experience was explored at the Hidden Hunger scientific Symposium:

solutions for America's aging population, March 23, 2018, also at the 4th International Congress Hidden Hunger in Germany, March 01 2019.

All these three kinds of hunger could be alleviated only with the help of three following pro-visions:

a) Availability of food in the market/shops: Making available food grains, pulses, oilseeds, vegetables and fruits is a function of 'production' locally or importation. Several unfortunate countries in the developing world do not produce enough to meet the local needs, and have no enough money either to import. Consequently, the people starve and in the recent years the print and electronic media have been effectively portraying the heart-breaking stories of hunger and deprivation hunger of the children orphaned by civil wars, terrorism etc.
b) Access: The markets/shops might be loaded with a variety of food items, but people may not have money to buy them. Lack of «purchasing power» in such situations as «abject poverty» denies access to food. In India, famine of rural livelihoods (i.e. inadequate jobs/means of livelihoods) is a major cause of food insecurity and hunger of millions of people in the rural households.
c) Absorption: Lack of clean drinking water in the rural areas of several developing countries leading to gastro-enteric diseases (e.g. dysentery, diarrhea etc.). This block the «absorption» of the ingested food; the net result is «leaky pot». What have been aforementioned bring out the complexity of the problem of achieving a truly «Zero Hunger» world proposed by the then Secretary – General of the United Nations at the «RIO + 20» Conference held in June 2012 in Johannesburg, South Africa.

India has a national food security program. The program provides food for 800 million citizens of the country by subsidizing the production and distribution of grain crops. The implementation of the program showed the existence of contradictions related to the production, purchase, storage, transportation of products, and distribution of subsidies along the logistics chain. The problems are the lack of integration of various participants in the supply process, shortcomings in the organization of the warehouse and transport facilities, losses during loading and unloading. Some inefficient production processes are subsidized. Labor relations should be regulated. Organizational reforms, improvement of inventory management policy and grain supply logistics system, use of cloud technologies for information support of the program management are recommended [7].

3 Ecotechnologies for Eco-agriculture and Eco-enterprises for Sustainable Food Security

During the epoch Holocene about 10,000 years ago, the Homo sapiens started farming and domestication of farm animals. Earlier, about 30 to 40 thousand years ago they transformed wild dogs as pet animals as partners in hunting and also as loyal companions. The then agriculture was largely ecofriendly except in the case of conversion of forestland into cropland. Things, however, rapidly changed soon after the «Industrial Revolution» that was ushered in with the invention of steam engine by James Watt in 1780. In the middle of the 19th century, Justus von Liebig in Germany discovered that nitrogen-containing chemical compounds added to the soil promoted luxurious growth and productivity of plants, particularly the crop plants. As to the means of commercial production of nitrates (plants uptake nitrogen in the form of nitrates), the Haber-Bosch Process was adopted. And then, in 1939 Paul Muller discovered the insecticidal properties of a chemical compound, DDT had been synthesized much earlier (in 1874) by Othmar Zeidler. The humankind was jubilant as it could get rid of every unwanted insect in the world. Then came the book, «Silent Spring» by Rachael Carson (1962) which described the mass destruction of non-target beneficial organisms by the DDT. The pesticides in general are carcinogens, endocrine disruptors and killers. More recently, Barbara Cohn et al. (2015) have shown that DDT, an endocrine disruptor induces breast cancer in women of about 52 years of age, following their exposure in utero in the 1960s. Their mothers who had high exposures to DDT in the 1960s showed high levels of DDT in their system. It is an amazing study of 54-year follow-up of mothers exposed to DDT in the 1960s, the consequent exposure of the female fetuses in such pregnant mothers, and these daughters exposed to high levels of DDT in their mothers' wombs developing breast cancers when they reach about 52 years of age. The study showed that women exposed to the higher levels of DDT in the womb had 3.7 times higher risk of breast cancer than those who had the lowest exposure to DDT. It is, therefore, highly inaccurate to conclude that short-term studies did not reveal adverse effects. Long-term studies for at least two years using chronic feeding of experiments rats/mice are essential to gather reliable data.

Today, the Earth's biosphere is accumulating at an accelerated rate several chemical pollutants derived from anthropogenic activities in agriculture, manufacturing industry, nuclear, space, nano- and biotechnologies. The «ecological footprint» (Wackernagel M. et al.) is overshooting while the «biocapacity», of the Earth to restore what has been degraded is concomitantly diminishing.

Notwithstanding the fact that most of the enlightened lay citizens of the world understand that the Planet Earth, and in particular, its biosphere are at a crossroad, development under the globalization goes on largely as «business-as-usual». This brings into discussion the Corporate (factory) Farming vis-a-vis Family Farming. Factory farming largely involves monocropping, that is just one high-yielding variety (HYV) of corn, wheat, soy, cotton, paddy or any other. Chemical fertilizers and pesticides (including herbicides) are used liberally. It also aims at having as few farm workers as possible, so that jobless economic growth increases profits. Consequently, automation adds to the numbers of the unemployed. On the environmental side, chemicals and over exploitation of groundwater lead to degradation of land, aquifers and biodiversity. More importantly, it is generally forgotten that food production in the world towards food «availability» is largely due to the toil of about 500 million family farmers throughout the world. The family farms in North America and Western Europe could be large (over 100 ha) whereas those in several developing countries are about 1 to 2 ha or even less. Labor for farm work involves farm animals for draught purpose. Further the members of the families and friends also contribute labor and resources. From the point of protecting farm-based livelihoods, studies show that corporate farming creates 9.44 jobs displacing 27.97. So, the number of unemployed people, particularly youth from the farming sector in the developing countries in-creases. Jobless economic growth breeds discontentment and violence in the hearts of people. In a nutshell, the shift from exploitative to sustainable agriculture needs to be in the direction of using biologically-derived inputs than chemically-synthesized inputs to augment yield increase. Further, focusing only on high yields or overcoming susceptibility to a pest or disease alone will not do; concurrent attention to soil health, quality fresh water, rich biodiversity, renewable energy including use of manual labor and farm animals is very essential to achieve what is called «farming with nature» and not «farming against nature».

Lessons learnt from the *«Green Revolution»* and *«Genetic Engineering»* technologies over the last several decades is that there is an urgent need to shift our strategies towards sustainable agriculture. With due recognition of the need for concurrent attention to soil health, fresh water, biodiversity, renewable (clean) energy etc., has elaborated his concept of «evergreen revolution» which he had proposed in 1996. He defines the «evergreen revolution» as achieving productivity in perpetuity without accompanying environmental and social harm. In today's parlance, it is the «systems approach» to crop and farm animals husbandry. Its basic tenets are that chemistry-based (chemical) inputs such as inorganic fertilizers and chemical pesticides are increasingly replaced with bio-derived (biological) *biofertilizers* and *biopesticides.* Biofertilizers are the various microbial organisms which atmospheric fix nitrogen as nitrate in the soil. An age-old method of traditional farming involved crop rotation in which naturally nitrogen-fixing leguminous crops are grown before a cereal crop. *Sesbania rostrata* has nitrogen-fixing bacteria in the nodules on leaf surface. Cultivating this species and ploughing it in the soil enriches the nitrogen-content (i.e. nitrate) by about 70 to 80 kg of N2 per hectare. The succeeding paddy crop therefore, would require substantially low level of inorganic nitrogen fertilizer addition. There are also fertilizer trees such as the African fertilizer tree (*Faidherbia albida*) which takes nitrogen from the air and fixes it as nitrates in its leaves, these leaves eventually fall and their nitrates are incorporated into the soil. *Azolla* is an alga that is cultured in special facilities and applied as nitrogen-containing biofertilizer.

As regards the biopesticides, there are several herbal formulations which do not leave toxic/carcinogenic residues. Thus, these are safe from health and environmental point of view. Even more spectacular is the ingenious use of the parasites and predators in the biological world to contain the serious insect pests of agricultural crops. An example is the use of egg parasitoid (*Trichogramma chilonis*) a tiny wasp that lays its eggs into the eggs of cotton bollworm and consequently the pest production is drastically curtailed. The ingenious use of insect parasites against insect pests is absolutely free from chemical toxicity in the environment. Natural parasites exist against cotton bollworm (*Helicoverpa armigera and Pectinophora gossypiella*) and fruit and shoot borer of brinjal (*Leucinodes orbonalis*).

Using the example of 707 French farms in 2008, a comparative analysis of the cost of growing arable crops at high and low concentrations of pesticides per hectare of crop area was performed. The results of the analysis showed that the use of lower concentrations of pesticides resulted in more competitive economic performance in terms of direct costs. Such results were statistically reliable regardless of crop type, and provide ecofriendly farming and crop productivity [8]. The introduction of eco-agricultural technologies should encourage farmers to restructure their activities. Governments should therefore provide appropriate compensation to farmers when needed. In China, a study was conducted on the willingness of farmers to respond to compensation in response to the introduction of environmentally friendly technologies. The study was conducted on the basis of the traditional food of the region a pair of rice – fish [9]. A by-product of animal production and urbanization are the various bio-waste, including feces. On farms and in cities, these by-products are produced in large quantities, which requires processing. At the same time, ammonia emissions into the atmosphere are a negative phenomenon. Biogas is produced from waste by anaerobic digestion, composting and chemical hydrolysis. Biogas can be used as a motor fuel. The article [10] presents the positive results of anaerobic digestion of manure on farms and biowaste in cities. Thus, damage to the environment is prevented and at the same time environmentally friendly fuel is obtained.

For many developing countries, rain-fed agriculture is important in providing crop products. To evaluate the productivity of dry land crops and forecasting yields it is necessary to use reliable and available indicators. Argentine scientists in semi-arid pampas conducted a study of the quality of various indicators to judge the presence in the soil of organic substances, phosphorus and nitrogen compounds, moisture reserves. The study of the correlation between yield and indicators was carried out using aerial photography differentiated for different areas of land used for growing corn. Aerial photography is a fast and effective non-contact way to obtain information about the state of rainfed lands on a large area. The results obtained made it possible to carry out targeted control actions on corn yield [11].

The spread of innovative technologies of agricultural production in the era of the knowledge economy requires the creation of effective systems of training. In agriculture are a large proportion of small businesses today. Such producers usually do not have the opportunity to break away from their production for a long time to gain new knowledge. Therefore, it is important to organize a system of remote self-education of farmers. Information support of such systems is provided by modern cloud technologies. The practical application of distance learning in agriculture based on cloud technologies is considered in the article [12].

Important is the transition of shift from the chemical to biological inputs. The latest production processes for the production of materials for biomedical and agricultural purposes are based on environmentally friendly «green» biotechnologies and nanosynthesis. A comprehensive review of various biosynthesis methods using fungi, yeast, algae, viruses and other biological forms is made in the analytical review [13].

Rational agricultural production is based on the use of healthy soil. Healthy soil keeps water in dry weather and drains water with its excess, minimizes nutrient consumption, allows you to maximize the number of days of its useful use in the year. Soil health is assessed by a set of physical, chemical and biological indicators. Biological indicators are the least studied. One indicator is the presence and diversity of earthworms, which are easy to observe [14].

Brazilian scientists have studied the quality of untreated wastewater flowing into rivers flowing into Guanabara Bay. Coprostanol, produced in the intestine and not assimilated by humans and animals, was used as an indicator to assess the state of water. The use of this indicator allows the monitoring of river waters for compliance with standards, and to determine the suitability of river waters for the organization of people's rest [15].

Smallholder family farms have an important role in world's food and nutrition security. According to agricultural censuses and studies, there are currently more than 570 million farms in the world. Most of these farms are small farms with 2 ha or less of land. Small firms account for about 12% of agricultural producers. About 75% of agricultural land belongs to family agriculture. In low-income countries, farm sizes have been declining over the past 40 years. In middle-income countries, farm sizes tend to increase. The food and agriculture organization of the United Nations (Food and Agriculture Organization of the United Nations) focuses on the study of the number and size of farms to establish monitoring through agricultural censuses to enable the formulation of sound policies for the development and promotion of agricultural production [16].

However, according to a study by Brazilian scientists, conventional agricultural censuses are not sufficiently informative. The study was conducted on a representative database of 4.7 million farms. Income, productivity and diversification of production in small farms were analyzed. This analysis revealed the characteristics of Brazil's most numerous participants in the agricultural business. Small agricultural producers are able to improve the profitability of the business, it is easier to carry out diversification of agricultural production and can increase productivity. The results of the analysis are used to develop sound and effective policy decisions to stimulate agricultural activities [17].

4 Conclusion

The question is as to how this goal could be achieved in the light of the fact that the science and technology of today are hardly eco-friendly as also without social and gender equity concerns. As Jeffrey Sachs Director of the Earth Institute at Columbia University, New York observed, «Even with all our technological wizardry, we have not yet conquered the Malthusian challenge since we have not yet adopted truly sustainable method of feeding the planet».

In modern society there is a deep transformation of socio-economic foundations of all forms of social relations. A post-industrial society is being formed, the foundations of which have been studied by Daniel Bell, Peter Ferdinand Drucker and their followers [18, 19]. In post-industrial society, physical labor is gradually mechanized and automated. Educated people become the main class, and science becomes a productive force. For example, in the most advanced economy of the modern world – in the United States, about 79% of jobs are related to the knowledge economy. To refer to this new intellectual class, Alvin Toffler proposed the use of the concept of "cognitariat" as an alternative to the dominant class of the "proletariat" in the twentieth century [20]. Therefore, it is important to organize a system of remote self-education of farmers. Information support of such systems is provided by modern cloud technologies. The introduction of eco-agricultural technologies should encourage farmers to restructure their activities. Governments should therefore provide appropriate compensation to farmers when needed.

The spread of education and the leading role of cognitariat promote deep understanding by most people of the necessity of harmonization of relations with nature [21].

Environmental knowledge is shaping the understanding of intelligent interaction with nature in new generations [22, 23]. Socio-economic relations are globalizing. Solving SD tasks depends on the quality of training of professionals who are currently receiving vocational education. One of the areas of SD is the participation of specialists from various fields of activity in solving environmental problems [24, 25]. Therefore, it is very important to pay close attention to the training of students in the field of ecological education.

In the era of the modern Anthropocene it is necessary to transform social relations and understanding of man's place in nature. At present, debates are being held to clarify the historical significance of the Anthropocene, various measures are being developed to overcome the crisis phenomena, and the psychology of people is being transformed in the direction of eco-consciousness [26].

References

1. Muster, S., Langer, M., Abnizova, A., Young, K.L., Boike, J.: Spatio-temporal sensitivity of MODIS land surface temperature anomalies indicates high potential for large-scale land cover change detection in arctic permafrost landscapes. Remote Sens. Environ. **168**, 1–12 (2015). https://doi.org/10.1016/j.rse.2015.06.017]
2. Luan, Y., Fischer, G., Wada, Y., Sun, L., Shi, P.J.: Quantifying the impact of diet quality on hunger and undernutrition. J. Clean. Prod. **205**, 432–446 (2018). https://doi.org/10.1016/j.jclepro.2018.09.064
3. Charbonnier, L., van der Laan, L.N., Viergever, M.A., Smeets, P.A.M.: Functional MRI of challenging food choices: forced choice between equally liked high- and low-calorie foods in the absence of hunger. Plos One **10**(7), e0131727 (2015). https://doi.org/10.1371/journal.pone.0131727
4. More, P., Fiszman, S.: Revisiting the role of protein-induced satiation and satiety. Food Hydrocolloids **68**(SI), 199–210 (2017). https://doi.org/10.1016/j.foodhyd.2016.08.003
5. Harding, K.L., Aguayo, V.M., Webb, P.: Hidden hunger in South Asia: a review of recent trends and persistent challenges. Public Health Nutr. **21**(4), 785–795 (2018). https://doi.org/10.1017/S1368980017003202
6. Eggersdorfer, M., et al.: Hidden hunger: solutions for America's aging populations. Nutrients **10**(9), 1210 (2018). https://doi.org/10.3390/nu10091210
7. Singha Mahapatra, M., Mahanty, B.: India's national food security programme: a strategic insight. Sādhanā **43**(12), 1–13 (2018). https://doi.org/10.1007/s12046-018-0947-2
8. Boussemart, J.-P., Leleu, H., Ojo, O.: Exploring cost dominance in crop farming systems between high and low pesticide use. J. Prod. Anal. **45**(2), 197–214 (2015). https://doi.org/10.1007/s11123-015-0443-1
9. Liu, M.-C., Xiong, Y., Yuan, Z., Min, Q.-W., Sun, Y.-H., Fuller, A.M.: Standards of ecological compensation for traditional eco-agriculture: taking rice-fish system in Hani terrace as an example. J. Mt. Sci. **11**(4), 1049–1059 (2014). https://doi.org/10.1007/s11629-013-2738-x
10. Riggio, V., Rosso, M., Comino, E., Biagini, D., Montoneri, E.: Ecofriendly manure anaerobic digestion assisted by soluble bio-based substances obtained from anaerobic digestion, composting and chemical hydrolysis of urban bio-wastes: a step toward the integration of urban and agriculture waste management. J. Chem. Technol. Biotechnol. **92**(5), 1111–1117 (2017). https://doi.org/10.1002/jctb.5106
11. Farrell, M., Gili, A., Noellemeyer, E.: Spectral indices from aerial images and their relationship with properties of a corn crop. Precision Agric. **19**(6), 1127–1137 (2018). https://doi.org/10.1007/s11119-018-9570-9

12. Jia, L.F.: Research on remote training system of eco-agriculture based on cloud computing. Agro Food Ind. Hi-Tech **28**(3), 2657–2660 (2017)
13. Saratale, R.G., et al.: A comprehensive review on green nanomaterials using biological systems: recent perception and their future applications. Coll. Surf. B-Biointerfaces **170**, 20–35 (2018). https://doi.org/10.1016/j.colsurfb.2018.05.045
14. Griffiths, B.S., Faber, J., Bloem, J.: Applying soil health indicators to encourage sustainable soil use: the transition from scientific study to practical application. Sustainability **10**(9), 3021 (2018). https://doi.org/10.3390/su10093021
15. Costa, L.A.D., Pessoa, D.M.M., Carreira, R.D.: Chemical and biological indicators of sewage river input to an urban tropical estuary (Guanabara Bay, Brazil). Ecol. Indic. **90**, 513–518 (2018). https://doi.org/10.1016/j.ecolind.2018.03.046
16. Lowder, S.K., Skoet, J., Raney, T.: The number, size, and distribution of farms, smallholder farms, and family farms worldwide. World Dev. **87**, 16–29 (2016). https://doi.org/10.1016/j.worlddev.2015.10.041
17. Herrera, G.P., et al.: Econometric analysis of income, productivity and diversification among smallholders in Brazil. Land Use Policy **76**, 455–459 (2018). https://doi.org/10.1016/j.landusepol.2018.02.025
18. Bell, D.: The Coming of Post-Industrial Society: A Venture of Social Fore-casting, vol. XIII, p. 507. Basic Books, Cop, New York (1973)
19. Drucker, P.F.: Managing in the Next Society, p. 336. Truman Talley Books/St. Martin's Press, New York (2007)
20. Toffler, A.: Powershift: Knowledge, Wealth and Violence at the Edge of the 21st Century, p. 506. Bantam Books, New York (1990)
21. Grishaeva, Y., Spirin, I.V., Matantseva, O.: Aspects of professional education in the higher school in the interests of the techno-sphere safety. Mod. Res. Social Prob. (Online Sci. J.) **9**(65), 5–18 (2016). https://doi.org/10.12731/2218-7405-2016-9-5-18
22. Grishaeva, Y.M., Wagner, I.V., Tkacheva, Z.N., Lugovskoy, A.M., Moro, P.N.: Education for sustainable development today: a problem area for overcoming difficulties of pedagogical adaptation (on the example of a higher school). South Russ. Ecol. Dev. **13**(3), 159–166 (2018). (In Russian), https://doi.org/10.18470/1992-1098-2018-3-159-166
23. Grishaeva, Y., Gagarin, A., Spirin, I., Evstafieva, N., Napolov, O.: Ecological culture of students in the trends of the Concept of sustainable development (2021). https://doi.org/10.1051/e3sconf/202126507003
24. Spirin, I.V., Grishaeva, Y.M., Matantseva, O.Y., Tkacheva, Z.N.: Human capital as a tool of ensuring sustainable development. In: Solovev, D.B., Savaley, V.V., Bekker, A.T., Petukhov, V.I. (eds.) Proceeding of the International Science and Technology Conference "FarEastCon 2020." SIST, vol. 227, pp. 709–714. Springer, Singapore (2021). https://doi.org/10.1007/978-981-16-0953-4_69
25. Grishaeva, Yu.M., Matantseva, O.Yu., Spirin, I.V., Savosina, M.I., Tkacheva, Z.N., Vasin, D.V.: Sustainable development of transportation in the cities of Russia: experience and priorities. South Russ. Ecol. Dev. **13**(4), 24–46 (2018). (In Russian), https://doi.org/10.18470/1992-1098-2018-4-24-46
26. Shim, Y.-S., Bellomy, D.C.: Thinking and acting systematically about the anthropocene. Syst. Pract. Act. Res. **31**(6), 599–615 (2018). https://doi.org/10.1007/s11213-018-9442-2

Sustainable Municipal Solid Waste Management with Zero Waste Approach

Serpil Öztaş and Nihal Bektaş(✉)

Environmental Engineering Department, Gebze Technical University, Gebze, Turkey
{soztas,nbektas}@gtu.edu.tr

Abstract. The world's resources are limited and threatened by numerous factors such as population growth, climate change and waste and pollution from municipal and industrial processes. Human beings use technologies and improvements to expand economic growth and enhance their lives. However, it is essential to be used these developments in a clever or sustainable way so that future generations can use the same resources. Also, great loss and collapse of biodiversity and ecosystems with irreversible consequences which leads to a severe depletion of resources and environment for humanity and industry. Therefore, sustainability is an important perception and Sustainable Waste management does not only generate local impacts but also would minimize environmental burden of the plant earth. Municipal Solid Waste (MSW) is recognized as a public service and should be managed using methods supported by different technologies within the framework of the rules determined by legal requirements. The aim of this presentation to show the importance of sustainable waste management options for our society through the "Zero Waste Regulation" since it can reduce or eliminate harmful environmental effects through proper initial municipal waste management applications.

Keywords: Sustainability · Waste management · Circular economy · Zero waste

1 Introduction

In original definition of sustainability can be described as meeting of present needs without compromising the ability of future generations to meet their own needs [1]. More recently sustainably definition has shifted to "A dynamic process which enables all people to realize their potential and to improve their quality of life in ways which simultaneously protect and enhance the earth's life support systems" (Forum for the Future). In this world, everything for our well-being and survival depends on our surrounding environment. Sustainability can be associated with every step from manufacturing to transformation to customer service to finally to disposal. the company can maximize the benefits from an environmental focus in the long-term. Earth can be protected for our environment, humanity and all living things by seeing the environment is an exhaustible resource [2, 3].

H. Gökçekuş and Y. Kassem (Eds.): NRSEM 2021, EESCI, pp. 306–312, 2022.
https://doi.org/10.1007/978-3-031-04375-8_34

However, protection of environment is not final edge of sustainability. The terms of economic development and social equity are also be covered by theory of sustainability. These three forms of sustainability are known as the three pillars of sustainability «people, planet and profits». Therefore, true sustainability can be reached by the equal harmony of economic, social and environmental sustainability. In a sustainable world, humans live in agreement with the natural environment, preserve natural resources for future generations and enjoy social justice and a high quality of life (Fig. 1) [3, 5].

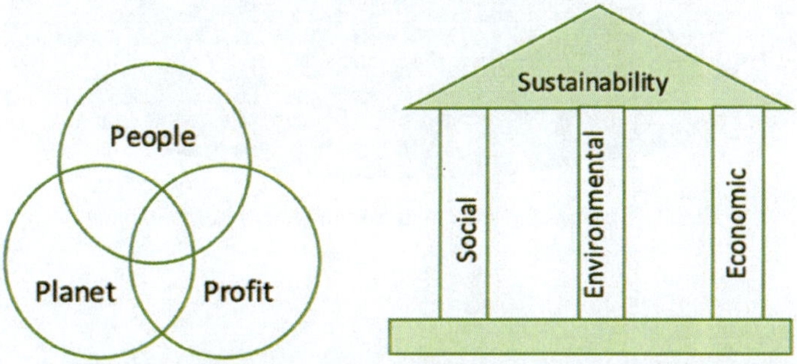

Fig. 1. Relationships with the dimensions of sustainability

2 Municipal Solid Waste Management in Sustainable World

Environmental engineering uses many different disciplines such as engineering, soil science, biology, and chemistry to improve ecosystem health, develop solutions to environmental problems and conserve natural resources [5–7]. However, in this century, the older style of engineering has changed to a new style that makes ecology a priority.

Human activities generate waste materials that are often discarded because they are considered useless. Solid waste can be defined as [11, 19] any unwanted-unused item or substance resulting from a human activity or process. Since these materials aren't functioning anymore can be classified as unwanted or un-useful material however it may be useful to other person/process. The concept of waste economy in the world and in Turkey has emerged not only to protect nature but also to contribute to the economic cycle creating a new economic sector related to waste. Solid Waste Management includes all administrative, economic, legal, design and engineering functions related to solid waste problem solution. Managing MSW is challenging because of its heterogeneous nature.

Solid Waste Management is an issue that needs to be addressed with a systems approach. Hence, there is a global urgent need for sustainable, cost-effective, as well as integrated approaches, to better manage risks and recover resources from solid waste. Managing of solid waste is not only a technical problem but also a problem of organization, education, industrial design and responsibility and a key steppingstone in moving from a throwaway society to a sustainable society. Increasing amount of MSW is one of the greatest concerns of the modern world. There are several strategies can be implemented to overcome and manage the ever-increasing solid wastes. However, there are various issues associated with these waste management strategies, resulting despoiled the quality of natural resources globally [5, 7–9] (Fig. 2).

Sustainable Solid Waste Management

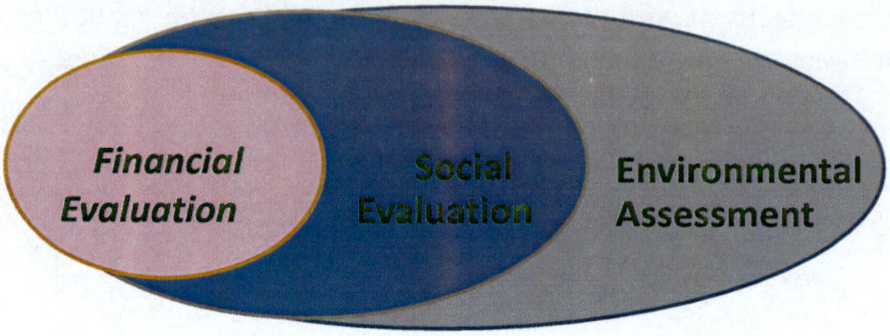

Fig. 2. Sustainable solid waste management dimensions

Financial Evaluation

The management of wastes as an important source of income that should be evaluated, while reducing the administrative expenses, also enables them to be evaluated as a surplus income item. The budget allocated by the cities for waste management decreases and provides an opportunity to transfer them to different sources.

Social Evaluation

Dissemination of wastes is prevented, and thus, infectious diseases are reduced. The factors that threaten the health of humans and other living things are reduced. Generations are raised, who approach the environment with a more sensitive consciousness and are aware of the future danger. In this way, the probability of people living together sociologically in prosperity and happiness increases.

Environmental Assessment

Reducing the damage to nature and the environment on a global scale, such as the reduction of chemical pollution, greenhouse gas emissions, photochemical ozone gas emissions, and locally, such as less acidification, less eco-toxicity.

3 Circular Economy

Efficient and successful waste management is needed to protect the environment and economic development and to also reduce potential impacts. Successful SWM should be both environmentally and economically sustainable and must reduce the environmental impacts and operate at acceptable cost to community. Therefore, sustainable solid waste management is vital to conserve valuable natural resources, public health and natural ecosystems and prevent the unnecessary emission of GHGs.

In this century the main issue was waste management to make sure remove waste without damaging to our health and the environment. Nowadays, Resource Management is key point so that future generations can use existing resources. Therefore, there is a need to turn our waste into resources with new perspectives and approaches. The bending circle of "production consumption-waste management" can be defines as circular economy. In this economic approach, the value of products, materials and resources is retained in the economy as long as possible, and the amount of waste is minimal (Fig. 3).

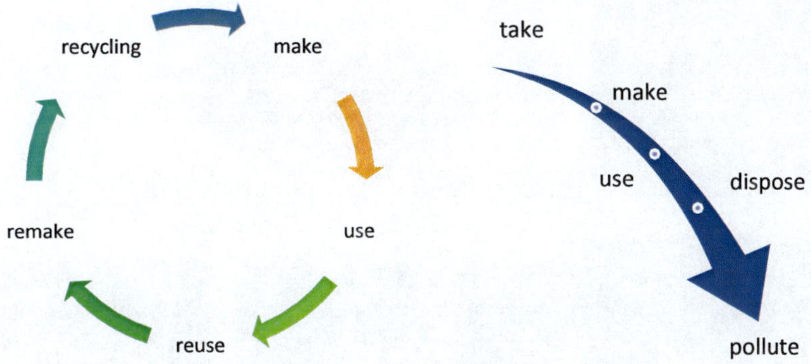

Fig. 3. Circular Economy and Linear Economy

There are some useful applications to support circular economy through the MSWM (Fig. 4). Separate collection system at the source should be established for recycling and evaluation of the organic content of the waste. The use of supplementary fuel (WDF) and pre-processing, biogas, gasification plants should be increased for waste economy. Industry and community need to use waste materials instead of natural resources. Moreover, products should be designed suitable for recycling or reusing. Societies should recognize the zero-waste philosophy as a lifestyle.

3.1 Contribution to Environment and Economy through the Circular Economy

- It can support to the need of raw materials/production
- Waste materials are brought back to the economy
- Reduce the cost of transportation for storage and disposal
- Reducing the waste quantity going to landfill.
- By turning wastes into raw materials, the need to be imported raw materials from abroad is also reduced.
- It can save energy in many ways.
- By increasing independence of energy and raw materials the competitiveness increases.
- Consumer satisfaction by producing new products from waste materials
- To prevent waste management issues, especially in crowded and densely populated areas
- Job opportunities increase through recycling industry.

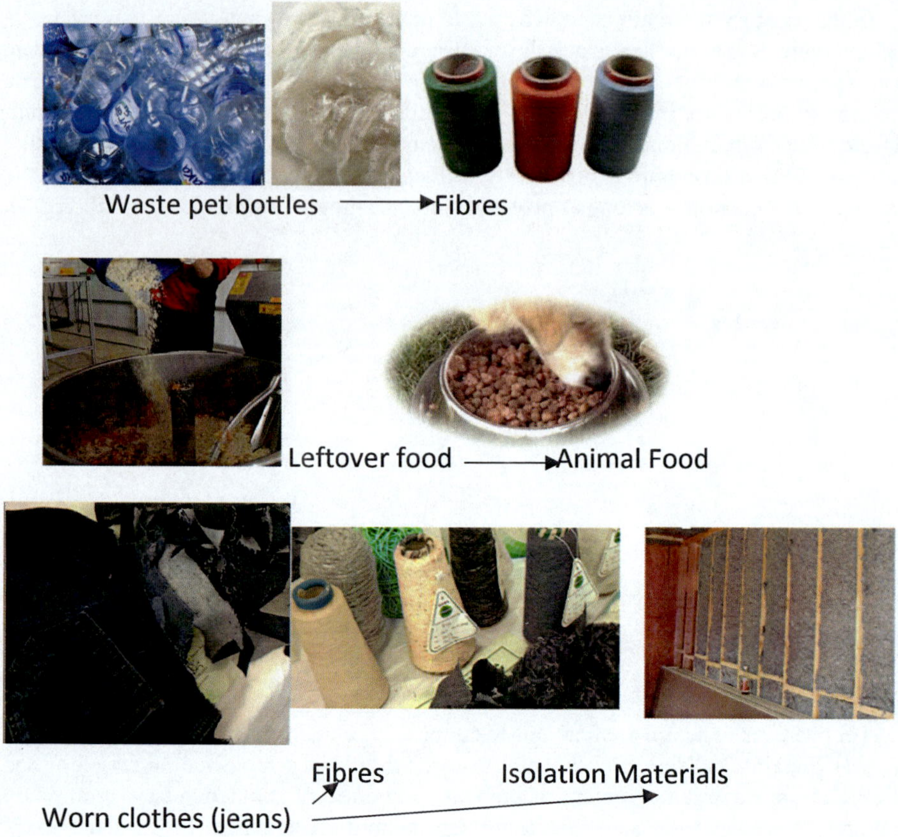

Fig. 4. The examples of circularity

4 Zero Waste Approach

Integrated municipal waste management includes different steps such as minimization of wastes, separate collection at the source, intermediate storage, establishment of transfer stations, transportation, recovery, disposal, operation of facilities, closing, monitoring and control of landfills etc. In 2017, Turkey initiated "Zero Waste Project" aims to control and upgrade the waste collection systems under sustainable development principles, leaving a clean Turkey and a liveable world to future generations. Zero Waste System is a seven-step roadmap developed by the Turkish Environment and Urban Planning Ministry, consisting of various steps that companies, institutions, or organizations should apply to be included in Zero Waste Movement. Current "Zero Waste Regulation" in our county aims to determine the general principles and implementation principles for the establishment of the zero-waste management system, and the adoption, implementation and dissemination of the zero-waste vision throughout the country. This legislation is forced public and government offices to make the separate collection of recyclable and organic waste and also create temporary storage areas to achieve zero waste target as well as given informative seminars and meetings on this manner. Zero waste is a waste management gaol that contains the prevention of waste at occurrence and the collection and recovery of source separated waste. The initial results showed that MSW recovery rate, which was 13% at the beginning of the project (2017), reached 19% and planning to increase the recycling rate 35% until 2023 in Turkey [4, 15, 16].

5 Conclusion and Recommendations

Municipal solid waste management is an issue that needs to be solved with a systems approach in a collaborative working environment. Sustainable waste management plans and solutions should consider the environment, society and economy as whole. Also, there is a need to develop high-efficiency processes, systems and technologies for reducing the environmental impacts e.g. developing a software model. Circular economy and zero waste approaches are tools for sustainable waste management. Contributing to the correct and effective use of resources by targeting the circular economy, not only seeing it as recycling and energy saving, but also creating new business opportunities and income sources, as well as leading to the spread of new financing model. Circular economy should be seen as an opportunity to protect a more liveable world for all of us.

References

1. Brundtland Commission: Report of the World Commission on Environment and Development: Our Common Future (1987)
2. André, P., Delisle, C.E., Revéret, J.-P.: Environmental Assessment for Sustainable Development: Processes, Actors and Practice: Presses inter Polytechnique (2004)
3. Sauvé, S., Bernard, S., Sloan, P.: Environmental sciences, sustainable development and circular economy: alternative concepts for trans-disciplinary research. Environ. Dev. **17**, 48–56 (2016)
4. T.C.Çevre ve Şehircilik Bakanlığı. n.d. Ulusal Atik Yönetimi ve Eylem Plani 2023. Ankara. https://webdosya.csb.gov.tr/db/cygm/haberler/ulusal_at-k_yonet-m--eylem_plan-20180328154824.pdf
5. Christensen, T.H., (ed.): Solid Waste Technology & Management, 1 & 2. Blackwell Publishing Ltd., Denmark (2011). https://doi.org/10.1002/9780470666883.ch1
6. IPCC: Climate Change 2007: Synthesis Report. Contribution of Working Groups I, II and III to the Fourth Assessment Report of the Intergovernmental Panel on Climate Change [Core Writing Team, Pachauri, R.K and Reisinger, A. (eds.)]. IPCC, Geneva, Switzerland (2007). 104 pp.
7. Pires, A., Martinho, G., Chang, N.-B.: Solid waste management in European countries: a review of systems analysis techniques. J. Environ. Manag. **92**(4), 1033–1050 (2011)
8. Wilson, D.C., Rodic, L., Cowing, M.J., et al.: Wasteware' benchmark indicators for integrated sustainable Waste management in cities. Waste Manage. **35**, 329–342 (2015)
9. Lehmann, S.: Optimizing urban material flows and waste streams in urban development through principles of zero waste and sustainable consumption. Sustainability **3**, 155–183 (2011)
10. Mauch, C.: Introduction: the call for zero waste. In: Mauch, C. (ed.) A Future Without Waste? Zero Waste in Theory and Practice, pp. 5–12. Munich: Rachel Carson Center Perspectives (2016)

11. Murray, R.: Zero waste. Greenpeace Environmental Trust, London (2002)
12. Nizar, M., Munir, E., Munavar, E., Irvan, M.: Implementation of zero waste concept in waste management of Banda Aceh City. J. Phys: Conf. Ser. **1116**, 1–12 (2018)
13. Simon, J. M.: A zero waste hierarchy for Europe new tools for new times: from waste management to resource management (2019). https://zerowasteeurope.eu/2019/05/a-zero-waste-hierarchy-for-europe/ (E. T. 26.08.2020)
14. Song, Q., Li, J., Zeng, X.: Minimizing the increasing solid waste through zero waste strategy. J. Clean. Prod. **104**, 199–210 (2015)
15. Uz Zaman, A.: Identification of key assessment indicators of the zero waste managament systems. Ecol. Ind. **36**, 682–693 (2014)
16. Uz Zaman, A.: A comprehensive review of the development of zero waste management: lessons learned and guidelines. J. Clean. Prod. **91**, 12–25 (2015)
17. Yaman, K., Olhan, E.: Atık yönetiminde sıfır atık yaklaşımı ve bu anlayışa küresel bir bakış. Biyoloji Bilimleri Araştırma Dergisi **3**(1), 53–57 (2010)
18. T.C.Çevre ve Şehircilik Bakanlığı. n.d. "Ulusal Atik Yönetimi ve Eylem Plani 2023." Ankara. https://webdosya.csb.gov.tr/db/cygm/haberler/ulusal_at-k_yonet-m--eylem_plan--20180328154824.pdf
19. Tchobanoglous, G., Theisen, H., Vigil, S.A.: Integrated Solid Waste Management: Engineering Principles and Management Issues. McGraw-Hill, New York (1993)
20. Coventry, Z.A., Tize, R., Karunanithi, A.T.: Comparative life cycle assessment of solid waste management strategies. Clean Technol. Environ. Policy **18**(5), 1515–1524 (2016). https://doi.org/10.1007/s10098-015-1086-7
21. Murray, A., Skene, K., Haynes, K.: The circular economy: an interdisciplinary exploration of the concept and application in a global context. J. Bus. Ethics **140**, 369–380 (2017)

The Inventory of Flood Disasters in Turkey

Ibrahim Gürer[1] (✉) and Ibrahim Uçar[2]

[1] Başkent University, Bağlıca, 06790 Ankara, Turkey
igurer@baskent.edu.tr
[2] FLOODIS Engineering, Çankaya, 06530 Ankara, Turkey
ucar@floodis.com

Abstract. Turkey is located both Europe and Asia, and covers an area of 780 580 km^2 including the lakes. Turkey undergoes different types of natural disasters because of its geographical location, geomorphology, and topography. Flooding is the second important natural hazard after earthquakes.

A flood inventory of the period extending from 1955 to 2020 having a total of about 3250 events was prepared using a simple computer program based on Excel for easy access to different geomorphologic parameters such as surface areas of river basins, slope, geological structure, vegetative cover, type of precipitation, and the elevation above mean sea level (a.m.s.l) and hydro-meteorological parameters. In the same inventory each flood has been defined with damages on human as lost and injured, size of flooded area and loss of wealth (not exact information for wealth). By categorization of the available data in hand, spatial and time distributions of past flood events were determined. In large basins, negative impacts are more closely related to climatic factors, but in small watersheds, the urbanization along the rivers, internal migration, regional planning, urban drainage infrastructure are more important on negative impacts.

In order to prevent the floods and minimize the adverse effects to property, both structural and non-structural solutions are applied in Turkey. Two case studies added to show the solutions.

Keywords: Flood disaster · Turkey · Flood inventory · Flood damage · Structural and nonstructural solution

1 Introduction

In designing the engineering structures on the rivers to cope with a flood of a certain magnitude, means a flood of a certain return period. At present due to economical restrictions it is not possible to design and build unnecessarily large and expensive water structures. The reliable estimate of the magnitude of the project flood becomes very important. This creates the necessity of having a flood inventory covering the whole Turkey. It is believed that such a document will also clarify some of the questions when new settlements will be planned at the flood prone areas and help the decision makers at the very early stage of the investment.

In this connection the floods of the period extending from 1955 to 2020, were analyzed from engineering and economic perspectives, by creating a data base having

H. Gökçekuş and Y. Kassem (Eds.): NRSEM 2021, EESCI, pp. 313–322, 2022.
https://doi.org/10.1007/978-3-031-04375-8_35

different parameters to define a single flood event. The archives of the Turkish State Hydraulic Works (DSI), the Turkish State Meteorological Service (MGM), the General Directorate of Disaster Affairs (AFAD) and local newspapers and TV's on different natural disasters such as earthquake, land slide, rock fall, snow avalanches and inundations were utilized and 3250 flood cases were selected (Table 1). With the data in hand it was possible to study the spatial (Fig. 1) and seasonal distribution of the flood events in Turkey (Table 2).

2 Factors Affecting Flooding

Besides parameters originating from the types of precipitation, there are other important factors, such as the prevailing hydro-meteorological conditions prior to flooding, topography, vegetation, and land use in the flood-prone area.

In the snow melt-produced floods of March-May period at Central Anatolia, it is not enough to have data of heavy snow accumulation on the ground; but also the previous autumn's meteorological conditions, and soil moisture conditions have important roles in producing flooding.

If the previous autumn had very little rainfall and a warm period before the permanent snow cover, a major portion of the snow melt water will be absorbed by the dry soil mantle, and surface flow will start only after the upper soil zone is completely saturated. But after a rainy and cool autumn almost all the water in the snow pack will be transformed to surface runoff and subsequently to floods. Of course other meteorological parameters such as the relative humidity, and wind velocity and geomorphologic parameters such as surface size, slope, geological structure, elevation above mean sea level (a.m.s.l) and existing vegetation have important roles in the formation and magnitude of floods. There is one more important factor to be considered for developing countries such as Turkey; the reduction in flood plain capacity. This causes a decrease in the available safe wetted

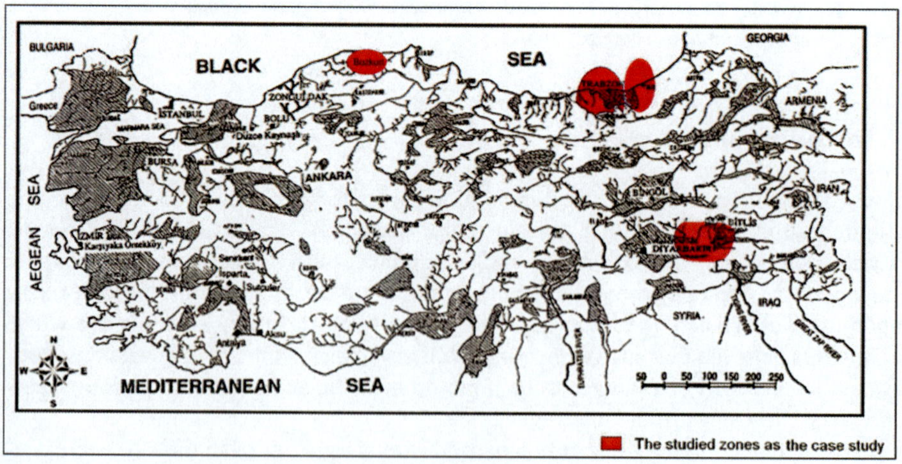

Fig. 1. Areal distribution of flood events in Turkey [1]

Table 1. Distribution of floods and damages caused by calendar year

Year	Number of flood events	Death Toll	Flooded Area (ha)
1955	27	–	86675
1956	22	90	178668
1957	22	13	49336
1958	49	190	61266
1959	50	17	21954
1960	98	1	78123
1961	42	24	45860
1962	117	15	94014
1963	122	25	191983
1964	69	32	5349
1965	63	3	96358
1966	90	25	137971
1967	81	4	5466
1968	170	24	170029
1969	30	6	125104
1970	8	3	18306
1971	31	6	4400
1972	109	33	21076
1973	19	22	44188
1974	35	73	2534
1975	62	8	36714
1976	29	5	22536
1977	27	11	3317
1978	21	0	13850
1979	21	61	40966
1980	44	6	83016
1981	16	2	58413
1982	10	0	784
1983	14	33	2113
1984	12	0	29140
1985	7	0	2318
1986	8	4	679

(*continued*)

Table 1. (*continued*)

Year	Number of flood events	Death Toll	Flooded Area (ha)
1987	7	0	564
1988	24	17	3910
1989	10	1	9500
1990	26	57	7450
1991	23	23	15770
1992	14	1	690
1993	2	0	60
1994	9	4	1680
1995	20	164	201100
1996	4	1	11000
1997	1	0	1390
1998	2	57	7000
1999	1	3	0
2000	4	0	8066
2001	42	8	43297
2002	27	27	510
2003	21	7	64200
2004	23	3	25750
2005	25	14	13855
2006	24	45	85810
2007	22	11	1050
2008	10	2	10
2009	84	59	3250
2010	110	25	44279
2011	56	13	202
2012	69	23	19685
2013	47	7	17569
2014	124	14	4455
2015	137	13	7985
2016	31	5	1100
2017	177	3	55813
2018	221	4	935

(*continued*)

Table 1. (*continued*)

Year	Number of flood events	Death Toll	Flooded Area (ha)
2019	169	19	3996
2020	157	24	3980
TOTAL	**3248**	**1390**	**2398417**
AVERAGE	**49**	**21**	**36340**

area and an increase in the devastating effect of floods. Unauthorized use of flood plains by the construction of local barriers and settlements, construction of inadequate bridges and culverts, and creating new agricultural plots are examples.

Continuous forest cutting during the last 40 years to obtain new plots for agricultural purposes, especially on the steep slopes of the northern Anatolian mountain range, and the clear cutting of shrub-size oaks in the Southeastern Anatolia for winter use in stoves have increased the possibility of landslides with flash floods in these parts of the country. Erosion of valuable surface soil and transportation of sediments first to rivers then to the sea is very common.

3 Floods Inventory in Turkey

According to statistics used on flood inventory for the period 1955–2020; the total number of floods analyzed for this period about 3250. There are 25 drainage basins in Turkey. The Susurluk Basin has the highest number of floods, the maximum number of floods (221) occurred in 2018, 4 people died and the size flooded was rather small. As shown also in Table 1, the most devastating year was 1995. In terms of seasonal distributions, 672 floods took place in summer; and as for monthly distributions, 309 floods took place in December. The areal distribution of floods occurred during the period 1955–2020 shows that Black Sea Region, Eastern Anatolia and Mediterranean Sea Region respectively had higher flood risk than the others It is a recorded information that the flood causing the maximum number of deaths in Turkey occurred on September 11, 1958 in Ankara, Central Anatolia. The number of 169 people living on the banks of Hatip Creek died of this event. After that flood event, Hatip Creek has been regulated by DSI, and now this river flows through the city of Ankara in an under ground culvert. During the period that flood inventory covers, about 50 floods have occurred and 21 people have lost their lives average per year. Besides this, the monetary cost of the economic losses per ysear on average has been over than $ 58 106. The cost of the flood disasters cannot be priced very exactly due to many unknowns in estimating the damage.

Although the Eastern Black Sea region has the highest long term mean annual precipitation (Fig. 3), the Susurluk Basin in the south of Marmara Sea experienced the highest number of floods and these floods in the Susurluk Basin during 66 years period caused 17 death and $ 65 10^6 damage (Fig. 2).

In Central Anatolia Kızılırmak Basin which has dry continental climate experiences snowmelt induced floods with lower peaks but longer flood periods and more than 100

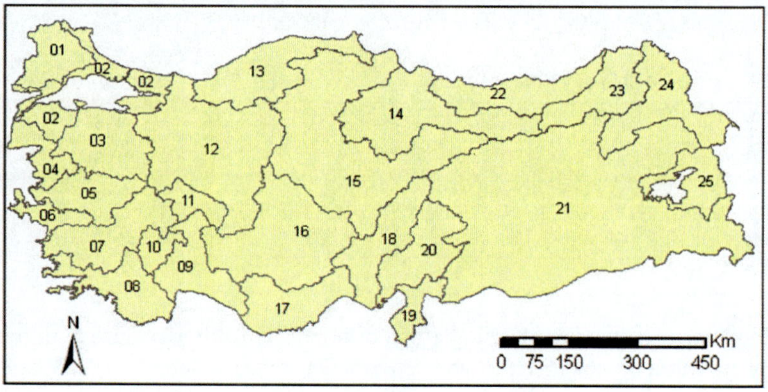

Fig. 2. Locations and numbers of watersheds in Turkey

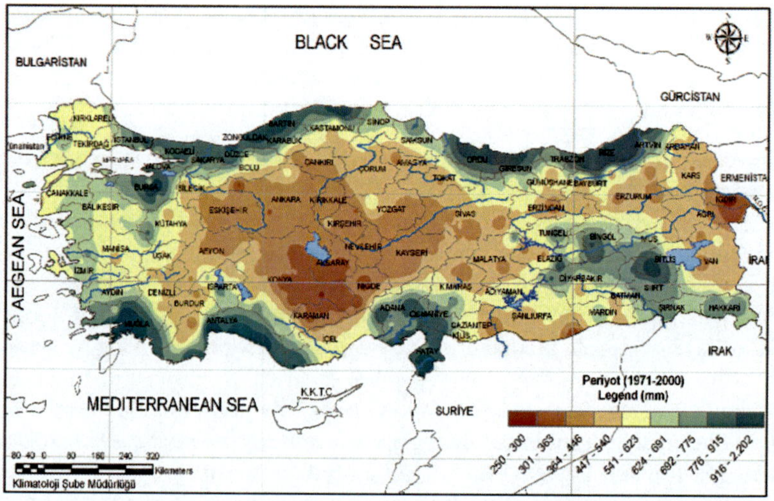

Fig. 3. The long term mean annual precipitation map of Turkey [2]

people died during 65 years. The Eastern Black Sea region has the less number of floods, but more destructive. Therefore, about 300 people have died in this region since 1955. Property damage totaled $ 359 185 738 but not exact.

In this study the floods and their damages for the last 65 years are studied and two extreme cases are presented as case studies. The first case is from Bozkurt town of Kastamonu, the flood which occurred on August 8, 2021 and this flood caused 81 deaths and 30 is still missing. The second case is about South Eastern Anatolia Flood on November 02, 2006 which caused 39 deaths.

4 Case Studies

The Mountains are generally parallel to the Black Sea coast. In the west, the mountains tend to be low, with elevations rarely exceeding 1500 m, but they rise in an easterly direction to heights greater than 3000 m south of Rize. Rivers in the Black Sea region flow through narrow valleys. Therefore, big flood disasters can happen without very high discharge.

4.1 Case 1. Flood Disaster of 2021 at Kastamonu, Bozkurt

After the heavy rain that affected the Western Black Sea Region on August 11, a flood disaster occurred in the region. One of the regions most affected by the flood was Kastamonu, Bozkurt (Fig. 4). AFAD said "As a result of the heavy rains that started in the Western Black Sea region on August 11, 2021, floods occurred in the cities of Bartın, Kastamonu and Sinop. Bartın province Ulus district, Kastamonu province Azdavay, İnebolu, Bozkurt, Küre and Pınarbaşı districts and Sinop province Ayancık districts were affected by the flood. The number of death was 81 and the number of injured was 7, the number of missing people was about 30 [3, 4]. After the flood, it was announced that 2380 people were evacuated from floods. The analysis of rainfall and floods and damages of Western Black Sea region floods continues by relevant state organizations, universities and NGOs.

Fig. 4. Flood and torrential flow at Bozkurt town on August 11, 2021 [3].

4.2 Case 2. Flood Disaster of 2006 at Southeastern Anatolia

Euphrates and Tigris Rivers' system experiences mainly snowmelt floods. Actually this is the largest drainage basin of Turkey and about 5 times larger than Eastern Black Sea region. In this basin, snowmelt induced floods occur mainly in April, May and even in June. The flood on November 2, 2006 at Southeastern Anatolia, chosen as the case study was due to the heavy rainfall in and around Batman (Table 2). In a similar way, duration, amount, intensity, and specific flow produced by this rainfall are presented during the rain storm in Table 7. In this flood 39 people lost their lives, 3800 houses damaged and 365000 ha of agricultural lands flooded (Fig. 5 and Fig. 6). The disaster has the biggest financial damage as the flood hazard up to the present in Turkey with a total loss of $ $300*10^6$ it was the biggest flood damage in Turkey.

Table 2. The characteristics of precipitations of the Batman Flood [2]

Station Name: Batman						
Date	Start	Finish	Duration (min)	Amount (mm)	Intensity (mm/min)	Specific Flow Produced (L/sec/ha)
1/11	19:37	21:37	120	26.6	0.222	36.9
1/11–2/11	17:54	5:54	720	54.8	0.076	12.7

Fig. 5. Batman City Center [5]

Fig. 6. Bitlis City (After The Flood) [5]

5 Structural and Non-structural Flood Protection Measures in Turkey

The structural flood protection measures in Turkey are classified as follows [6]:

- Construction of water impoundment structures (dams, artificial lakes, etc.) mainly in upstream parts of drainage areas.
- Construction of transverse structures such as weirs, drop structures, chutes, etc. to dissipate the kinetic energy of the supercritical flow of mountain rivers, most of which are torrential in character.
- Construction of longitudinal structures to define the flow path, and to protect river banks and areas on the banks of rivers.

- Stream bed modification such as change in the existing river bed and setting up new diversion structures. Dikes and groins are economical, but they are temporary solutions.
- Reforestation and improvement of land use techniques, such as terracing, strip cropping and enrichment of the existing vegetation, will decrease the velocity of the sheet flow into the river channel and minimize erosion and the amount of sediment transported into rivers and to the sea.

The non-structural flood protection measures can be summarized as:

- Education of the local people about the reasons of flooding, and the probable magnitude of damage.
- Convincing the people of using the flood plains for recreational purpose during dry seasons.
- Setting up first-aid teams made up of local people to help the victims in remote locations immediately where state help may come very late due to damaged roads is necessary in remote but flood-prone areas.

6 Conclusion

An inventory of about 3250 floods throughout Turkey was prepared using available data for the period 1955 to 2020. Floods in western Turkey and in the coastal zones mainly are produced by heavy rainfall in combination with geomorphologic features; however, in the central and eastern parts of Anatolia snow accumulation has an important role on spring floods. The adverse effects of floods on communities and other types of flood damage can be minimized by a comprehensive flood management program for each newly proposed development zone. It is also important to have good coordination among the stakeholders, the state, local people and non-governmental organizations.

In the Black Sea region, people settle on very narrow flood plains of rivers having torrential flows, because the mountains generally run parallel to the Black Sea coast and the agricultural plots are very limited due to the narrow valley.

In Turkey, the approach used to tackle with the natural disasters is "relocation of the disaster-hit people", it is hardly possible to answer the demand of the people at once. The local people expect the state to take into consideration their traditional way of living when providing new settlements [6].

References

1. Gurer, I.: Flood disasters and preventative measures in turkey. J. Nat. Disaster Sci. **20**(1), 1–9 (1998)
2. The Turkish State Meteorological Service (MGM), 2009–2020 data logs
3. Kronos. Kastamonu'da sel felaketi: Yüzlerce kişi evlerinde yardım bekliyor. https://kronos34.news/tr/kastamonuda-buyuk-sel-felaketi-yuzlerce-kisi-evlerinde-yardim-bekliyor
4. Euronews, "Karadeniz'de sel: Hayatını kaybedenlerin sayısı 78'e yükseldi" https://tr.euronews.com/2021/08/16/karadeniz-de-sel-hayat-n-kaybedenlerin-say-s-74-e-yukseldi

5. Balikci, F., Aslan, F., Duru, M., Arslan, A., Ciris, Ö.: The Archive of Hürriyet Gazette November 2 (2006)
6. Gurer, I., Ucar I.: Flood Disasters Inventory in Turkey, Eleventh International Symposium on Water Management and Hydraulic Engineering, Ohrid, Macedonia (2009)

The Sensitivity Analysis and Performance of SWAT+ in Simulation of Stream Flow in a Mountainous Catchment

Soghra Andaryani[1,4](✉), Farnaz Ershadfath[2], and Vahid Nourani[1,3]

[1] Center of Excellence in Hydroinformatics, Faculty of Civil Engineering, University of Tabriz, Tabriz, Iran
s.andaryani@gmail.com

[2] Water Engineering Department, Sari Agricultural Sciences and Natural Resources University, Sari, Iran
farnaz@agro.au.dk

[3] Faculty of Civil and Environmental Engineering, Near East University, Mersin, Turkey

[4] Geological Survey of Denmark and Greenland, GEUS, Øster Voldgade 10, 1350 Copenhagen K, Denmark

Abstract. In the context of floods, water stress or crises, climate change planning, and human activities, the application of ecological models for environmental management is critical. In this study, we used the Soil and Water Assessment Tool Plus (SWAT+), an upgraded version of SWAT, to route runoff in a mountainous catchment, Bostanabad, which is one of the headwaters of Ajichay (connected to the Urmia Lake basin), at monthly scale. At first the most sensitive parameters were selected by sensitivity analysis, and then the model was calibrated and validated during different periodss of time. We employed gridded temperature and precipitation produced based on elevation and data from existing gauges to overcome the model calibration tool's (SWAT+ toolbox) limitations in mountainous areas. The results indicated that the model is sensitive to the parameters of snowmelt-min, CN2 and CN3-swf, snomelt_min (table of hru), revap-min, bf-max and alpha, flo_min (table of aqu), bd and awc (table of soil) and bd (table of rte). The gridded approach for simulation of discharge resulted in calibration performance based on R2, NSE, BAIS of 0.71, 0.68 and −2.81, respectively. This approach can be used to simulate discharge in mountainous catchments.

Keywords: SWAT · Water Assessment Tool plus · Mountainous catchments

1 Introduction

The Soil and Water Assessment Tool (SWAT) [1] is considered as one of the most commonly used hydrologic models, having been utilized in a variety of basins around the world for more than two decades. It has been demonstrated that this model can accurately simulate streamflow and pollution transport at various scales and circumstances [2, 3]. This model has several advantages, including the ability to combine plant growth and

H. Gökçekuş and Y. Kassem (Eds.): NRSEM 2021, EESCI, pp. 323–329, 2022.
https://doi.org/10.1007/978-3-031-04375-8_36

land management with channel processes, as well as the ability to combine various environmental processes and contaminants. The model's codes are also open source, allowing users to improve the model based on their requirements [4]. Several limitations and drawbacks in this model have been identified by various scholars. Simulating all hydrologic processes using hydrologic response units (HRUs), which are areas with a unique combination of land use, soil type, and slope, for example, makes it difficult to pinpoint crucial source locations and execute conservation measures within a subbasin. Furthermore, the total amount of water and pollutant in each subbasin is equal to the total amount of water and pollutant in all HRUs in that subbasin. Furthermore, the landscape's transport and sedimentation processes are not appropriately considered [5]. Meeting the demand for increasing spatial complexity while maintaining model usability will be a major challenge in the model's future development [4]. Therefore, in order to overcome the limitations of the SWAT model, a completely reconstructed SWAT+ model was created [6, 7]. Even though the core methods for calculating the processes in the model are same, the structure and arrangement of the input code and files has changed dramatically [8]. Model maintenance and future code revisions should be easier as a result of the core code enhancements [9]. In SWAT+, HRUs, aquifers, canals, ponds, reservoirs, point sources, and inlets have all been regarded as independent modules. In addition, aquifers were linked to HRUs, and the accompanying aquifer borders can now be established more freely without being constrained by HRU constraints in SWAT$^+$ [6].

The objective of this study is to evaluate the SWAT+ model's application for streamflow simulation in mountainous areas, where there are currently restrictions due to the model's lack of consideration of laps rate.

2 Method and Materials

2.1 Study Area

Bostan Abad watershed is one of the catchments of Lake Urmia, which is located in northwestern Iran and in East Azarbaijan province (Fig. 1). This watershed covers an area of 584 km^2, which is limited to Kordkandi from the north, to the slopes of Sahand mountain from the southwest, to Tabriz from the west, and to Sarab from the east. Ojan River is the only surface water resource in this area which originates from Biokdagh heights in the west and flows along the west to the east. It drains into the Ajichai river after receiving water from many branches and passing through Bostan Abad city. According to the Amberge climate index, the region is dry and cold. The average annual precipitation, temperature, and evaporation are 276 mm, 8 °C, and 1802.3 mm, respectively, and the average annual discharge of the Ojan River is 138.8 m^3/s.

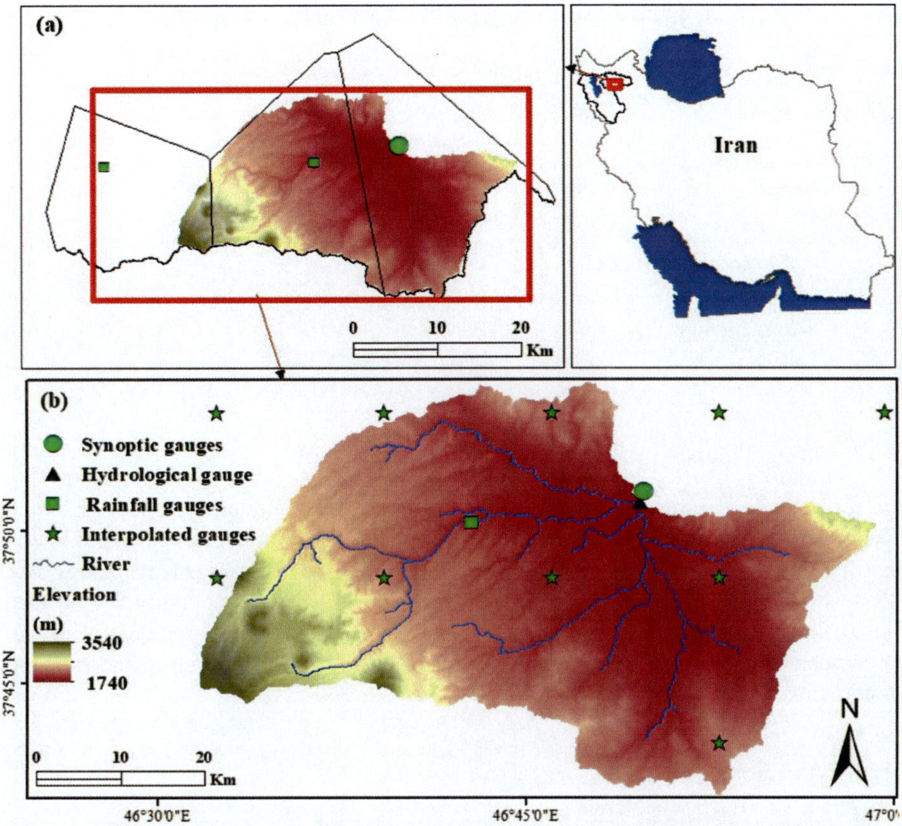

Fig. 1. Location of Bostan Abad in Iran, (a) the location of rainfall gauges in Thiessen polygons. (b) Bostan Abad River and its basin, and the location of the interpolated rainfall gauges in the basin.

2.2 Data Collection

Table 1 lists the data sets used in this investigation as well as their sources. To delineate the study region, digital elevation model (DEM) was obtained from USGS at a resolution of 30 m. The HRUs (Hydrologic Response Units) were created using land use and soil data. The Iranian Meteorological Organization (IRIMO) and East Azerbaijan Water Company (EAWC) provided meteorological (e.g., precipitation and temperature) and stream flow data from 2002 to 2016.

Table 1. Data used (inputs) for SWAT+ model development.

Data	Resolution	Source
DEM	1:30 m	USGS https://earthexplorer.usgs.gov/
Soil	1:1 million	FAO
Land use	1:250,000	Iranian ministry of Jahad agriculture
Meteorological Data	Daily	IRIMO https://www.irimo.ir
Stream flow gages	Daily	EAWC http://www.azarwater.ir

2.3 Creating Thiessien Polygon

Due to the lack of raps rate consideration in the SWAT+ model, we used Thiessien polygons to construct a link between height and precipitation and temperature. It was identified from which station each region's data is received (see Fig. 1(a)). Using Arc GIS software, one station was established for every 10 km of the study region (see Fig. 1(b)). The lapse rate was calculated using DEM data (elevation) and data from rain and temperature gauges, assuming a linear relationship between elevation and rain and temperature change.

2.4 SWAT⁺ Setup

The delineated river basin was divided into 7 subbasins. The subbasins were further subdivided into upland areas and floodplains. Slope Position method was applied for delineating the Land Scape Units (LSUs). By superimposing the soil, land use, and slope maps, HRUs are defined. 5 classes of slope include 0–10%, 10–20%, 20–30%, 30–60% and above 60%. Totally 30 LSUs and 427 HRUs were created. The surface runoff is computed using the SCS curve number (CN) method, and the runoff is routed from the sub-basin through the river network and to the main basin outlet using the variable storage method. Hargreaves equation was applied for potential evapotranspiration calculation as well. The discharge data set was separated into two time periods: 2007–2011 (calibration period) and 2012–2016 (validation period to mitigate the effects of unknown initial conditions, the year 2002–2006 was chosen as a warm-up period, and as a result, this time period was eliminated from the analysis.

3 Results and Discussion

The model was calibrated and evaluated against the observed stream flow data (source: USGS). Among different algorithms for sensitivity analysis such as Sobol, Fourier Amplitude, Random Balance Design Fourier Amplitude, Delta Moment- Independent

Table 2. The sensitive parameters

No.	Parameter		Max	Min	Final
1	cn2	percent	−10	20	13
2	snomelt_min	absolute	−1.09	1.7	0.12
3	cn3_swf	absolute	0.1	0.9	0.2
4	revap_min	absolute	2	6	3
5	bf_max	absolute	0.1	0.6	0.22
6	alpha	absolute	0.1	0.8	0.4
7	flo_min	absolute	2	15	6
8	awc	percent	−10	10	−6
9	bd	percent	−10	10	2.5

Measure we selected SOBOL algorithm due to its high usage which estimate the statistical significance of a parameter is based on t-stat and p-value. The most sensitive parameters are listed in Table 2.

The performance of SWAT+ model was evaluated by using the goodness-of-fit criteria, such as, coefficient of determination (R2), Nash-Sutcliffe Efficiency (NSE) (Nash and Sutcliff, 1970). The model calibration and validation results are shown in Table 3. Figure 2 demonstrates the time series of modeled and observed discharge during the calibration and validation period. Our results fell in the ranges of good and very good based on the Moriasis' findings [10]. According to the literature [11] SWAT model has some limitations in the mountainous area due to not considering the snowmelt. The above-mentioned references showed poor performance in the mountainous areas, however the model performances in the plain areas are much more satisfying [12]. Thus, calculating labs rate using Thiessen gridding and building extra stations can be useful for more detailed discharge simulating.

Table 3. Goodness of fit statistics between modeled and observed streamflow for the study area

	R^2	NSE	MSE	RMSE	PBIAS
Calibration (2007–2011)	0.71	0.68	0.47	0.68	−2.81
Validation (2012–2016)	0.68	0.6	0.23	0.48	9.0

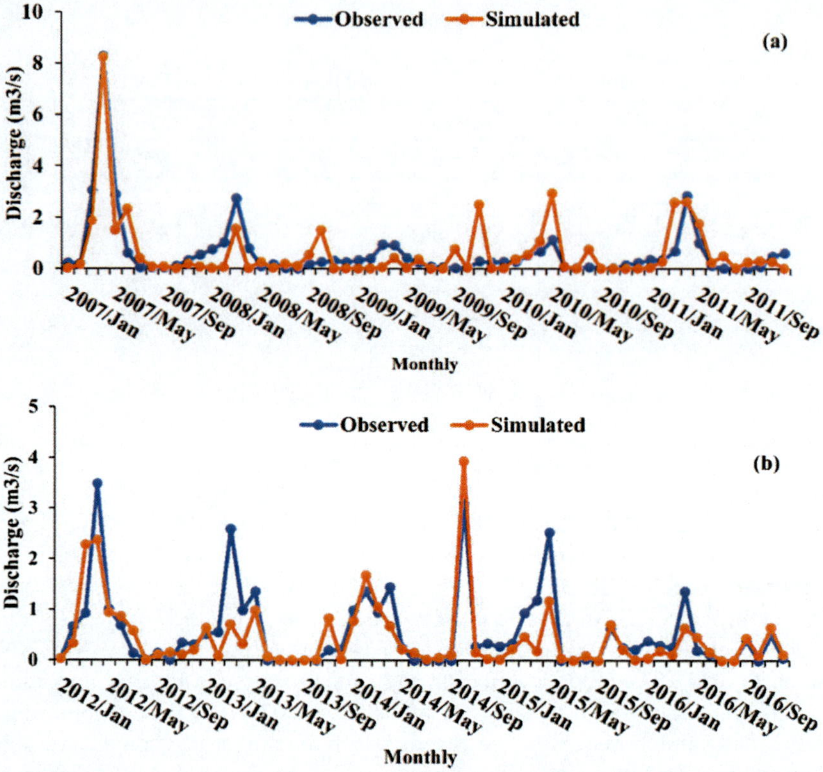

Fig. 2. Monthly modeled and observed stream flow in the study area.

4 Conclusion

In this study, the SWAT+ model was used to simulate discharge in a mountainous watershed. Due to the restricted number of meteorological stations in the area (i.e., two stations in a 584 km^2 area), Thiessen polygons were created and climatic data for ten new stations were interpolated in order to analyze the effects of height variations on climatic parameters. The Thiessen interpolation approach, according to the findings, can overcome the constraint of the SWAT+ model's lack of laps rate consideration and improve the model's efficiency.

Acknowledgement. This work was supported by Iran's National Elites Foundation (INEF) and the University of Tabriz [grant agreement 15/7806] to SA.

References

1. Arnold, J.G., Srinivasan, R., Muttiah, R.S., Williams, J.R.: Large area hydrologic modeling and assessment Part I: model development. J. Am. Water Resour. Assoc. **34**(1), 73–89 (1998). https://doi.org/10.1111/j.1752-1688.1998.tb05961.x

2. Krysanova, V., Arnold, J.G.: Advances in Ecohydrological modelling with SWAT—a review. Hydrol. Sci. J. **53**(5), 939–947 (2008). https://doi.org/10.1623/hysj.53.5.939
3. Douglas-Mankin, K.R., Srinivasan, R., Arnold, J.G.: Soil and Water Assessment Tool (SWAT) model: current developments and applications. Trans. ASABE **53**(5), 1423–1431 (2010). https://doi.org/10.13031/2013.34915
4. Gassman, P.W., Reyes, M.R., Green, C.H., Arnold, J.G.: The soil and water assessment tool: historical development, applications, and future research directions. Trans. ASABE **50**(4), 1211–1250 (2007). https://doi.org/10.13031/2013.23637
5. Arnold, J.G., Fohrer, N.: Current capabilities and research opportunities in applied watershed modelling. Hydrol. Process. **19**, 563–572 (2005). https://doi.org/10.1002/hyp.5611
6. Bieger, K., et al.: Introduction to SWAT+, a completely restructured version of the soil and water assessment tool. J. Am. Water Resour. Assoc. **53**(1), 115–130 (2017). https://doi.org/10.1111/1752-1688.12482
7. Arnold, J.G., Bieger, K., White, M.J., Srinivasan, R., Dunbar, I.J.A., Allen, P.M.: Use of decision tables to simulate management in SWAT+. Water **10**(6), 713 (2018). https://doi.org/10.3390/w10060713
8. Bieger, K., Arnold, J.G., Rathjens, H., White, M.J., Bosch, D.D., Allen, P.M.: Representing the connectivity of upland areas to floodplains and streams in SWAT+. J. Am. Water Resour. Assoc. **55**(3), 578–590 (2019). https://doi.org/10.1111/1752-1688.12728
9. Wu, J., et al.: Development of reservoir operation functions in SWAT+ for national environmental assessments. J. Hydrol. **583**, 124556 (2020). https://doi.org/10.1016/j.jhydrol.2020.124556
10. Moriasi, D.N., Arnold, J.G., Van-Liew, M.N., Bingner, R.L., Harmel, R.D., Veith, T.L.: Model evaluation guidelines for systematic quantification of accuracy in watershed simulations. Trans. ASABE **50**(3), 885–900 (2007)
11. Andaryani, S., Nourani, V., Ball, J., Janbakhsh Asl, S., Keshtkar, H., Trolle, D.: A comparison of frameworks for separating the impacts of human activities and climate change on river flow in existing records and different near-future scenarios. Hydrol. Process. **35**(7), e14301 (2021). https://doi.org/10.1002/hyp.14301
12. Andaryani, S., Trolle, D., Nikjoo, M.R., Moghadam, M.H.R., Mokhtari, D.: Forecasting near-future impacts of land use and climate change on the Zilbier river hydrological regime, northwestern Iran. Environ. Earth Sci. **78**(6), 1–14 (2019). https://doi.org/10.1007/s12665-019-8193-4

The Use of Modified Drastic Method (DRASTIC-LU) for Assessment of Groundwater Vulnerability to Pollution at the Palas Basin/Turkey

Mehmet Soylu[1], Ugur Bozdoganlio[2], and Filiz Dadaser-Celik[3](✉)

[1] Cappadocia University, 50400 Ürgüp/Nevsehir, Turkey
mehmet.soylu@kapadokya.edu.tr
[2] Erciyes University, 38280 Talas/Kayseri, Turkey
[3] Erciyes University, 38280 Talas/Kayseri, Turkey
fdadaser@erciyes.edu.tr

Abstract. Groundwater is one of our major water resources, which is affected by hydrological, climatic, and anthropogenic factors. DRASTIC model is a widely-used index-based method that can be used to evaluate the vulnerability of the groundwater systems to pollution. DRASTIC uses depth to water (D), net recharge (R), aquifer media (A), soil media (S), topography (T), impact of vadose zone (I), and hydraulic conductivity (C) information to create a GIS-based vulnerability map. In this study, we used a modified version of DRASTIC model (DRASTIC-LU) for vulnerability assessment at the Palas Basin/Turkey. DRASTIC-LU uses land use (LU) information in addition to the hydrogeological parameters. The measured nitrate concentrations in the Palas Basin were used to test the original and modified DRASTIC models. The results of the study revealed that the DRASTIC-LU improved predictions of nitrate concentrations. The results of the vulnerability assessment in this study can be used for groundwater management and planning at the Palas Basin.

Keywords: Groundwater · Vulnerability assessment · DRASTIC · DRASTIC-LU · Palas basin

1 Introduction

Groundwater forms by infiltration of precipitation and is renewed much more slowly than surface water. Therefore, it is more vulnerable to overexploitation and pollution [1].

The pressure on groundwater is increasing day by day due to population growth and anthropogenic impacts such as agricultural intensification and industrial development [2]. The use of fertilizers and pesticides in agriculture [3] and nitrogen and phosphorus compounds from the livestock sector threaten groundwater [4, 5]. Heavy metal pollution originating from mining sites and organic pollutants from various industrial sectors

H. Gökçekuş and Y. Kassem (Eds.): NRSEM 2021, EESCI, pp. 330–338, 2022.
https://doi.org/10.1007/978-3-031-04375-8_37

pose risks. Some elements or compounds originating from natural processes can also interfere with groundwater [5]. Treatment of groundwater in case of contamination is more difficult and costlier than other water sources [6]. Therefore, prevention of groundwater contamination is important for sustainability of these resources [7].

In the study, we examine groundwater contamination potential or aquifer vulnerability at a semi-arid basin in Turkey (Palas Basin) (Fig. 1). Aquifer vulnerability is defined as "the susceptibility of the aquifer to an imposed contaminant load caused by the human and/or natural impacts" [8, 9]. DRASTIC model and modified version of DRASTIC model (DRASTIC-LU) were used in the analyses. Measured nitrate values were used in order to evaluate the success of vulnerability assessment.

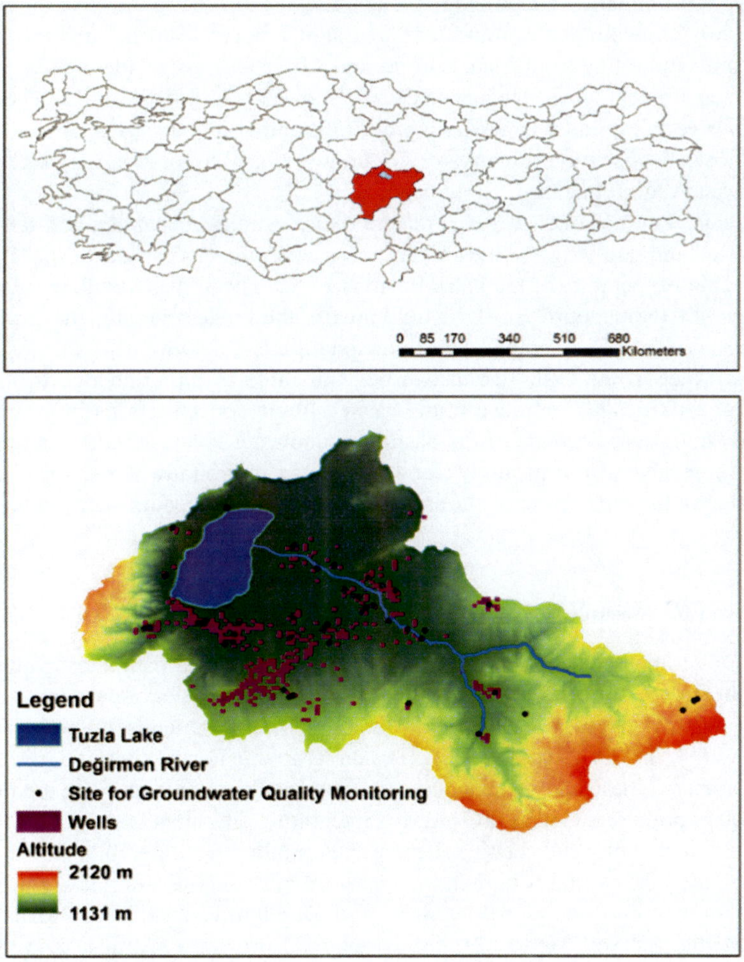

Fig. 1. Study area

2 Methods

In the study, the vulnerability of groundwater system to pollution at the Palas Basin was investigated. Below, we provide information about study area. Then, we explain DRASTIC and DRASTIC-LU models and data used in the analyses.

2.1 Study Area

This study was carried out at the Palas Basin (Fig. 1). Palas Basin is an agricultural basin, located in Kayseri, in the Central Anatolia Region of Turkey. The basin is a hydrologically closed basin. The altitude of the region ranges from 1131 to 2120 m, and this area is approximately 100 km^2 [10]. Tuzla Lake is located to the west of the basin. Tuzla Lake is an ecologically important area as it is located in the junction point of routes of birds migrating from Asia, Europe, and Africa and hosts endemic plant species [11]. A small stream, named Değirmen Stream, flows towards Tuzla Lake. The flow in the stream is very low and it is mostly dry during summer months. Therefore, the major water source in the basin is groundwater. Groundwater is used for meeting irrigation and drinking water requirements.

Economic activities in the region are agriculture, animal husbandry, and salt extraction. Irrigated and non-irrigated agriculture areas cover majority of the basin. The average annual air temperature of the Palas Basin is 11 °C. The hottest month is July, where the average air temperature is 20 °C. In January, the coldest month, the average air temperature is −2.5 °C. Average annual precipitation is 402 mm [10].

Groundwater in the Palas Basin can be vulnerable to pollution due to intensive agricultural activities taking place in the basin. Nitrate concentrations in groundwater are already high in some parts of the basin. Groundwater levels are also decreasing as a result of intensive use of groundwater. The change in groundwater salinity can be a problem due to the interaction of the saline Tuzla Lake with groundwater in the region [10].

2.2 DRASTIC Method

DRASTIC is a method developed by Aller et al. [12], which estimates groundwater vulnerability to pollution based on hydrogeological characteristics of the area of interest. In this approach, depth to water (D), net recharge (R), aquifer media (A), soil media (S), topography (T), impact of vadose zone (I), and hydraulic conductivity (C) are used as input parameters. These parameters are rated (r) and weighted (w) based on their relative importance to pollution. DRASTIC index is integrating all inputs based on Eq. 1. Final DRASTIC index shows the relative degree of groundwater vulnerability of the study area. When the index value is higher, the possibility of contamination is higher. In this study, the following weighting values were used: Depth to Water [5], Net Recharge [4], Aquifer Media [3], Soil Media [2], Topography [1], Impact of Vadose Zone [5] and Hydraulic Conductivity [3]. The input data are also shown in Table 1 and Fig. 2 and the rating values were listed in Table 2.

$$DRASTIC\ Index = D_R * D_w + R_R * R_w + A_R * A_w + S_R * S_w + T_R * T_w + I_R * I_w + C_R * C_w \quad (1)$$

Table 1. Data used in this study

Layers	Raw data source	Data adjustments
Depth to water	Log data for different location at basin	* Coordinate transformation * GRID installation (100 m × 100 m) * Classification
Net recharge	SWAT model	* Coordinate transformation * GRID installation (100 m × 100 m) * Classification
Aquifer media	State hydraulic works	* Coordinate transformation * GRID installation (100 m × 100 m) * Classification
Soil media	FAO world soil map	* Image clip process * Coordinate transformation * GRID installation (100 m × 100 m) * Classification
Topography	Using DEM image by SRTM satellite	* Image clip process * Coordinate transformation * GRID installation (100 m × 100 m) * Slope Analysis * Classification
Impact of vadose zone	FAO world soil map	* Image clip process * Coordinate transformation * GRID installation (100 m × 100 m) * Classification
Hyraulic conductivity	SPAW hydrology programme	* Hyraulic conductivity values search and input for soil types * Coordinate transformation * GRID installation (100 m × 100 m) * Classification

2.3 Modified DRASTIC Method (DRASTIC-LU)

Classical DRASTIC index does not consider land use information. However, land use can affect contamination possibility greatly and inclusion of this parameter can improve vulnerability predictions [13–16]. In this study, we integrated land use information based on Eq. 2.

$$DRASTIC-LU = D_R * D_w + R_R * R_w + A_R * A_w + S_R * S_w + T_R * T_w + I_R * I_w + C_R * C_w + L_R * L_w \quad (2)$$

According to Eq. 2, different types of land use can affect the transport of pollution to groundwater. For example, the pollution potential in areas used for agricultural purposes and covered with forests are different. In the land use layers, there are artificial areas such as settlements, industrial activity areas, airports, and areas where agriculture and livestock activities are carried out. There are also areas of low pollution potential, such as forests, wetlands, lakes and streams.

Table 2. Rating, weighting and classes information for each parameter

Parameter	Classes	Rating	Weight
Depth to water (m)	0–1.5	10	5
	1.5–4.6	9	
	4.6–9.1	7	
	9.1–15.2	5	
	15.2–22.9	3	
	22.9–30.5	2	
	>30.5	1	
Net recharge (m/year)	0–141	1	4
	141–282	3	
	282–494	6	
	494–705	8	
	>705	9	
Aquifer media	Silty clay, Sand, Gravel	4	3
	Sandstone	6	
	Siltstone, sandstone, clay limestone	6	
	Silty clay	7	
	Broken cracked rock	3	

(*continued*)

Table 2. *(continued)*

Parameter	Classes	Rating	Weight
	Impervious tuffs	9	
Soil media	Sandy clay loam	2	2
	Clay silty	3	
	Silty clay loam, sandy clay loam, clay loam	4	
	Sandy clayey gravelly	6	
	Fine sandy loam - sandy loam	8	
	Clay - silty clay - sandy clay	2	
Topography (percent)	0–2	10	1
	2–6	9	
	6–12	5	
	12–18	3	
	>18	1	
Impact of vadose zone	Sandy clay loam	2	5
	Clay silty	3	
	Silty clay loam, sandy clay loam, clay loam	4	
	Sandy clayey gravelly	6	
	Fine sandy loam - sandy loam	8	
	Clay - silty clay - sandy clay	2	
Hyraulic conductivity (m/day)	0,19	1	3
	0,21	1	
	0,18	1	
	0,23	1	
	0,24	1	
	0,27	1	

2.4 Nitrate Data Comparison

The results from the DRASTIC and DRASTIC-LU models were compared using groundwater quality data collected at 54 groundwater wells in the basin. Nitrate values were examined and the average of these values was calculated. The nitrate values detected in the monitoring wells were in the range of 0.97–53.52 mg/L NO_3^-. The average of the nitrate data in the study area was 20 mg/L. In the study, these nitrate values expressed were used to evaluate the relationships of water quality with the DRASTIC index values.

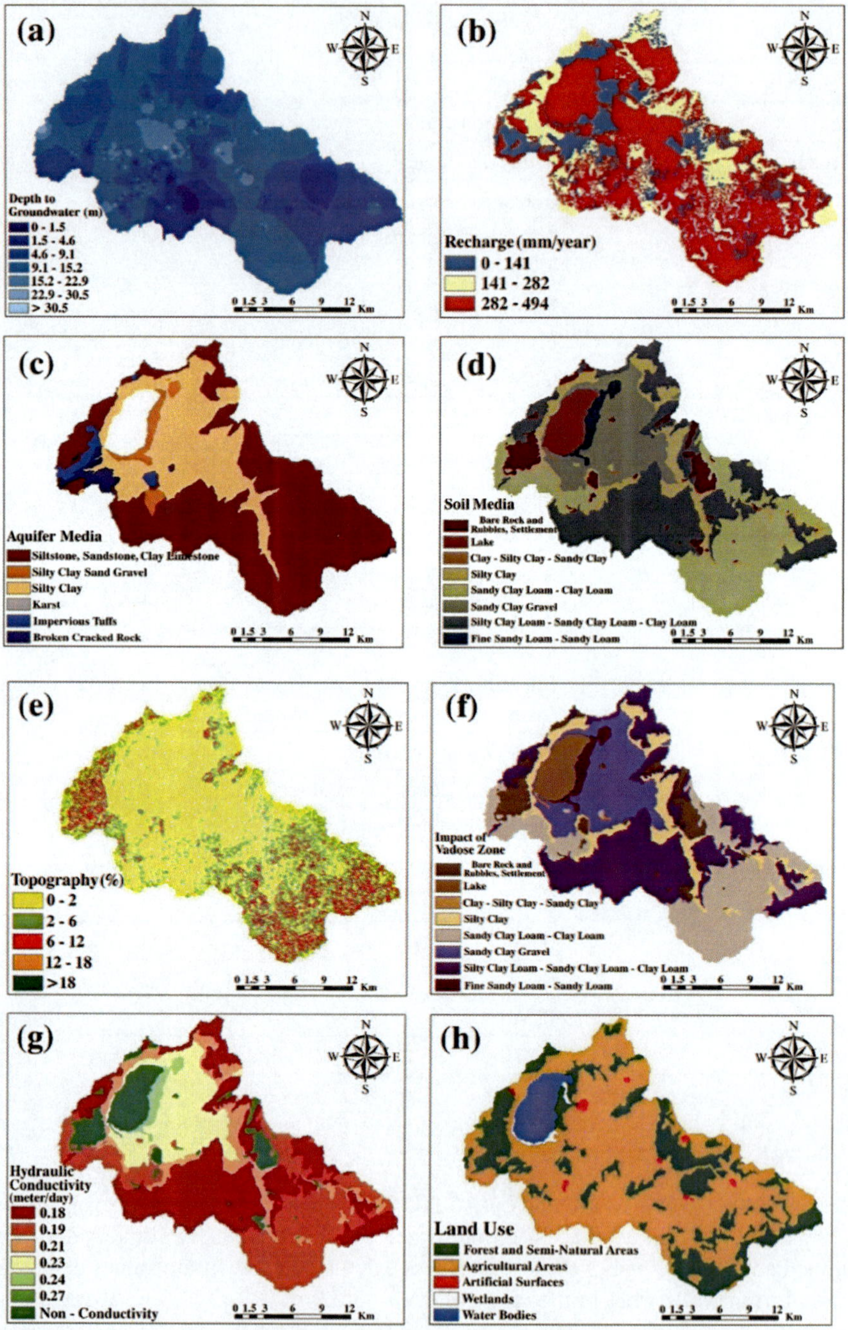

Fig. 2. Classified maps for each layer

3 Results

DRASTIC Index map was created by combining seven separate layers (Fig. 3). In this map, a numerical output value was created for each grid value and these values were divided into three classes as low, moderate and high vulnerability zones. Both classic DRASTIC and DRASTIC-LU index values were classified into three classes as low medium, and high. The majority of the basin was covered by moderate vulnerability class (60%). Low vulnerability areas covered 17% and high vulnerability areas covered 23% in the DRASTIC vulnerability map. However, it was seen that this situation changed in the DRASTIC-LU map. On the DRASTIC-LU map, the areal coverage of low vulnerability areas increased to 27% and the areal coverage of high areas increased to 27%, while the areal coverage of moderate vulnerability areas decreased to 46%. One of the main reasons for this change is the presence of low pollutant profile areas of land use density and artificial areas with high pollutant profile. The variation between maps is shown in Fig. 3.

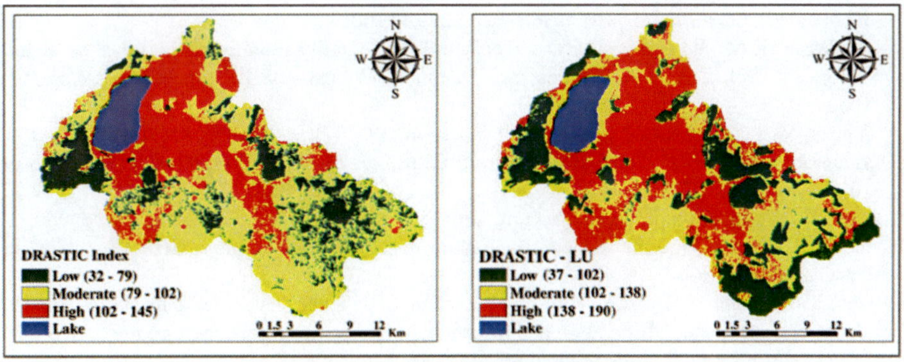

Fig. 3. Drastic index and drastic-LU maps

We examined the relationships of index values of DRASTIC and DRASTIC-LU maps with nitrate values measured in the basin. We found that the 48% of the nitrate values over 20 mg/L were located on the high vulnerable zone on the DRASTIC index map. However, this ratio was found to be 60% on the DRASTIC-LU map. When both maps are compared, nitrate values above the average were represented better by the DRASTIC-LU map.

4 Conclusion

In this study, we evaluated the vulnerability of groundwater to pollution at the Palas Basin. By applying DRASTIC and DRASTIC-LU models, the basin was classified to low, moderate, and high vulnerability regions. DRASTIC-LU model, which considers land use as input, over performed the classical DRASTIC model in estimating measured nitrate concentrations in the basin.

Acknowledgements. This study was supported by the Scientific and Technological Research Council of Turkey (TUBITAK) (Project No: 118Y178). The second author received funding from the 2247-C STAR-Intern Researcher Scholarship Programme provided by TUBITAK.

References

1. Freeze, R.A., Cherry, J.A.: Groundwater. Prentice-Hall, New Jersey (1979)
2. Famiglietti, J.S.: The global groundwater crisis. Nat. Clim. Chang. **4**, 945–948 (2014). https://doi.org/10.1038/nclimate2425
3. Foster, S.S.D., Chilton, P.J., Stuart, M.E.: Mechanisms of groundwater pollution by pesticides. Water Environ. J. **5**, 186–193 (1991). https://doi.org/10.1111/j.1747-6593.1991.tb00606.x
4. Zhao, Y.Y., Pei, Y.S.: Risk evaluation of groundwater pollution by pesticides in China: a short review. Procedia Environ. Sci. **13**, 1739–1747 (2012). https://doi.org/10.1016/j.proenv.2012.01.167
5. Kurwadkar, S.: Groundwater pollution and vulnerability assessment. Water Environ. Res. **89**, 1561–1577 (2017). https://doi.org/10.2175/106143017X15023776270584
6. Talabi, A.O., Kayode, T.J.: Groundwater pollution and remediation. J. Water Resour. Prot. **11**, 1 (2019). https://doi.org/10.4236/jwarp.2019.111001
7. Balderacchi, M., et al.: Groundwater pollution and quality monitoring approaches at the European level. Crit. Rev. Environ. Sci. Technol. **43**, 323–408 (2013). https://doi.org/10.1080/10643389.2011.604259
8. Jenifer, M.A., Jha, M.K.: Comparative evaluation of GIS-based models for mapping aquifer vulnerability in hard-rock terrains. Environ. Earth Sci. **77**(19), 1–26 (2018). https://doi.org/10.1007/s12665-018-7821-8
9. Zaporozec, A.: Groundwater contamination inventory: a methodological guide with a model legend for groundwater contamination inventory and risk maps. IHP-VI, Series on Groundwater, 2 (2004)
10. Cengiz, E., Dadaser-Celik, F.: Hydrological changes at Tuzla (Palas) Lake in Turkey. In: 5th Conference on Water Observation and Information System for Decision Support (BALWOIS 2012), Ohrid, Republic of Macedonia (2012)
11. Dadaser-Celik, F., Celik, M.: Modelling surface water-groundwater interactions at the Palas Basin (Turkey) using FREEWAT. Acque Sotterranee-Italian Journal of Groundwater **6**, 53–70 (2017). https://doi.org/10.7343/as-2017-288
12. Aller, L., Bennett, T., Lehr, J.H., Petty, R.: Drastic: A Standardized System for Evaluating Ground Water Pollution Potential using Hydrogeologic Settings. U.S. Environmental Protection Agency, Washington, D.C., EPA/600/2–85/018 (1986)
13. Alam, F., Umar, R., Ahmed, S., Dar, F.A.: A new model (DRASTIC-LU) for evaluating groundwater vulnerability in parts of central Ganga Plain, India. Arab. J. Geosci. **7**(3), 927–937 (2012). https://doi.org/10.1007/s12517-012-0796-y
14. Sinha, M.K., Verma, M.K., Ahmad, I., Baier, K., Jha, R., Azzam, R.: Assessment of groundwater vulnerability using modified DRASTIC model in Kharun Basin, Chhattisgarh, India. Arab. J. Geosci. **9**(2), 1–22 (2016). https://doi.org/10.1007/s12517-015-2180-1
15. Zafane, D., Gharbi, B., Douaoui, A.A.: New model (DRASTIC-LU) for evaluating groundwater vulnerability in alluvial aquifer of upper Cheliff (Algeria). In: Euro-Mediterranean Conference for Environmental Integration, pp. 615–617. Springer, Cham. (2017). https://doi.org/10.1007/978-3-319-70548-4_185
16. Kumar, A., Pramod Krishna, A.: Groundwater vulnerability and contamination risk assessment using GIS-based modified DRASTIC-LU model in hard rock aquifer system in India. Geocarto Int. **35**, 1149–1178 (2020). https://doi.org/10.1080/10106049.2018.1557259

Thornthwaite's Method for the Computation of the Water Balance

Selmin Burak[1(✉)], Ayşe Hümeyra Bilge[2], and Duygu Ülker[1]

[1] Institute of Marine Sciences and Management, Istanbul University, 34134 Istanbul, Turkey
{sburak,duygu.ulker}@istanbul.edu.tr
[2] Faculty of Engineering and Natural Science, Kadir Has University, 34083 Istanbul, Turkey
ayse.bilge@khas.edu.tr

Abstract. Evapotranspiration is one of the crucial processes on water balance under the effect of meteorological and environmental factors and soil and vegetation characteristics. Thornhtwaite's method for the evaluation of the water balance at a given place is related to the temperature, precipitation data at monthly or daily intervals and on the information on the water holding capacity of the soil at that place. The method is based on equating the sum of evapotranspiration and runoff to the sum of precipitation and soil moisture. In Thornthwaite's method, most steps of the calculation are expressed by mathematical formulas but the values of the amount of water retained in the soil as a function of potential evapotranspiration (PE), for each value of the water holding capacity (WHC) of the soil are given by tables. In this paper we present an overview of the method and we present mathematical formulas that fit the data of the of these tables, to express the relation between water retained in soil and potential evapotranspiration (PE) for different water holding capacity (WHC).

Keywords: Evapotranspiration · Runoff · Soil moisture retention · Thornthwaite's method · Water balance

1 Introduction

Evapotranspiration is one of the crucial processes on water balance under the effect of meteorological and environmental factors and soil and vegetation characteristics. Quantification of real evapotranspiration (ETr) is a significant part to be considered for water resources management, irrigation management, agricultural planning, modelling of climate change effects etc. [1, 2]. However, the computation of ETr under actual field conditions has difficulties. For this purpose, estimation of potential evapotranspiration (PE) is required to estimate actual evapotranspiration in consideration of water balance method.

The climatic water balance was first introduced by Thornthwaite in "the Report of the Committee on Transpiration and Evaporation" in 1944 [3]. On this basis, the concept of potential evapotranspiration is introduced by [4] with the improved classification of climate approach. After the development studies on water balance, Thornthwaite and Mather published PE calculation tables in 1955 and its revised edition published in 1957

H. Gökçekuş and Y. Kassem (Eds.): NRSEM 2021, EESCI, pp. 339–347, 2022.
https://doi.org/10.1007/978-3-031-04375-8_38

[5, 6]. Drought indexes and water-balance models underlying Thornthwaite's method are widely used today [7–10].

The mechanisms on the occurrence of the runoff on the basis of Thornthwaite's water balance approach are shown in Fig. 1. The model estimates the water amount of the several components of the hydrological cycle. Thornthwaite's water balance method that is based on equation between the sum of evapotranspiration and runoff, and the sum of precipitation and soil moisture. Necessary information to compute water balance for a region are; (i) Mean monthly or daily air temperatures,(ii) Mean monthly or daily precipitation (iii) Necessary conversion and computational tables, (iv) Information on the water holding capacity of the depth of soil for which the balance is to be computed [5].

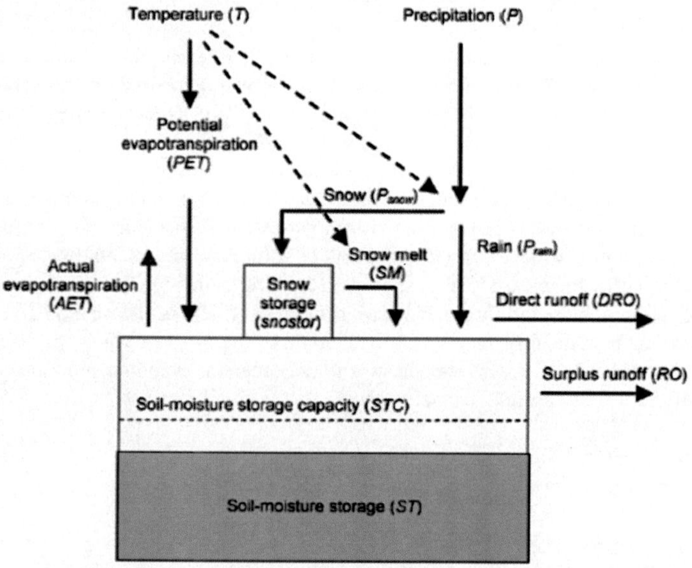

Fig. 1. Thornthwaite's water balance diagram (adapted from [11])

In this study, we aimed to particularize the Thornthwaite's method step by step to show the each stages of the method and to avoid missing and incorrect implementation of the method taking into account the widespread use of the method in the literature [8, 12–15]. Additionally, 11 tables presented by Thornthwaite and Mather 1957 are derived as a graph to show the relation between water retained in soil and potential evapotranspiration (PE) for different water holding capacity (WHC). The function fitting of the tables and implementation in a catchment area are aimed for the next study purpose to show the easier implementation of the Thornthwaite and Mather method in the literature.

2 Literature Review

The hydrological balance equation is a beneficial method in many disciplines related to water resources management. Water balance method is based on the equation between entering water from precipitation (rain and snow) and the outflow of water by evapotranspiration, groundwater recharge, and runoff [16]. Even though such models may use hourly or daily data, these models are however more data intensive and have more parameters than do the corresponding monthly models [17]. On the other hand a another monthly water balance calculation methods are accessible in the literature [18–20], Thornthwaite and Mather method is widely used and accepted method in many different fields [5] such as hydrology, hydraulic engineering, agricultural management, ecology with easily accessible data required which are long-term monthly mean temperature, mean precipitation and latitude. For instance apply to compute PE and provide input on COSERO Hydrological modelling of the Danube basin upstream of Vienna [21], to compute monthly water balance in semi-arid region in Texas to observe basin scale risk assessment from climate change [22], to compute PE for observing Dune Mobility Index [23], to determine of planting time and planting area for agricultural management [24], to compute runoff at seven watersheds and evaluate water demand in Istanbul [25, 26], to assess surface runoff changes in the Mountainous basin [27], to compute PE in the non-irrigated areas for the purpose of observing climate change and its impacts on coastal aquifers [28], to compute potential and actual evapotranspiration for the purpose of observing patterns and ecological implications [29].

Additionally, computer programs and models based on Thornthwaite and Mather method have been developed since 1980s. For instance; WTRBLN calculates water balance based on the basis of long-term average monthly precipitation [16], USGS is developed a software which is "A Monthly Water-Balance Model Driven By a Graphical User Interface" [11]. However, it calculates the PET using the Hamon's equation [30]. The STREAM s a grid-based spatially distributed hydrological balance model developed by [31] and has been applied in different studies. For instance; applied for modelling of runoff to observe climate and land-use changes in the Belgian Dijle catchment by [32] and applied for hydrological modelling to identify priority areas in terms of flood regulation services in Europe [33]. The WebApp based WaterbalANce software written in Phyton which is recently introduced by [34] applying in two watersheds in Europe.

3 Computation of Monthly Runoff with Thornthwaite Method

The sum of actual evapotranspiration, the soil moisture increase, infiltration, and runoff is equal to precipitation in the water balance method [6]. The relation between precipitation and runoff varies not only on the soil type and evaporation characteristics, but also on the strength and the duration of the precipitation and it involves delays due to snow melting effects, hence it is far from being linear. Nevertheless, the ratio of runoff to precipitation, denoted as the runoff coefficient, is a generally accepted measure for estimating runoff. Computation of ETr provides an indirect assessment of the runoff via water balance [5].

In this section we describe the method pre as given in [5]. Data needed for this computation are:

- Monthly tempertures (T_m), in degree celsius.
- Monthly precipitation (P_m), in milimeters.
- The water holding capacity of the root zone of the soil (ST_{max}) for the location under consideration, in milimeters.

- The latitude λ of the location and the mean sunlight hours for each month.

The quantities described below are arranges in the rows of a table and displayed as monthly values of the corresponding variables.

3.1 Stages of the Method

Step 1 "Monthly Temperatures (T_m)". Obtaining of the monthly temperatures data is needed.

Step 2 "The Heat Index I = I(T)". The values for the heat index for each month are given by the formula below. If T_m is negative, the heat index is zero when the mean temperature is 0 °C or below.

$$I_m = \left(\frac{T_m}{5}\right)^{1.514} \text{ if T} > 0 \text{ and } I_m = 0 \text{ if } T \leq 0 \quad (1)$$

Step 3 "The Unadjusted Potential Evapotranspiration (PE_m)". In the earlier work of Thornthwaite[4] these values are computed in terms of monthly temperature and the heat index as

$$PE_m = 1.6\left[10\frac{T_m}{I}\right]^a \quad (2)$$

$$a = 6.75 \times 10^{-7} \times I^3 - 7.71 \times 10^{-5} \times I^2 + 1.792 \times 10^{-2} \times I + 0.49239^7 \quad (3)$$

Step 4 "The Adjusted Potential Evapotranspiration λm PEm". The values for potential evaporation are adjusted for the daylight hours and for the duration of each month. The correction factors are introduced by Thornthwaite 1948 which the unadjusted potential evapotranspiration of each month must be multiplied (Thornthwaite 1948).

Step 5 "Monthly Precipitation (P_m)". Obtaining of the monthly precipitation data is needed.

Step 6 "Difference between Monthly Precipitation and Potential Evapotranspiration, (P_m- $\lambda_m PE_m$)". Computation of P-PE value is necessary to determine periods of moisture excess and deficiency.

Step 7 "Accumulated Potential Water Loss (WL_m)". The values in are computed as follows.

- The values of accumulated water loss, WL is zero for months for which P_m- PE_m is positive.
- If the sum of P-PE is positive, for months during which P_m- PE_m is negative, accumulated potential water loss is defined cumulatively, as the sum of the potential water loss during the previous month and of the difference P_m-PE_m. Then, the amount of water lost from ground due to moisture reduction can be expressed in mathematical terms as

$$WL_m = 0 \text{ if } P_m - PE_m \text{ is positive} \quad (4)$$

$$WL_m = WL_{m-1} + P_m - PE_m \text{ if } P_m - PE_m \text{ is positive} \quad (5)$$

- If the sum of P-PE is negative, and month j is the first month at which P-PE is negative, then an initial value for WLm is estimated by using successive approximations as described above, and this value WL_0 is added to the difference P-PE at month j. Accumulated potential water loss is defined then cumulatively, as above.

Step 8 "Soil Moisture Storage (ST_m)".

- If the accumulated water loss at month m is zero and $T \geq 1$ °C, then the value ST is equal to the maximum storage capacity ST_{max}. Therefore, the value reported in line 8, is equal to ST_{max} for all consecutive months for which

$$P_m - PE_m \text{ is positive and } T \geq 1\,^\circ C \tag{6}$$

- If the accumulated water loss at month m is zero but $T \leq 1$ °C, then the value ST can exceed the maximum storage capacity ST_{max}. Therefore, the value reported in line 8, is equal to

$$ST_m = ST_{m-1} + P_m - PE_m, \text{ if } P_j - PE_j > 0 \text{ for all and } T \leq 1\,^\circ C \tag{7}$$

- For the next group of consecutive monts during which P_m- PE_m is negative, the value reported in line 8, ST_m:, is obtained from the Tables 11–33, corresponding to the appropriate value of water holding capacity ST_{max}. The formulas that fit to the values given in these tables are given in the previous section.

$$ST_m = f(WL_m), \text{ if } P_j - PE_j < 0 \tag{8}$$

- For the next group of consecutive months during which P_m- PE_m is again positive, the value P_m- PE_m is added to the value of stored water in the previous month, but it cannot exceed ST_{max} if $T \geq 1$ °C.

$$ST_m = max\{ST_{max,} ST_{m-1} + P_m - PE_m\}, \text{ if } P_j - PE_j > 0 \tag{9}$$

Step 9 "Change in the Soil Moisture Storage (ΔST_m)". The value reported in line 9 is the difference in the soil moisture storage between one month and next month. It is assumed that $\Delta ST_m = 0$ if value in the storage line is above the water holding capacity

$$\Delta\, ST_m = ST_m - ST_{m-1} \tag{10}$$

The first value is zero, assuming the the value of ST is equal to ST_{max} at the last month.

Step 10 "Actual evapotranspiration (AE)". When P_m- PE_m is postive, actual evapotranspiration is equal to the potential evapotranspiration. On the contrary, if P_m- PE_m is negative, actual evapotranspiration is equal to the sum of precipitation and the absolute value of ΔST_m.

$$AE_m = PE \text{ if } P_j - PE_j > 0 \tag{11}$$

$$AE_m = P_m + |\Delta STm| \text{ if } P_j - PE_j < 0 \tag{12}$$

Step 11 "Moisture deficit (D)". The moisture deficit is defined as the difference between actual and potential evapotranspiration.

Step 12 "Moisture surplus S for T ≥ 1 °C". If the temperature is below -1 °C, the precipitation is in the form of snow and there is no moisture surplus. If the temperature is above -1 °C., and the soil moisture surplus reaches its maximum capacity, then any excess precipitation is subject to runoff.

The moisture surplus is zero if $ST_m < ST_{max}$. If $ST_m = ST_{max}$, $ST_{m\text{-}1} = ST_{max}$, then the moisture surplus is the difference between precipitation and and actual evapotranspiration. But if $ST_{m\text{-}1} < ST_{max}$, the precipitation is also used to saturate the soil. Therefore,

$$S_m = 0, \text{ if } \mathrm{ST_m} < \mathrm{ST_{max}} \tag{13}$$

$$\mathrm{S}_m = \mathrm{P}_m - AE_m, \text{ if } \mathrm{ST_m} = \mathrm{ST_{max}},\ \mathrm{ST_{m-1}} = \mathrm{ST_{max}} \tag{14}$$

$$S_m = P_m - AE_m - (S_{max} - ST_{m-1}) \text{ if } \mathrm{ST_m} = \mathrm{ST_{max}},\ \mathrm{ST_{m-1}} < \mathrm{ST_{max}} \tag{15}$$

Step 13 "Runoff (RO)". The conversion of the moisture surplus is converted into runoff obeys an exponential decay. A certain percentage of the moisture surplus of a given month is converted to runoff. The same percentage of the remaining amount is converted to runoff next month and so on. This percentage is usually assumed to be 50%. The exponential decay formula is presented in Eq. 16.

$$RO(t) = S \times \exp(-\ln(2)t) \tag{16}$$

Step 14 "Snow melt runoff (SMRO)".

Step 15 "Total runoff (TotRO)". Total runoff is the sum of runoff and snow melt runoff.

Step 16 "Total moisture detention (DT)". The moisture detention is the total of the water stored within the soil, the snow remaining on the surface and the surplus water in the process of running off which has been detained for a month.

4 Results and Discussion

Thornthwaite and Mather method is a traditional and widely used method to compute water budget with hydrological balance equation as shown in the literature [34]. It is easily accessible data requirement makes the method more common in the hydrological modelling studies comparing the other methods. Thornthwaite and Mather method requires monthly data and equation unit is in the monthly basis. Even though such models may use hourly or daily data, these models are however more data intensive and have more parameters than do the corresponding monthly model [17].

The 16 stages of the method are defined in the Sect. 2 to particularize of each stage of the method to avoid missing and incorrect implementation of the method in the literature. Tables of soil moisture retention depend on PE to compute the water balance are given by [5] for different values of WHC between 25 mm and 400 mm. Derivation of the tables

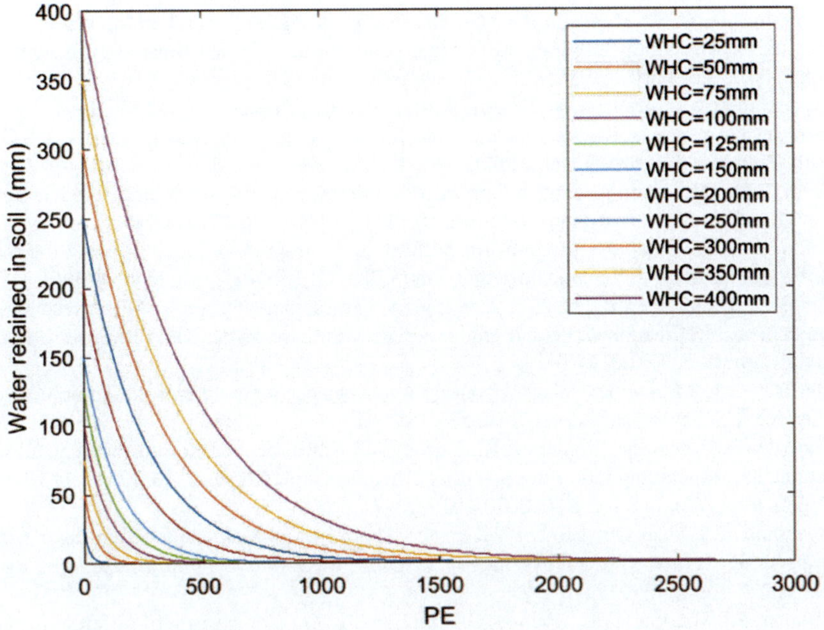

Fig. 2. Water retained in soil depends on PE for different value of WHC.

is presented in Fig. 2. Increase of PE depending of WHC is shown in exponential decay in Fig. 2.

The function fitting from the tables according to different WHC and implementation of the method in a catchment area are in the scope of the next study to show the easier and clear implementation of the method.

References

1. Özhan, S., Gökbulak, F., Serengil, Y., Özcan, M.: Evapotranspiration from a mixed deciduous forest ecosystem. Water Resour. Manag. **24**, 2353–2363 (2010). https://doi.org/10.1016/0022-2836(81)90087-5
2. Pandey, P.K., Dabral, P.P., Pandey, V.: Evaluation of reference evapotranspiration methods for the northeastern region of India. Int Soil Water Conserv. Res. **4**, 52–63 (2016). https://doi.org/10.1016/j.iswcr.2016.02.003
3. Wilm, H.G., Thornthwaite, C.W., Colman, E.A., Cummings, N.W., Croft, A.R., Gisborne, H.T., et al.: Report of the Committee on Transpiration and Evaporation, 1943–1944. Eos Trans. Am. Geophys. Union. **25**, 683–693 (1944)
4. Thornthwaite, C.W.: An approach toward a rational classification of climate. Geogr. Rev. **38**, 55–94 (1948). https://doi.org/10.2307/210739
5. Thornthwaite, C.W., Mather, J.R.: Instructions and tables for computing potential evapotranspiration and water balance. Drexel Institute of Technology, Laboratory of Climatology, Publications in Climatology, New Jersey, vol. 10, pp.185–231 (1957)

6. Thornthwaite, C.W., Mather, J.R.: The water balance. Publ. Climatol. New Jersey: Drexel Institute Of Technology.Laboratory of Climatology, Publications in Climatology, New Jersey, vol. 8, no. 1, pp. 1–104 (1955)
7. Westenbroek, M.S., Kelson, V., Dripps, W.R., Hunt, R.J., Bradbury, K.R.: SWB—a modified thornthwaite-mather soil-water- balance code for estimating groundwater recharge. US Geol. Surv. Tech Methods 6(A31), 60 (2010)
8. Keim, B.D.: The lasting scientific impact of the thornthwaite water-balance model. Geogr. Rev. **100**, 295–300 (2010). https://doi.org/10.1111/j.1931-0846.2010.00035.x
9. Sellinger, C.E.: Computer program for performing hydrograph separation using the rating curve method. NOAA Technical Memorandum ERL GLERL-100, Michigan (1996)
10. Toure, A., Diekkrüger, B., Mariko, A., Cissé, A.S.: Assessment of groundwater resources in the context of climate change and population growth: case of the Klela Basin in Southern Mali. Climate. **5**, 45 (2017)
11. McCabe, G.J., Markstrom, S.L.: A monthly water-balance model driven by a graphical user interface. US Geological Survey Reston, VA (2007)
12. Trajkovic, S., Gocic, M., Pongracz, R., Bartholy, J.: Adjustment of thornthwaite equation for estimating evapotranspiration in vojvodina. Theoret. Appl. Climatol. **138**(3–4), 1231–1240 (2019). https://doi.org/10.1007/s00704-019-02873-1
13. Steenhuis, T.S., Van Der Molen, W.H.: The Thornthwaite-Mather procedure as a simple engineering method to predict recharge. J. Hydrol. **84**, 221–229 (1986). https://doi.org/10.1016/0022-1694(86)90124-1
14. Gudulas, K., Voudouris, K., Soulios, G., Dimopoulos, G.: Comparison of different methods to estimate actual evapotranspiration and hydrologic balance. Desalin Water Treat. **51**, 2945–2954 (2013). https://doi.org/10.1080/19443994.2012.748443
15. Calvo-Alvarado, J.C.: An evaluation of thornthwaite's water balance technique in predicting stream runoff in Costa Rica. Hydrol. Sci. J. **31**, 51–60 (1986). https://doi.org/10.1080/02626668609491027
16. Donker, N.H.W.: WTRBLN: a computer program to calculate water balance. Comput. Geosci. **13**, 95–122 (1987). https://doi.org/10.1016/0098-3004(87)90002-1
17. Xu, C.-Y., Singh, V.P.: A review on monthly water balance models for water resources investigations. Water Res. Manag. **12**, 20–50 (1998). https://doi.org/10.1023/A:1007916816469
18. Snyder, F.F.: A water yield model derived from monthly runoff data. Int. Assoc. Sci. Hydrol. Publ. **63**, 18–30 (1963)
19. Kuczera, G.: Improved parameter inference in catchment models: 1. Evaluating parameter uncertainty. Water Res. Res **19**, 1151–1162 (1983). https://doi.org/10.1029/WR019i005p01151
20. Gabos, A., Gasparri, L.: Monthly runoff model for regional planning. Water Int. **8**, 42–45 (1983). https://doi.org/10.1080/02508068308686000
21. Santos, I.M., Herrnegger, M., Holzmann, H.: Seasonal discharge forecasting for the Upper Danube. J. Hydrol. Reg. Stud. **37**, 1–18 (2021). https://doi.org/10.1016/j.ejrh.2021.100905
22. Martin, N.: Watershed-scale, probabilistic risk assessment of water resources impacts from climate change. Water **13**, 40 (2021). https://doi.org/10.3390/w13010040
23. Santos, L.A., Latrubesse, E.M.: Aeolian mobility in the Middle São Francisco Dune Field, Northeast Brazil, as a response to Caatinga's droughts and land-use changes. Geomorphology **393**, 1–9 (2021). https://doi.org/10.1016/j.geomorph.2021.107940
24. Apriyana, Y., Surmaini, E., Estiningtyas, W., Pramudia, A., Ramadhani, F., Suciantini, S., et al.: The integrated cropping calendar information system: a coping mechanism to climate variability for sustainable agriculture in Indonesia. Sustainability. **13**, 1–22 (2021). https://doi.org/10.3390/su13116495

25. Burak, S., Bilge, A.H., Ülker, D.: Assessment and simulation of water transfer for the megacity Istanbul. Phys. Geogr., 1–25 (2021). https://doi.org/10.1080/02723646.2021.1904698
26. Burak, S., Bilge, A.H., Ulker, D.: Computation of monthly runoff coefficients for Istanbul (Turkey). Therm. Sci. **25**, 147 (2021). https://doi.org/10.2298/TSCI191102147B
27. Stefanidis, S., Stathis, D.: Assessing Surface Runoff in Future Climate Conditions in Mountainous Catchments (Case Study: Klinovitikos Torrent, Central Greece). In: HAICTA 2020, Thessalonki, Greece, pp. 425–435 (2020). http://ceur-ws.org/Vol-2761/HAICTA_2020_paper62.pdf
28. Stigter, T.Y., et al.: Comparative assessment of climate change and its impacts on three coastal aquifers in the Mediterranean. Reg. Environ. Change **14**(1), 41–56 (2012). https://doi.org/10.1007/s10113-012-0377-3
29. Frank, D.A., Inouye, R.S.: Temporal variation in actual evapotranspiration of terrestrial ecosystems: patterns and ecological implications. J. Biogeogr. **21**, 401–411 (1994). https://doi.org/10.2307/2845758
30. Hamon, W.R.: Estimating potential evapotranspiration. J. Hydraul. Div. **87**, 107–120 (1961). https://doi.org/10.1061/JYCEAJ.0000599
31. Aerts, J., Kriek, M., Schepel, M.: STREAM (Spatial tools for river basins and environment and analysis of management options):'set up and requirements.' Phys. Chem. Earth Part B Hydrol. Ocean Atmos. **24**, 591–595 (1999). https://doi.org/10.1016/S1464-1909(99)00049-0
32. Notebaert, B., Verstraeten, G., Ward, P., Renssen, H., Van, Rompaey, A.: Modeling the sensitivity of sediment and water runoff dynamics to Holocene climate and land use changes at the catchment scale. Geomorphology **126**, 18–31 (2011). https://doi.org/10.1016/j.geomorph.2010.08.016
33. Stürck, J., Poortinga, A., Verburg, P.H.: Mapping ecosystem services: the supply and demand of flood regulation services in Europe. Ecol. Indic. **38**, 198–211 (2014). https://doi.org/10.1016/j.ecolind.2013.11.010
34. Mammoliti, E., Fronzi, D., Mancini, A., Valigi. D., Tazioli, A.: WaterbalANce, a WebApp for thornthwaite–mather water balance computation: comparison of applications in two European watersheds. Hydrology **8,** 34 (2021). https://doi.org/10.3390/hydrology8010034

Urban Heat Island Effects of Pavements

Gokhan Calis[1,2(✉)], Sadik Alper Yildizel[1,2], and Ulku Sultan Keskin[1,2]

[1] Karamanoglu Mehmetbey University, 70100 Karaman, Turkey
gokhancalis@kmu.edu, sayildize@springer.com
[2] Konya Technical University, 42000 Konya, Turkey
uskeskin@ktun.com

Abstract. Urban heat island effect is one of the major issues and to be addressed in order to achieve sustainability. Human population in the cities has been increasing tremendously. As a result, natural areas are turned into buildings, roads etc. Scholars from various disciplines are now focused on the causes of Urban Heat Island effect and mitigation strategies. Considering the areas covered by the road pavements in the cities, its share on Urban Heat Island Effect is indisputable. Within the scope of this study a sample area was selected and modelled by utilizing various pavement materials. These areas have been investigated in terms of temperature, reflectance, view factor. The results are promising and likely to change the common beliefs. Pavements with higher albedos are helpful to reduce heat island effect of the non-natural materials in the environments. The temperature difference can be 0.5 °C to 2 °C.

Keywords: Urban heat island · Concrete pavement · Asphalt pavement · Cool pavement

1 Introduction

The urbanization leads an increase in the human population and built environment within the cities. This development has significant impact on the environment. With the increase of non-natural covering materials in the cities, the soil which was previously reflective loses its reflectivity and absorbs sun light. Consequently, the temperature in the urban areas become higher than the surrounding rural areas. This is the known phenomenon called "Urban Heat Island".

There are two type of factors that has direct impact over generation of Urban Heat Island effect. The first one contains environmental, and natural factors that cannot be changed by human beings. The second type is human related factors namely; decrease of heat flux due to vegetation lose [1], large amount of solar heat absorbed and re-emitted by the unnatural surfaces (low albedo) [2], reduce ventilation and airflow due to high density of buildings [3], high heat flux [4]. The increase in the building stock brought about by urbanization and road coverings mean the reduction of natural vegetation and the placement of materials that absorb heat and cause less reflection than natural vegetation [5–11]. The causes of urban heat island formation are determined by the scholars. These causes are; pavement materials [12–14], human population increase [15–17], increase of

H. Gökçekuş and Y. Kassem (Eds.): NRSEM 2021, EESCI, pp. 348–357, 2022.
https://doi.org/10.1007/978-3-031-04375-8_39

energy consumption, reduce of natural soils and green areas, increase solid surfaces in the city [5–11]. Furthermore, Nwakaire has classified UHI causes under 3 title namely; anthropogenic, structural and climatic [18]. Anthropogenic causes contain metabolism, heating & air conditioning, transportation and manufacturing. These are the natural results of increase in human population and almost inevitable.

The temperature increase in an urban area might be 0.5 °C up to 12 °C [19, 20].With the temperature increase urban heat island (UHI) causes some important issues need to be addressed properly. These issues are; increase in energy need, carbon emission, death rate or high temperature related diseases and global warming [21]. This has negative impacts on human being health and other ecosystems. In 2013, more than 5000 patients were diagnosed with heat related sickness in China [22–24]. These cases mostly located in the dense areas whereas high level of urban heat island impact exists. This is concrete data proves that UHI forms solid danger for public health. In addition, UHI along with global warming causes chronic diseases such as cardiovascular, mental disorders [25]. Susca (2012) carried out a study investigating a life cycle analysis of albedo changes in order to reduce the UHI in New York City and determined that increasing all albedo value of the roofs from 0.32 to 0.90 would lead 0.5 °C temperature decrease. It is predicted that this air temperature decrease would be a way to avoid nearly 750 disability-adjusted life years (DALY-defined by World Health Organization as the lost years of health life [26]. In three metropolitan areas, combining vegetation (roof greening and street trees) with albedo increase to 0.9 would reduce heat related deaths by 40% to 90% [27]. Another study indicates that if reflectance of surfaces is increased 20%, heat related mortality can be reduced up to 20% within 10 years [28]. Netherlands reveals how ambient temperature affected the number of deaths between 1979 and 1997. Based on the increase of the weather temperature the total mortality rate determined to be increased by 2.72% for each degree increase in air temperature. Furthermore, the other study that was carried out in Canada indicates that relative mortality rate increases by 2.3% for each degree increase in air temperature [29].

Scholars expect that %20 of world crowded cities will suffer temperature increase from 4 °C up to 6 °C in the next 3 and 8 decades [30]. UHI is considered as one of the biggest problems that human beings face with in the 21st century [31]. Hence, scholars and professionals must pay attention and focus on preventing or reducing urban heat island formation as early as possible. The fundamental problem is the lack of understanding and awareness of UHI [14, 32]. Engineers, urban developers, and architects do not take in to account urban heat island mitigation strategies in. In fact it was determined that in local authorities and professionals do not have sufficient information and awareness of UHI [33]. It was also pointed out that some mitigation strategies might be useful in one season such as summer, and not have the same impact on the winter. Therefore, while mitigation strategies are determined the geographical location of the country and other variables should be considered in deep [34]. This can also be interpreted as one mitigation strategy for a country can increase urban heat island impact in another. In fact, this could be valid for two cities in the same country too.

Academics have investigated various mitigation strategies to address urban heat island effect properly [19, 35–42]. Recent study concluded that best solution against UHI in New York City is to increase reflectivity of roofs and increasing trees in the streets [43]. In their study, the surfaces have been investigated and potential mitigation strategies were performed. In terms of heat flux, wind speed change, air temperature, increasing reflectivity of roofs and vegetation at street level was the most effective method. Increasing albedo of a surface may result 10 °C decrease in the surface temperature [27]. Cool roof application is an effective way to decrease office buildings energy consumption for cooling, and can reduce air temperature 1 °C in the summer season [44].

Green roof is another roof related solution based on planting vegetations in the roof areas. The fundamental effect of greening the area is creating shade and protections the building against solar energy. As a results up to 30% decrease in the air temperature can be obtained in the summer. More dense vegetation will increase evatranspiration and relative humidity [26]. Another study carried out in Texas, indicated that by changing roof material of the simulation area can reduce air temperature nearly 2 °C [45].

Buildings façades in the urban areas receive significant amount of solar radiation, and depending upon the type and properties of this façade material urban heat island forms namely air temperature increases, wind speed reduces, relative humidity decreases [37]. In the recent research the impact of retro reflective facades has been investigated under various angles. When the solar radiation reaches the retro reflective facades with high angles, it is reflected symmetrically [36]. Hence, the reflectivity of these facades is dependent upon the incoming angle of solar radiation, the location of building, and building density in the area.

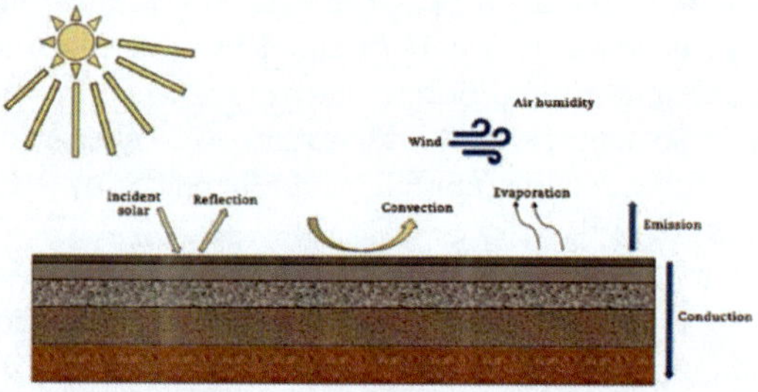

Fig. 1. Heat transfer in road pavements [46].

Cool pavements can be classified under 3 titles namely; evaporative, heat storage modified and reflective pavements [47]. Reflective pavements have higher albedo value than traditional asphalt pavement. With various types of cool pavements and various properties, they are capable to reflect solar radiation more than black color asphalt pavements [48]. Heat transfer mechanism in road pavements is presented in Fig. 1. Increasing albedo of pavements is considered one of the most effective mitigation strategies against urban heat island. It has been predicted that changing albedo from 0.1 to 0.35 can reduce the air temperature 0.6 °C in Los Angeles [49]. In the recent study solar reflective coating with functional gradient multilayer structure was performed on asphalt pavement. Different coatings were prepared, and their cooling impacts were investigated. Proposed solar reflective coating contains 3 layers with pigments. The increase of pigment dosage results increases in reflectivity, reduce urban heat island impact. However, in this multi-layer coating each layer has its own effects and the most successful combination was able to reduce the temperature °C [50]. Yinfei et. al [51] examined the benefits of replacing limestone mineral filler with hollow glass microsphere. The developed material with hollow glass microsphere has shown 40% less thermal conductivity, and 60% higher near infrared radiation reflectance than the control sample. Asphalt pavements albedo value is at the range of 0.05 to 0.20 while concrete pavements have brighter surface with an albedo of 0.35. Therefore, it is expected to be a better paving material in terms of reflectance (albedo) and urban heat island forming effect. In this study, it is aimed to quantify results of utilization of concrete pavement as an alternative paving material to the traditional asphalt pavement.

Urban heat island (UHI) has become a global issue that each state must deal with both individually and with other studies. Considering the total amount of concrete use and existing asphalt pavements in the world, they both have effect of urban heat island formation. Application of pavements and their effects concern civil engineering and environmental engineering disciplines. In this study the objective is to investigate which type of pavement has greater effect in terms of urban heat island parameters.

Urban heat island (UHI) has become a global issue that each state must deal with both individually and with other studies. Considering the total amount of concrete use and existing asphalt pavements in the world, they both have effect of urban heat island formation. Application of pavements and their effects concern civil engineering and environmental engineering disciplines. In this study the objective is to investigate which type of pavement has greater effect in terms of urban heat island parameters.

2 Simulation and Results

Within the scope this research, an area of 400 m^2 land (20 m × 20 m) with two buildings, tree and pavement was investigated. The model can be seen in the Fig. 2. Envi-Met software lite version was utilized to analyze the area and pavements There are two buildings positioned next to each other. Black area represents the pavement and orange represents natural soil. In the second case only black is changed to concrete and simulation was performed.

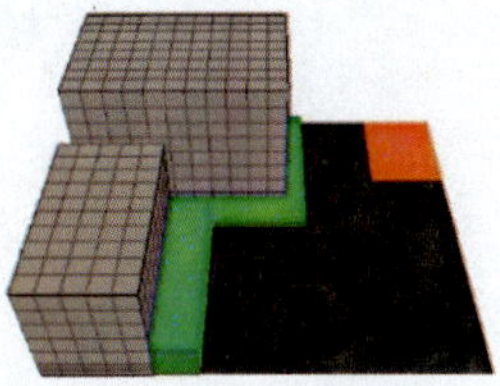

Fig. 2. Simulation model 3d

Potential air temperature is given in the Fig. 3. The one on the left hang side represent the temperature of asphalt pavement scenario and the one on the left represents concrete pavement. It was determined that the temperature of area covered with light color is significantly lower than the area covered with asphalt. In the area where concrete pavement used, the minimum temperature is 29.52 °C while the maximum is 29.74 °C. In the asphalt pavement area the maximum temperature is 31.49 °C and 30.11 °C. It can be concluded that concrete pavement can reduce the temperature 0.59 °C–1.75 °C.

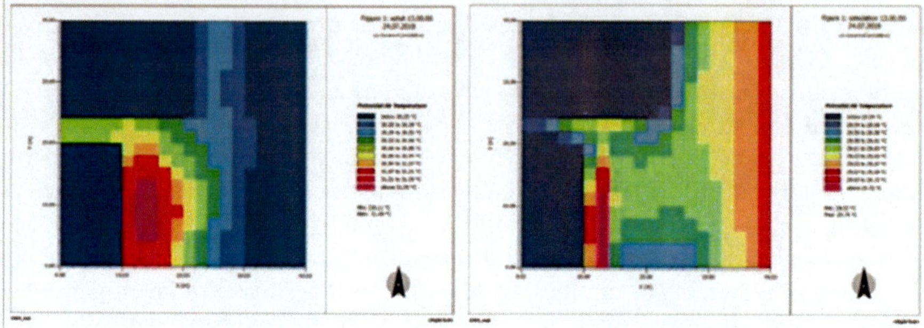

Fig. 3. Potential air temperature comparison of asphalt and concrete

Concrete pavement has significantly difference level of reflectance in comparison to asphalt pavement. Concrete pavement's surface has 617.04 w/m^2 reflectance capacity while asphalt surface has 192.92 w/m^2.Therefore, it can be concluded that concrete pavement is 3 times stronger than asphalt pavement in terms of reflecting the solar radiation. This can be seen in Fig. 4.

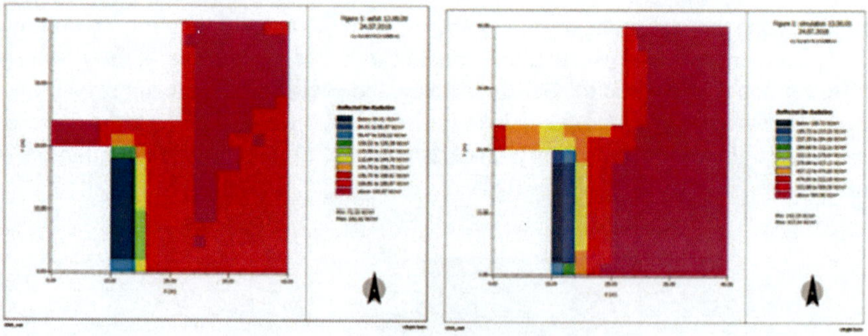

Fig. 4. Reflected SW radiation (atmosphere) of Asphalt and Concrete

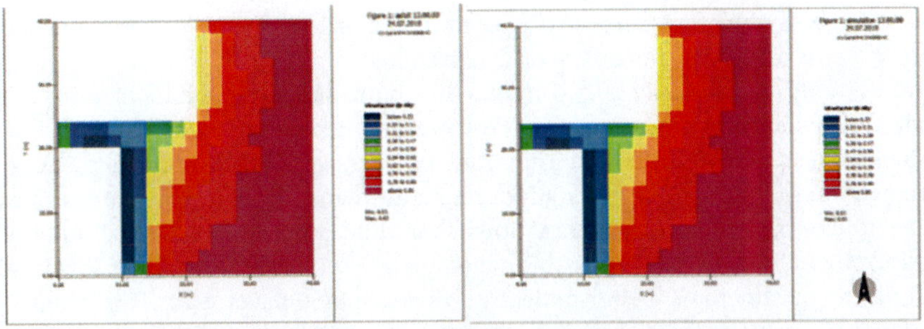

Fig. 5. View factor comparison of Asphalt and Concrete

View factor impact of asphalt is presented in Fig. 5 on the left side and concrete is on the right-hand side. As it can be seen from Fig. 57, both pavements can provide the same view factor.

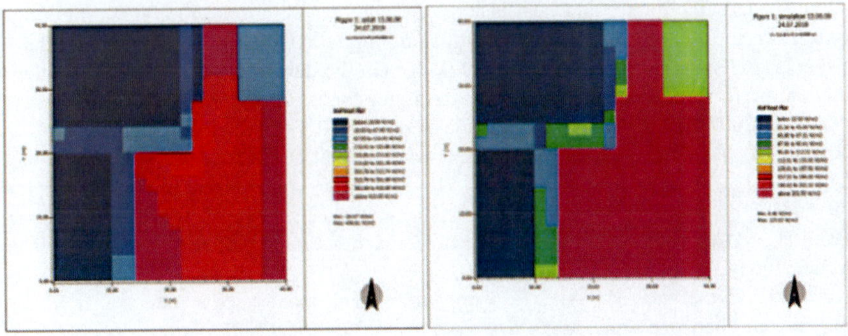

Fig. 6. Soil heat flux comparison of Asphalt and Concrete

Soil heat flux comparison of the pavements is shown in the Fig. 6 on the left side and concrete is on the right hand-side. Asphalt pavement casted area receives more energy per square meter than concrete pave casted area. This can be seen as the reason for the high temperature of the asphalt surface.

3 Discussion

The materials that replace the natural soil has significant effect on formation of urban heat island effect even the area is small as it is in this case study. Especially open large spaces such as parking lots, building front/back yards are vitally important as the volume of incoming solar radiation is high.

If the pavement material is altered, urban heat island formation can be easily changed. At this point scholars are advised to focus on developing cool pavement and cool construction materials that has high albedo.

View factor is not affected by the type of pavement in this study. Therefore, pavement type does not provide any advantage or disadvantages.

Natural soils are fundamentally different than human made materials in terms of soil heat flux, reflectance as it is expected. Furthermore, lack of green areas plays crucial role in urban heat island formation. Therefore, cities must develop and enlarge not only in terms of number of buildings, length of road pavements but also green areas as well. Needles to emphasize that the natural soils should be protected and, local authorities should afforest as much areas as possible not only in the rural areas but also in downtowns.

Concrete pavement is better reflective than asphalt pavement. This should be considered by municipalities especially the ones in the hot regions.

For the future studies the differences caused by the height and position of buildings must be investigated. In a largescale simulation urban sprawl and organized urban development can be compared in terms of wind speed, air temperature increase etc.

References

1. Taha, H.: Urban climates and heat islands: Albedo, evapotranspiration, and anthropogenic heat. Energy Build **25**, 99–103 (1997). https://doi.org/10.1016/s0378-7788(96)00999-1
2. Synnefa, A., Santamouris, M., Apostolakis, K.: On the development, optical properties and thermal performance of cool colored coatings for the urban environment. Sol. Energy **81**, 488–497 (2007). https://doi.org/10.1016/j.solener.2006.08.005
3. Tayanç, M., Toros, H.: Urbanization effects on regional climate change in the case of four large cities of Turkey. Clim Change **35**, 501–524 (1997). https://doi.org/10.1023/A:1005357915441
4. Bahi, H., Mastouri, H., Radoine, H.: Review of methods for retrieving urban heat islands. Mater. Today Proc. **27**, 3004–3009 (2020)
5. Akbari, H., Pomerantz, M., Taha, H.: Cool surfaces and shade trees to reduce energy use and improve air quality in urban areas. Sol. Energy **70**, 295–310 (2001). https://doi.org/10.1016/S0038-092X(00)00089-X
6. Doulos, L., Santamouris, M., Livada, I.: Passive cooling of outdoor urban spaces. The role of materials. Sol. Energy **77**, 231–249 (2004). https://doi.org/10.1016/j.solener.2004.04.005
7. Cotana, F., Rossi, F., Filipponi, M., et al.: Albedo control as an effective strategy to tackle global warming: a case study. Appl. Energy **130**, 641–647 (2014). https://doi.org/10.1016/j.apenergy.2014.02.065
8. Santamouris, M., Yun, G.Y.: Recent development and research priorities on cool and super cool materials to mitigate urban heat island. Renew. Energy **161**, 792–807 (2020). https://doi.org/10.1016/j.renene.2020.07.109
9. Jusuf, S.K., Wong, N.H., Hagen, E., et al.: The influence of land use on the urban heat island in Singapore. Habitat Int **31**, 232–242 (2007). https://doi.org/10.1016/j.habitatint.2007.02.006
10. O'Malley, C., Piroozfar, P., Farr, E.R.P., Pomponi, F.: Urban Heat Island (UHI) mitigating strategies: a case-based comparative analysis. Sustain. Cities Soc. **19**, 222–235 (2015). https://doi.org/10.1016/j.scs.2015.05.009
11. Smith, C., Levermore, G.: Designing urban spaces and buildings to improve sustainability and quality of life in a warmer world. Energy Policy **36**, 4558–4562 (2008). https://doi.org/10.1016/j.enpol.2008.09.011
12. Takebayashi, H., Moriyama M.: Study on surface heat budget of varioud pavements for urban heat island mitigation. Adv. Mater. Sci. Eng. 2012 (2012). https://doi.org/10.1155/2012/523051

13. Golden, J.S., Kaloush, K.E.: Mesoscale and microscale evaluation of surface pavement impacts on the urban heat island effects.pdf. Int. J. Pavement Eng. (2006). https://doi.org/10.1080/10298430500505325
14. Phelan, P.E., Kaloush, K., Miner, M., et al.: Urban heat island: mechanisms, implications, and possible remedies. Annu. Rev. Environ. Resour. **40**, 285–307 (2015). https://doi.org/10.1146/annurev-environ-102014-021155
15. Form, U.: Urban form and extreme heat events : are sprawling cities more vulnerable to climate change than compact cities ? 118:1425–1428 (2010). https://doi.org/10.1289/ehp.0901879
16. Singh, R., Kalota, D.: Urban sprawl and its impact on generation of urban heat island: a case study of Ludhiana City. J. Indian Soc. Remote Sens. **47**(9), 1567–1576 (2019). https://doi.org/10.1007/s12524-019-00994-8
17. Cai, H., Qiao, Z.: Impacts of park landscape structure on thermal environment using QuickBird and landsat images (2016). https://doi.org/10.1007/s11769-017-0910-x
18. Nwakaire, C.M., Onn, C.C., Yap, S.P., et al.: Urban heat island studies with emphasis on urban pavements: a review. Sustain. Cities Soc. **63**, 102476 (2020). https://doi.org/10.1016/j.scs.2020.102476
19. Mohajerani, A., Bakaric, J., Jeffrey-Bailey, T.: The urban heat island effect, its causes, and mitigation, with reference to the thermal properties of asphalt concrete. J. Environ. Manage. **197**, 522–538 (2017). https://doi.org/10.1016/j.jenvman.2017.03.095
20. Giridharan, R., Kolokotroni, M.: Urban heat island characteristics in London during winter. Sol. Energy (2009). https://doi.org/10.1016/j.solener.2009.06.007
21. Kleerekoper, L., Van Esch, M., Salcedo, T.B.: How to make a city climate-proof, addressing the urban heat island effect. Resour. Conserv. Recycl. **64**, 30–38 (2012). https://doi.org/10.1016/j.resconrec.2011.06.004
22. Lee, W.V.: Historical global analysis of occurrences and human casualty of extreme temperature events (ETEs). Nat. Hazards **70**(2), 1453–1505 (2013). https://doi.org/10.1007/s11069-013-0884-7
23. Zuo, J., Pullen, S., Palmer, J., et al.: Impacts of heat waves and corresponding measures: a review. J. Clean. Prod. **92**, 1–12 (2015). https://doi.org/10.1016/j.jclepro.2014.12.078
24. Gu, S., Huang, C., Bai, L., Chu, C., Liu, Q.: Heat-related illness in China, summer of 2013. Int. J. Biometeorol. **60**(1), 131–137 (2015). https://doi.org/10.1007/s00484-015-1011-0
25. Gabriel, K.M.A., Endlicher, W.R.: Urban and rural mortality rates during heat waves in Berlin and Brandenburg, Germany. Environ. Pollut. **159**, 2044–2050 (2011). https://doi.org/10.1016/j.envpol.2011.01.016
26. Arabi, R., Shahidan, M., Kamal, M.S., et al.: Mitigating urban heat island through green roofs. Curr. World Environ. **10**, 918–927 (2015). https://doi.org/10.12944/cwe.10.special-issue1.111
27. Macintyre, H.L., Heaviside, C.: Potential benefits of cool roofs in reducing heat-related mortality during heatwaves in a European city. Environ. Int. **127**, 430–441 (2019). https://doi.org/10.1016/j.envint.2019.02.065
28. Vanos, J., Shickman, K., Alliance, G.C.C., Sheridan, S.: Assessing the health impacts of urban heat island reduction strategies in the cities of Baltimore, Los Angeles, and New York (2014)
29. Fahed, J., Kinab, E., Ginestet, S., Adolphe, L.: Impact of urban heat island mitigation measures on microclimate and pedestrian comfort in a dense urban district of Lebanon. Sustain. Cities Soc. **61**, 102375 (2020)
30. Estrada, F., Botzen, W.J.W., Tol, R.S.J.: A global economic assessment of city policies to reduce climate change impacts. Nat. Clim. Chang **7**, 403–406 (2017). https://doi.org/10.1038/nclimate3301
31. Rizwan, A.M., Dennis, L.Y.C., Liu, C.: A review on the generation, determination and mitigation of Urban Heat Island. J. Environ. Sci. **20**, 120–128 (2008). https://doi.org/10.1016/S1001-0742(08)60019-4

32. Wong, J.K.W., Lau, L.S.K.: From the "urban heat island" to the "green island"? A preliminary investigation into the potential of retrofitting green roofs in Mongkok district of Hong Kong. Habitat Int. **39**, 25–35 (2013). https://doi.org/10.1016/j.habitatint.2012.10.005
33. Ramakreshnan, L., Aghamohammadi, N., Fong, C.S., et al.: A qualitative exploration on the awareness and knowledge of stakeholders towards Urban Heat Island phenomenon in Greater Kuala Lumpur: critical insights for urban policy implications. Habitat Int. **86**, 28–37 (2019). https://doi.org/10.1016/j.habitatint.2019.02.007
34. Debbage, N., Shepherd, J.M.: The urban heat island effect and city contiguity. Comput. Environ. Urban Syst. **54**, 181–194 (2015). https://doi.org/10.1016/j.compenvurbsys.2015.08.002
35. Heaviside, C., Macintyre, H., Vardoulakis, S.: The urban heat island: implications for health in a changing environment. Current Environ. Health Rep. **4**(3), 296–305 (2017). https://doi.org/10.1007/s40572-017-0150-3
36. Rossi, F., Castellani, B., Presciutti, A., et al.: Retroreflective façades for urban heat island mitigation: experimental investigation and energy evaluations. Appl Energy **145**, 8–20 (2015). https://doi.org/10.1016/j.apenergy.2015.01.129
37. Aflaki, A., Mirnezhad, M., Ghaffarianhoseini, A., et al.: Urban heat island mitigation strategies: a state-of-the-art review on Kuala Lumpur, Singapore and Hong Kong. Cities **62**, 131–145 (2017). https://doi.org/10.1016/j.cities.2016.09.003
38. Li, D., Bou-Zeid, E., Oppenheimer, M.: The effectiveness of cool and green roofs as urban heat island mitigation strategies. Environ. Res. Lett. **9** (2014). https://doi.org/10.1088/1748-9326/9/5/055002
39. Zhou, Y., Shepherd, J.M.: Atlanta's urban heat island under extreme heat conditions and potential mitigation strategies. Nat. Hazards **52**, 639–668 (2010). https://doi.org/10.1007/s11069-009-9406-z
40. Corburn, J.: Cities, climate change and urban heat island mitigation: localising global environmental science. Urban Stud. **46**, 413–427 (2009). https://doi.org/10.1177/0042098008099361
41. Deilami, K., Kamruzzaman, M., Liu, Y.: Urban heat island effect: a systematic review of spatio-temporal factors, data, methods, and mitigation measures. Int. J. Appl. Earth Obs. Geoinf. **67**, 30–42 (2018). https://doi.org/10.1016/j.jag.2017.12.009
42. Nuruzzaman, M.: Urban heat island: causes, effects and mitigation measures - a review. Int. J. Environ. Monit. Anal. **3**, 67 (2015). https://doi.org/10.11648/j.ijema.20150302.15
43. Lynn, B.H., Carlson, T.N., Rosenzweig, C., et al.: A modification to the NOAH LSM to simulate heat mitigation strategies in the New York City metropolitan area. J. Appl. Meteorol. Climatol. **48**, 199–216 (2009). https://doi.org/10.1175/2008JAMC1774.1
44. Kolokotroni, M., Gowreesunker, B.L., Giridharan, R.: Cool roof technology in London: an experimental and modelling study. Energy Build. **67**, 658–667 (2013). https://doi.org/10.1016/j.enbuild.2011.07.011
45. Asadi, A., Arefi, H., Fathipoor, H.: Simulation of green roofs and their potential mitigating effects on the urban heat island using an artificial neural network: a case study in Austin, Texas. Adv. Sp. Res. **66**, 1846–1862 (2020). https://doi.org/10.1016/j.asr.2020.06.039
46. Aletba, S.R.O., Hassan, N.A., Jaya, R.P., et al.: Thermal performance of cooling strategies for asphalt pavement: a state-of-the-art review. J. Traffic Transp. Eng. (English Ed **8**, 356–373 (2021). https://doi.org/10.1016/j.jtte.2021.02.001
47. Anupam, B.R., Sahoo, U.C., Chandrappa, A.K., Rath, P.: Emerging technologies in cool pavements: a review. Constr. Build Mater. **299**, 123892 (2021). https://doi.org/10.1016/j.conbuildmat.2021.123892
48. Li, H.: Evaluation of Cool Pavement Strategies for Heat Island Mitigation Evaluation of Cool Pavement Strategies for Heat Island Mitigation. University of California (2012)

49. Rosado, P., Gilbert, H., Pomerantz, M., et al.: Cool pavement demonstration and study. In: Third International Conference on Countermeasures to Urban Heat Islands, pp. 815–826 (2014)
50. Jiang, L., Wang, L., Wang, S.: A novel solar reflective coating with functional gradient multilayer structure for cooling asphalt pavements. Constr. Build. Mater. **210**, 13–21 (2019). https://doi.org/10.1016/j.conbuildmat.2019.03.180
51. Yinfei, D., Mingxin, D., Haibin, D., et al.: Incorporating hollow glass microsphere to cool asphalt pavement: preliminary evaluation of asphalt mastic. Constr. Build Mater. **244**, 118380 (2020). https://doi.org/10.1016/j.conbuildmat.2020.118380

Impact of Sucralose on Environmental Bacteria: Mechanistic Insights from Molecular Modeling

Victor Markus(✉)

Department of Medical Biochemistry, Near East University, Mersin 10, 99138 Nicosia/TRNC, Turkey
victor.markus@neu.edu.tr

Abstract. As governments around the world tighten regulations on sugar, companies look for alternatives to sweeten their products. One of the alternatives to sugar commonly used today is artificial sweeteners. Of the six artificial sweeteners approved by the Food and Drug Administration of United States (FDA), sucralose is one of the most widely used and the only sweetener produced from natural sugar. Sucralose is reported to be nonmetabolizable in the gastrointestinal tract (GIT) and thus passes out of the body unaffected via urine and faces reaching the environment through wastewater. Sucralose can withstand drastic heat and pH and is resistant to wastewater treatment. While studies have suggested the impart of sucralose on environmental bacteria, this synthetic agent's mode of action is not completely understood. Molecular modeling is a cost-effective and convenient computational method used to assess and accurately predict the reaction mechanism of proteins with ligands. In this study, molecular modeling was employed to evaluate the impact of sucralose on environmental bacteria. Findings reveal pivotal interactions between sucralose and key bacterial proteins in sucrose metabolic cascade. This study provides evidence at the molecular level on the reported bacteriostatic effects of sucralose. Additionally, the consequential binding sites residues and characteristic binding of the synthetic sweetener could be employed for site-directed mutagenic studies in bioremediation applications.

Keywords: Sucralose · Artificial sweeteners · Sucrose metabolism · Invertase · Sucrose permease · Molecular modeling

1 Introduction

Artificial sweeteners otherwise referred to as non-nutritive sweeteners (NNS), low-caloric sweeteners (LCS), or non-caloric sweeteners (NCS), are sugar substitutes found today in a variety of foods, beverages, and pharmaceutical products. These synthetic sweeteners are currently drawing considerable attention in the scientific community due to their unexpected contamination of aquatic and coastal habitats [1]. Most of the studies conducted on artificial sweeteners have been directly connected to human health. Studies on the impact of these synthetic sweeteners on the environment are quite lacking. Although generally regarded as beneficial and safe to humans, available scientific

H. Gökçekuş and Y. Kassem (Eds.): NRSEM 2021, EESCI, pp. 358–368, 2022.
https://doi.org/10.1007/978-3-031-04375-8_40

data in the literature suggest that the safety status of artificial sweeteners is inconclusive. Evidence from most studies conducted using rodent models indicates that artificial sweeteners have deleterious consequences on metabolic health [2, 3]. In humans randomized controlled trials (RCT), findings show conflicting evidence concerning the impact of synthetic chemicals on metabolic health [4] and bodyweight [5]. Recently, another angle of debate was sparked when artificial sweeteners were considered emerging pollutants owing to their persistence and chemical stability in the environment [6].

Of the six artificial sweeteners approved by the Food and Drug Administration of United States (FDA), sucralose is currently the most extensively used in food and pharmaceutical products [7] and the only one produced from natural sugar. Although it was discovered in 1976, it was not approved for usage in the United States until 1998 [8]. Sucralose, referred scientifically as 4,1′,6′-trichloro-4,1′,6′-trideoxy-galacto-sucrose or 1,6-dichloro-1, 6-dideoxy-β-d-fructofuranosyl 4-chloro-4-deoxy-α-d-galacto-pyranoside, was produced by replacing three hydroxyl groups in sucrose at positions 4, 1′, and 6′ with chlorines [9]. Because sucralose is hundreds of times (600 times) sweeter than sucrose [8] and has enormous advantages, including solubility in water (280 g L^{-1} sucralose solution in water at 20 °C is achievable), stability under a wide range of temperature, acidic and alkaline conditions [9], many companies use it in their products. Although resistant to photolytic [10] and microbial [11] degradation, sucralose can decompose into hazardous polychlorinated chemicals [7] such as polychlorinated dibenzo-*p*-dioxins and dibenzofurans [12] at high temperatures (around 125 °C or more). Sucralose is not degraded in the gastrointestinal tract (GIT) [3], passing out of the body unaffected. Because sucralose consumed in foods and beverages is not metabolized, it is excreted in feces and urine, reaching the environment through wastewater [13]. Additionally, due to the resistance of this agent to wastewater treatment [14], it is continuously introduced into the environment. Several artificial sweeteners, including sucralose, are found in air, soil, groundwater, lakes, surface water, seawater, and the water people drink in various places around the world [6, 15]. A 2012 study on water treatment plants in the United States from 19 states that serve more than 28 million people found sucralose in 15 of the 19 water sources of the treatment plants (47–2,900 ng L^{-1}), and 13 of the 17 “finished” water of the treatment plants (49–2,400 ng L^{-1}) and 8 of the 12 distribution system of the water treatment plants (48–2,400 ng L^{-1}) [16].

Molecular modeling is a cost-effective and convenient computational method used to assess and accurately predict the reaction mechanism of proteins with ligands. Although molecular modeling has been widely used for drug discovery, its use in environmental research to assess the mechanism of action of pollutants and predict targets for bioremediation applications holds promises [17]. This study evaluates the mechanism of action of sucralose in the bacteria and provides evidence at the molecular level on the reported bacteriostatic effects of the sweetener.

2 Methods

2.1 Selection and Preparation of Ligands

The canonical SMILES notation of the ligands, sucrose (PubChem compound identifier: 5988) and sucralose (PubChem compound identifier: 71485) obtained from the PubChem (https://pubchem.ncbi.nlm.nih.gov/) were entered into an automatic, fast, and powerful three-dimensional (3D) Generator of Structures, CORINA Classic Version 4.3.0 [18], that generates 3D coordinates for the structure of the ligands in Mol2 or PDB format. Utilizing the default settings, a low-energy and high-quality conformation of the 3D molecular models of ligands were generated and downloaded.

2.2 Selection and Preparation of Target

The Crystallographic structures of alkaline invertase A (InvA) from Anabaena sp. PCC 7120 bound to its native substrate sucrose (PDB ID: 5GOP, chain –A; Resolution: 2.35 Å; R-Value Free: 0.235; R-Value Work: 0.207) and alkaline invertase B (InvB) from Anabaena sp. PCC 7120 bound to its native substrate sucrose (PDB ID: 5Z74, chain –B; Resolution: 1.95 Å; R-Value Free: 0.239; R-Value Work: 0.208) were obtained from the RCSB Protein Data Bank (PDB) [19]. The structure from PDB was fed as an input to the UCSF Chimera (Version 1.11.2) Dock Prep tool [20] for preparation. In the preparation of the structures, water molecules were removed; atomic charges, protonation states, and hydrogen atoms were added to the crystallographic structures. The crystallographic structures were then energy minimized, first using the steepest descent algorithm and next by conjugate gradient algorithm provided by Molecular Modelling Toolkit (MMTK) included in UCSF Chimera software [20].

2.3 Homology Modeling

The amino acid sequence of the protein from *Escherichia coli* (1 to 415 residues) was retrieved from NBCI, GenBank reference: BCA74826.1 available at https://www.ncbi.nlm.nih.gov/protein/BCA74826.1, and fed as the Query sequence on HHpred [21]. From the HHpred search results (number of hits: 210), the first and the second hits on the list were selected as follow:

a) The crystallographic structure of Lactose permease bound to thiodigalactoside and nanobody 9043 (PDB ID: 6VBG, chain A) with probability of 100%, E-value of 1.5 e^{-36}, raw score of 270.2, secondary structure scores of 50.8, matched region lengths of 407, and template lengths of 417.
b) The crystallographic structure of Drug: Proton Antiporter-1 (DHA1) Family SotB, in the inward conformation (H115N mutant) (PDB ID: 6KKL, chain A) with probability of 100%, E-value of 3.9 e^{-36}, raw score of 268.38, secondary structure scores of 44.7, matched region lengths of 379, and template lengths of 423.

On the HHpred Results page, the above-selected templates were selected and forwarded to MODELLER, a popular homology modeling program (Sali & Blundell, 1993). Using MODELLA, a model was built. PROCHECK was used to assess the model protein's stereochemical quality [22] and validated using ProSA, a web server for protein structure validations [23]. Additionally, the ligand-binding pocket of the model was predicted using a web server, 3DLigandSite [24].

2.4 Docking Studies

The docking studies were performed by CB-Dock, a cavity detection-guided webserver and AutoDock Vina-based blind docking algorithm [25], available at http://cao.labshare.cn/cb-dock/. In the webserver form, the cavities number for the docking process was 5. The best protein-ligand complexes were selected from the resulting docking outcomes based on their binding energies (in kcal mol^{-1}).

2.5 Protein-Ligand Interaction Profiling

The enumeration of the favorable non-covalent bonds between the targets and the docked substrates was performed by PLIP, a Web server for identifying interatomic proteins-ligand interactions [26] available at https://plip-tool.biotec.tu-dresden.de/plip-web/plip/index. The inspection of the predicted interactions between the proteins and ligands, and the processing of the image was done using the PyMOL Version 2.5 (Schrödinger, LLC, Portland, OR, USA).

3 Results

To assess the ability of molecular docking to correctly replicate the experimentally determined protein-ligand complexes, redocking studies were conducted with the native ligand, sucrose. Generally, a root-mean-square (RMS) deviation cut-off of <2 Å is considered as a criterion for redocking accuracy [27]. The results suggest that CB-Dock faithfully generates the near-native binding poses of sucrose within InvA and InvB binding cavities, with RMS of deviations of 0.338 Å (across all 435 atom pairs) and 0.433 Å (across all 420 atom pairs), respectively (Fig. 1).

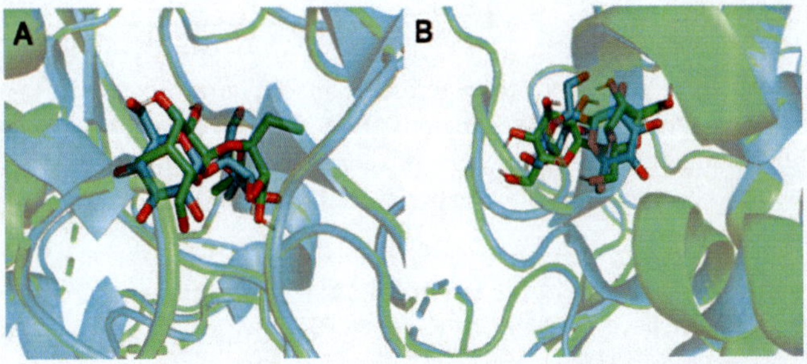

Fig. 1. Close-up view of the superposed atomic coordinates of co-crystallized (cyan) and redocked (green) ligands in the ligand-binding sites of proteins (A) Sucrose in ligand-binding site of InvA (PDB ID: 5GOP, chain –A); (B) Sucrose in ligand-binding site of InvB (PDB ID: 5Z74, chain –B). Images were processed using PyMOL, Version 2.5 (Schrodinger, LLC, Portland, OR, USA).

For lack of the crystal structure of sucrose permease, a model structure of the protein was built. HHpred, a web server that uses a multi-template approach for homology modeling and structural prediction of proteins based on HMM-to-HMM comparison, was used for the homology modeling. Suitable templates for the building of the model were obtained by feeding the amino acid sequence as the Query sequence on HHpred [21]. From the HHpred search results (Number of Hits: 210), the first and the second hits on the hitlist (prioritized based on HHpred score) were selected for the building of the model. High-scoring templates with lengths similar to the query and few gapped columns in the pairwise query-template alignment are typically considered to be the best [28]. Using PROCHECK [22], the stereo-chemical quality of the model protein was checked. The majority of the residues (96.4%) were found in the Ramachandran Plot's most favored regions, indicating a high-quality model. (Fig. 2A). A good quality model would have over 90% residues in the most favored areas [22]. Using ProSA, a web server for protein structure validations [23], the z-score plot was generated. The model protein's z-score (−3.85) was within the range of native conformation (Fig. 2B), depicting a good model. Using 3DLigandSite [24], the ligand-binding site of the sucrase permease-built 3D model was predicted by superimposing ligands (or substrates) from similar structures onto the built model. Mapping residue conservation onto the surface of the protein, the 3DLigandSite web server makes predictions of the ligand-binding site having Ile292, Ile296, Ile322, Glu323, and Lue326.

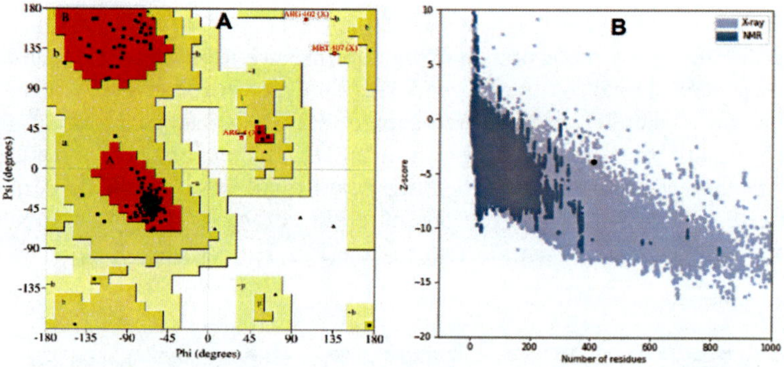

Fig. 2. (A) Ramachandran plot of sucrose permease model showing 96.4% residues in most favored regions, 2.8% in additional allowed regions, 0.6% in generously allowed regions, and 0.3% residues in disallowed regions. (B) The z-score of sucrose permease model (−3.85), is predicted to be within the scores observed by similar experimentally determined protein structures.

Protein structures from important environmental representative bacteria *Anabaena sp.* were selected. The role of *Anabaena sp.* in the environment has been well reported [29]. *Anabaena sp.* has two invertases, InvA and InvB [30, 31]. A previous report shows InvB is neutral invertase [32], however, a more recent study shows that the enzyme is alkaline, like InvA [30]. Although InvA shows a tight substrate specificity for sucrose's α1,2-glycosidic bond [31], InvB was reported to have higher catalytic efficiency and affinity for sucrose, and therefore, pivotal for nitrogen fixation [30]. Generally, invertases are classified into two: acid invertases (Ac-Invs) with pH 4.0–5.5 and alkaline/neutral invertases (A/N-Invs) with pH 6.5–8.0 [30]. Ac-Invs are β-fructofuranosidases localized in the cell wall or vacuole and A/N-Invs are of a novel glucosidase family localized in organelles or cytosol [33]. A/N-Invs have been reported to have a cyanobacterial origin [34], thus informing the choice of *Anabaena sp.* in this study as a representative of important environmental bacteria. Analysis of the substrate-binding cavity of InvA (a hexamer with each subunit binding a sucrose molecule) implicated Asp188 and Glu414 as the catalytic residues [31]. Although InvA and InvB share overall structure, InvB differs from InvA by containing the loop$_{\beta2-\beta3}$ near the binding cavity of the substrate and the residue Arg430 which enhances InvB high catalytic efficiency [30]. In this study, the cross-docking studies predicted the binding pose and interactions of sucralose in the substrate-binding cavities of InvA and InvB. The favorable non-covalent interactions between sucralose and the invertases are shown in Fig. 3. The binding energies of sucralose to InvA and InvB were estimated as –6.7 and –6.4 kcal mol^{-1} respectively. Within the binding site of InvA, sucralose engaged in a hydrogen-bonding interactions with Asp54, Asp188, His371, and Gln432 via the hydroxyl groups of 4-Chloro-4-deoxygalactose unit, and with Asn46, Asp188, Arg189, and Lys364 via the hydroxyl groups of 1,6-Dichloro-1,6-dideoxyfructose unit. Also, sucralose seemed to have also formed hydrophobic interactions with Leu45, Tyr47, Ala121, Lys364 through the chloro and methyl groups of 1,6-Dichloro-1,6-dideoxyfructose unit, and a salt bridge with Lys364 via 4-Chloro-4-deoxygalactose unit (all within 4.8 Å) (Fig. 3B). Within the substrate-binding pocket of InvB, sucralose was able to form hydrogen-bonding interactions with Leu308 and Leu380 via the hydroxyl groups of 4-Chloro-4-deoxygalactose unit. Additionally, it established hydrophobic interactions with Leu308 and Ile315 via the methyl groups of 4-Chloro-4-deoxygalactose unit, and with Arg54, Val57, Trp134, Leu308, Val379, Trp382, Trp435 via the chloro and methyl groups of 1,6-Dichloro-1,6-dideoxyfructose unit (all within 3.99 Å) (Fig. 3D).

Within the predicted substrate-binding pocket of the built-3D model of sucrose permease, it appears sucralose formed hydrogen-bonding interactions with Asn234 via the hydroxyl group of 4-Chloro-4-deoxygalactose unit, and with Gln353 and Asn234 via the hydroxyl groups of 1,6-Dichloro-1,6-dideoxyfructose unit. Additionally, the sweetener established hydrophobic interactions with Trp30 via the chloro group of 4-Chloro-4-deoxygalactose unit, and with Leu122, Asn234, Ile354, and Ile360 via the chloro and methyl groups of 1,6-Dichloro-1,6-dideoxyfructose unit (Fig. 4).

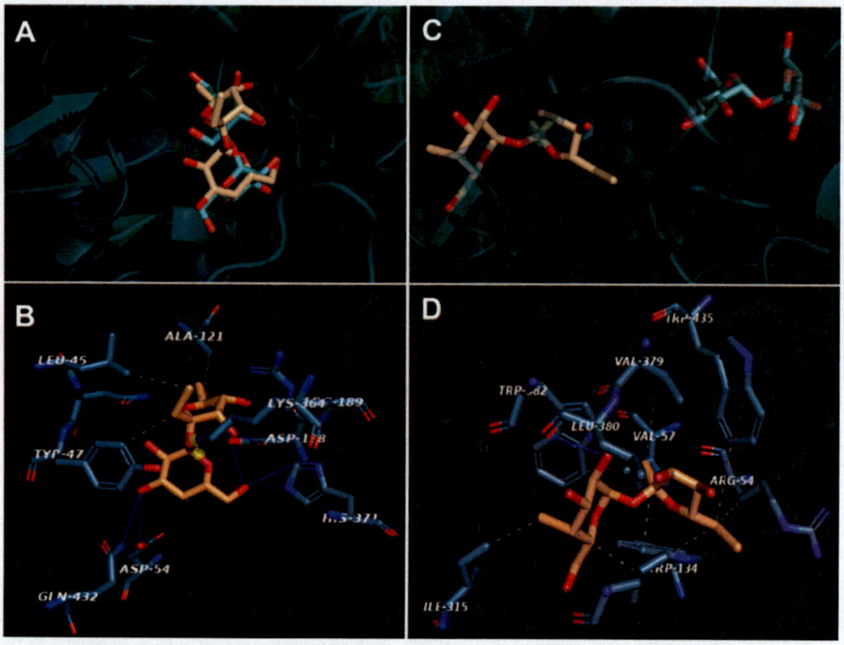

Fig. 3. Molecular docking and interaction profiling of InvA and InvB with sucralose. (A) Transparent cartoon close view of superimposed crystal structures of InvA (Cyan cartoon, PDB ID: 5GOP, chain –A) and best docking result (sky blue cartoon), with sucrose and sucralose represented as sticks with carbon atom coloured cyan and light pink respectively; (B) The interacting residues of InvA (Leu45, Asn46, Tyr47,Asp54, Ala121, Asp188, Arg189, Lys364, His371, Gln432) shown as sticks with the carbon atoms colored sky blue and sucralose represented as sticks with the carbon atoms colored light pink; (C) Transparent cartoon close view of superimposed crystal structures of InvB (Cyan cartoon, PDB ID: 5Z74, chain –B) and best docking result (sky blue cartoon), with sucrose and sucralose depicted as sticks with carbon atom coloured cyan and light pink respectively; (D) The interacting residues of InvB (Arg54, Val57, Trp134, Leu308, Ile315, Val379, Leu380, Trp382, Trp435) shown as sticks with the carbon atoms colored sky blue and sucralose depicted as sticks with the carbon atoms colored light pink. Solid blue line depicts hydrogen-bonding interactions, the gray dashed line depicts the hydrophobic interactions, and the yellow dashed line depicts the salt bridge. Images were processed by PyMOL, Version 1.8 (Schrodinger, LLC, Portland, OR, USA).

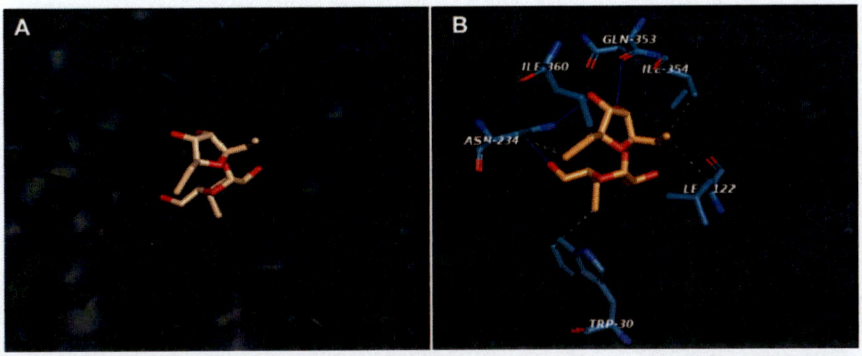

Fig. 4. Molecular docking and interaction profiling of sucrose permease with sucralose. (A) Transparent cartoon close view of the best pose of sucralose (sticks with carbon atom-light pink) docked on sucrose permease-built 3D model (sky blue cartoon); (B) The interacting residues of the built 3D model of sucrose permease (Trp30, Leu122, Asn234, and Gln353, Ile354, Ile360) shown as sticks with the carbon atoms colored sky blue and sucralose depicted as sticks with the carbon atoms colored light pink. Solid blue line depicts hydrogen-bonding interactions, and the gray dashed line depicts the hydrophobic interactions. Images were processed by PyMOL, Version 1.8 (Schrodinger, LLC, Portland, OR, USA).

4 Discussion

The universe harbors an estimated 10^{30} microorganisms, representing the most abundant living things [35]. These microorganisms play pivotal roles in several environmental processes. Any interference, therefore, in their activity may have far-reaching consequences. The use of artificial sweeteners in pharmaceuticals, beverages, and foods has increased in recent years because they are considered to increase the palatability of products with no increase in caloric content [36]. However, the issue with these synthetic agents seems beyond just "sweetness with no calories." Although approved by regulatory agencies such as FDA [8] and European Food Safety Authority [37], the extent of their impact on the environment is still largely unknown. In this study, molecular modeling was employed to glean insights into the mechanism of action sucralose.

Previous experimental data [11, 38] suggested that sucralose disrupts sucrose metabolism. In these molecular docking studies, sucralose appears to be housed within the ligand-binding cavities of the proteins although failing to establish interactions with the critical residues that are responsible for stabilizing the protein-ligand complexes. The localization of sucralose in the ligand-binding cavities of the proteins may induce structural instabilities, thereby disrupting the correct folding of the protein and eventually making them ineffective or dysfunctional. In the event where InvA, InvB, and sucrose permease become dysfunctional due to the impact of sucralose, the sucrose metabolic pathway of the affected bacteria may be disrupted. Because invertase is conserved in the microbial world, the impact of sucralose may be far-reaching. Sucralose is an analog of sucrose, indicating that the mode of inhibition of sucralose could be competitive and its impact could be dependent on concentration. The likely competitive binding of sucralose on the ligand-binding cavity may explain the reported bacteriostatic effect of the sweetener, resulting in reduced uptake and breakdown of sucrose [38]. This perhaps could justify the reported interference of oral bacteria proliferation by sucralose, preventing the formation of dental cavities [39]. This evidence further suggests that bacteria could not utilize sucralose as a source of carbon. Additionally, the interference in sucrose metabolism by sucralose may induce stress in bacteria that may trigger survival reactions. Studies show sucralose increases multidrug-resistant bacteria in the gut [40]. More recently, sucralose and other artificial sweeteners were demonstrated to enhance the spread of antibiotic-resistant genes among bacteria strains [41]. These impacts of sucralose on the gut microbial community suggest the sweetener as well could have the same effect on environmental bacteria.

With the great number of bacteria in environments, it may seem the current highest concentrations of sucralose reported in the environment (300 ng L^{-1}) may not affect bacterial growth [11]. However, the resistance of sucralose to wastewater treatment and its ability to withstand a wide range of pH, heat, and degradation, suggests it may accumulate in the environment to toxic levels, especially in areas that have limited water volumes. The findings in this present study show vital interactions between sucralose and important bacterial proteins in sucrose metabolic cascade. This study provides evidence at the molecular level on the reported bacteriostatic effects of sucralose. It is expected that this characterization of the binding poses of sucralose could contribute to designing site-directed mutagenic studies for bioremediation applications.

References

1. Lillicrap, A., Langford, K., Tollefsen, K.E.: Bioconcentration of the intense sweetener sucralose in a multitrophic battery of aquatic organisms. Environ. Toxicol. Chem. **30**, 673–681 (2011). https://doi.org/10.1002/etc.433
2. Suez, J., Korem, T., Zilberman-Schapira, G., Segal, E., Elinav, E.: Non-caloric artificial sweeteners and the microbiome: findings and challenges. Gut Microb. **6**, 149–155. https://doi.org/10.1080/19490976.2015.1017700
3. Suez, J., et al.: Artificial sweeteners induce glucose intolerance by altering the gut microbiota. Nature **514**, 181–186 (2014). https://doi.org/10.1038/nature13793
4. Ahmad, S.Y., Friel, J.K., MacKay, D.S.: Effect of sucralose and aspartame on glucose metabolism and gut hormones. Nutr. Rev. **78**, 725–746 (2020). https://doi.org/10.1093/nutrit/nuz099
5. Rogers, P.J., Appleton, K.M.: The effects of low-calorie sweeteners on energy intake and body weight: a systematic review and meta-analyses of sustained intervention studies. Int. J. Obes. **45**, 464–478 (2021). https://doi.org/10.1038/s41366-020-00704-2
6. Praveena, S.M., Cheema, M.S., Guo, H.R.: Non-nutritive artificial sweeteners as an emerging contaminant in environment: a global review and risks perspectives. Ecotoxicol. Environ. Saf. **170**, 699–707 (2019). https://doi.org/10.1016/j.ecoenv.2018.12.048
7. de Oliveira, D.N., de Menezes, M., Catharino, R.R.: Thermal degradation of sucralose: a combination of analytical methods to determine stability and chlorinated byproducts. Sci. Rep. **5**, 9598 (2015). https://doi.org/10.1038/srep09598
8. FDA Additional Information about High-Intensity Sweeteners Permitted for Use in Food in the United States. US Food Drug Adm, pp. 1–7 2018 (1981)
9. Glória, M.B.A.: Encyclopedia of food sciences and nutrition. Encycl. Food Sci. Nutr. **2003**, 5695–5702 (2003)

10. Perkola, N., Vaalgamaa, S., Jernberg, J., Vähätalo, A.V.: Degradation of artificial sweeteners via direct and indirect photochemical reactions. Environ. Sci. Pollut. Res. **23**(13), 13288–13297 (2016). https://doi.org/10.1007/s11356-016-6489-4
11. Omran, A., Baker, R., Coughlin, C.: Differential bacteriostatic effects of sucralose on various species of environmental bacteria. ISRN Toxicol. **2013**, 1–6 (2013). https://doi.org/10.1155/2013/415070
12. Dong, S., Liu, G., Hu, J., Zheng, M.: Polychlorinated dibenzo-p-dioxins and dibenzofurans formed from sucralose at high temperatures. Sci. Rep. **3**, 4–7 (2013). https://doi.org/10.1038/srep02946
13. Buerge, I.J., Buser, H.R., Kahle, M., Müller, M.D., Poiger, T.: Ubiquitous occurrence of the artificial sweetener acesulfame in the aquatic environment: an ideal chemical marker of domestic wastewater in groundwater. Environ. Sci. Technol. **43**, 4381–4385 (2009). https://doi.org/10.1021/es900126x
14. Sang, Z., Jiang, Y., Tsoi, Y.K., Leung, K.S.Y.: Evaluating the environmental impact of artificial sweeteners: a study of their distributions, photodegradation and toxicities. Water Res. **52**, 260–274 (2014). https://doi.org/10.1016/j.watres.2013.11.002
15. Lange, F.T., Scheurer, M., Brauch, H.J.: Artificial sweeteners-a recently recognized class of emerging environmental contaminants: a review. Anal. Bioanal. Chem. **403**, 2503–2518 (2012). https://doi.org/10.1007/s00216-012-5892-z
16. Mawhinney, D.B., Young, R.B., Vanderford, B.J., Borch, T., Snyder, S.A.: Artificial sweetener sucralose in U.S. drinking water systems. Environ. Sci. Technol. **45**, 8716–8722 (2011). https://doi.org/10.1021/es202404c
17. Srinivasan, S., Sadasivam, S.K., Gunalan, S., Shanmugam, G., Kothandan, G.: Application of docking and active site analysis for enzyme linked biodegradation of textile dyes. Environ. Pollut. **248**, 599–608 (2019). https://doi.org/10.1016/j.envpol.2019.02.080
18. Schwab, C.H.: Conformations and 3D pharmacophore searching. Drug Discov. Today Technol. **7**, e245–e253 (2010). https://doi.org/10.1016/j.ddtec.2010.10.003
19. Berman, H.M., et al.: The protein data bank. Acta Crystallogr. Sect. D Biol. Crystallogr. **58**, 899–907 (2002). https://doi.org/10.1107/S0907444902003451
20. Pettersen, E.F., et al.: UCSF Chimera - a visualization system for exploratory research and analysis. J. Comput. Chem. **25**, 1605–1612 (2004). https://doi.org/10.1002/jcc.20084
21. Zimmermann, L., et al.: A completely reimplemented MPI bioinformatics toolkit with a new HHpred server at its core. J. Mol. Biol. **430**, 2237–2243 (2018). https://doi.org/10.1016/j.jmb.2017.12.007
22. Laskowski, R.A., MacArthur, M.W., Moss, D.S., Thornton, J.M.: PROCHECK: a program to check the stereochemical quality of protein structures. J. Appl. Crystallogr. **26**, 283–291 (1993). https://doi.org/10.1107/s0021889892009944
23. Wiederstein, M., Sippl, M.J.: ProSA-web: Interactive web service for the recognition of errors in three-dimensional structures of proteins. Nucleic Acids Res. **35**, 407–410 (2007). https://doi.org/10.1093/nar/gkm290
24. Wass, M.N., Kelley, L.A., Sternberg, M.J.E.: 3DLigandSite: predicting ligand-binding sites using similar structures. Nucleic Acids Res. **38**, 469–473 (2010). https://doi.org/10.1093/nar/gkq406
25. Liu, Y., Grimm, M., Dai, W.T., Hou, M.C., Xiao, Z.X., Cao, Y.: CB-Dock: a web server for cavity detection-guided protein–ligand blind docking. Acta Pharmacol. Sin. **41**, 138–144 (2020). https://doi.org/10.1038/s41401-019-0228-6
26. Salentin, S., Schreiber, S., Haupt, V.J., Adasme, M.F., Schroeder, M.: PLIP: fully automated protein-ligand interaction profiler. Nucleic Acids Res. **43**, W443–W447 (2015). https://doi.org/10.1093/nar/gkv315

27. Bursulaya, B.D., Totrov, M., Abagyan, R., Brooks, C.L.: Comparative study of several algorithms for flexible ligand docking. J. Comput. Aided. Mol. Des. **17**, 755–763 (2003). https://doi.org/10.1023/B:JCAM.0000017496.76572.6f
28. Gabler, F., et al.: Protein sequence analysis using the MPI bioinformatics toolkit. Curr. Protoc. Bioinforma. **72**, 1–30 (2020). https://doi.org/10.1002/cpbi.108
29. Garlapati, D., Chandrasekaran, M., Devanesan, A., Mathimani, T., Pugazhendhi, A.: Role of cyanobacteria in agricultural and industrial sectors: an outlook on economically important by products. Appl. Microbiol. Biotechnol. **103**(12), 4709–4721 (2019). https://doi.org/10.1007/s00253-019-09811-1
30. Xie, J., et al.: Structural and enzymatic analyses of Anabaena heterocyst-specific alkaline invertase InvB. FEBS Lett. **592**, 1589–1601 (2018). https://doi.org/10.1002/1873-3468.13041
31. Xie, J., et al.: Structural analysis of the catalytic mechanism and substrate specificity of Anabaena alkaline invertase InvA reveals a novel glucosidase. J. Biol. Chem. **291**, 25667–25677 (2016). https://doi.org/10.1074/jbc.M116.759290
32. Vargas, W., Cumino, A., Salerno, G.L.: Cyanobacterial alkaline/neutral invertases. Origin of sucrose hydrolysis in the plant cytosol? Planta **216**, 951–960 (2003).https://doi.org/10.1007/s00425-002-0943-x
33. Shen, L.B., et al.: Genome-wide identification, expression, and functional analysis of the alkaline/neutral invertase gene family in pepper. Int. J. Mol. Sci. **19**, 1–14 (2018). https://doi.org/10.3390/ijms19010224
34. Vargas, W.A., Salerno, G.L.: The Cinderella story of sucrose hydrolysis: Alkaline/neutral invertases, from cyanobacteria to unforeseen roles in plant cytosol and organelles. Plant Sci. **178**, 1–8 (2010). https://doi.org/10.1016/j.plantsci.2009.09.015
35. Turnbaugh, P.J., Gordon, J.I.: An Invitation to the marriage of metagenomics and metabolomics. Cell **134**, 708–713 (2008). https://doi.org/10.1016/j.cell.2008.08.025
36. Baker-Smith, C.M., de Ferranti, S.D., Cochran, W.J.: The use of nonnutritive sweeteners in children. Pediatrics **144**, e20192765 (2019). https://doi.org/10.1542/peds.2019-2765
37. Mortensen, A.: Sweeteners permitted in the European Union: safety aspects. Scand. J. Food Nutr. **50**, 104–116 (2006). https://doi.org/10.1080/17482970600982719
38. Omran, A., Ahearn, G., Bowers, D., Swenson, J., Coughlin, C.: Metabolic effects of sucralose on environmental bacteria. J. Toxicol. **2013**, 1–6 (2013). https://doi.org/10.1155/2013/372986
39. Bowen, W.H., Young, D.A., Pearson, S.K.: The effects of sucralose on coronal and root-surface caries. J. Dent. Res. **69**, 1485–1487 (1990). https://doi.org/10.1177/00220345900690080701
40. Bian, X., Chi, L., Gao, B., Tu, P., Ru, H., Lu, K.: Gut microbiome response to sucralose and its potential role in inducing liver inflammation in mice. Front. Physiol. **8**, 1–13 (2017). https://doi.org/10.3389/fphys.2017.00487
41. Yu, Z., Wang, Y., Lu, J., Bond, P.L., Guo, J.: Nonnutritive sweeteners can promote the dissemination of antibiotic resistance through conjugative gene transfer. ISME J. **15**, 2117–2130 (2021). https://doi.org/10.1038/s41396-021-00909-x

Author Index

H. Gökçekuş and Y. Kassem (Eds.): NRSEM 2021, EESCI, pp. 369–370, 2022.
https://doi.org/10.1007/978-3-031-04375-8